Equilibrium Between Phases of Matter

Equilibrium Between Phases of Matter

Supplemental Text for Materials Science and High-Pressure Geophysics

M.H.G. Jacobs
Technical University, Clausthal, Germany

and

H.A.J. Oonk
Utrecht University, The Netherlands

 Springer

M.H.G. Jacobs
Institute of Metallurgy
Technical University
Clausthal-Zellerfeld
Germany

H.A.J. Oonk
Department of Earth Sciences
Utrecht University
Utrecht
The Netherlands

ISBN 978-94-007-1947-7   e-ISBN 978-94-007-1948-4
DOI 10.1007/978-94-007-1948-4
Springer Dordrecht Heidelberg London New York

Library of Congress Control Number: 2012932632

Printed on acid-free paper

Springer is part of Springer Science+Business Media (www.springer.com)

TABLE OF CONTENTS

INTRODUCTION

background

"Equilibrium between Phases of Matter - Supplemental Text for Materials Science and High-Pressure Geophysics" completes the project of writing two textbooks in the field of heterogeneous equilibrium and thermodynamic properties of materials. The first book of the project, by H.A.J. Oonk and M.T. Calvet, appeared a few years ago. The first book has an undergraduate character and it consists of three parts - referred to as 'level 0', 'level 1', and 'level 2'.

Level 0 is an introduction to phase diagrams, based on the principles of equilibrium, be it without the use of thermodynamics.
Level 1 corresponds to an introduction to classical thermodynamics and its basis of homogeneous chemical equilibrium and equilibrium between phases.
Level 2 can be characterized as a description of the road from thermodynamic properties - the Gibbs energy in particular - to phase diagrams; and the road back from phase diagrams to Gibbs energies.

level 3

The present book can be looked upon as an additional, an advanced level - level 3, so to say. This explains why its sections have been denoted as § 301 through § 308. The book has a postgraduate, a professional character; and compared with level 2, where the accent is on isobaric binary equilibrium, it introduces a number of new elements.
In sections 302 and 303 the theory on binary systems is extended to include *i*) the characteristics of slopes of equilibrium curves in phase diagrams, and *ii*) the background of retrograde behaviour.
In section 305 a move is made to ternary systems. From section 306 on, pressure makes its appearance as a leading experimental variable. And last but not least, in section 307 the mathematical, classical description makes place for a physical, statistical thermodynamics treatment.

account of research

From 304 on, the contents of the sections are intimately related to the scientific work of the authors. As such, the underlying work is a representative account of the research carried out by the authors and their colleagues. The opening sections of this book have been written from a materials science point of view; the three final chapters reflect the authors' interest in the phase behaviour and dynamics of materials in planetary interiors.

style

We have adhered to the style of the first book in that the sections have an *a priori* and an *a posteriori*, in which the contents of a section are announced, and summarized, respectively. In addition, a number of exercises have been added to the sections. The exercises that pertain to the sections 306-308 are assembled at the end of section 308. Some of the exercises are quite laborious; for each of the exercises the solutions are given. Section 301, with its numerous exercises, serves as a bridge between the first book and the new elements of the present book.

acknowledgements

We should not like to end this Introduction without an acknowledgement of the friendly contacts and stimulating discussions we have enjoyed with our colleagues, friends and students - at Utrecht University and elsewhere. In particular we are indebted to Arie van den Berg, Bernard de Jong, Teresa Calvet, Miquel Àngel Cuevas Diarte, Yvette Haget, Denise Mondieig, Josep Lluìs Tamarit and Nguyen-Ba-Chanh. Special thanks are due to Frouke Kuijer for fine artwork, and to Bettina Nelemans for a valuable suggestion.

Clausthal, Germany   Michael H.G. Jacobs
Utrecht, The Netherlands   Harry A.J. Oonk

Autumn 2011

LIST OF FREQUENTLY USED SYMBOLS

Latin letters

A	first component in binary or ternary system
A	Helmholtz energy; magnitude parameter of $AB\Theta$ model
a	activity; system-dependent parameter; anharmonicity parameter
B	general for substance; second component in binary or ternary system
B	asymmetry parameter of $AB\Theta$ model
b	system-dependent parameter
C	third component in ternary system
C_P	heat capacity at constant pressure
C_V	heat capacity at constant volume
c	number of components; system-dependent parameter
d	ordinary differential
f	variance, number of degrees of freedom
G	Gibbs (free) energy
G_{sh}	shear modulus
g	parameter in excess Gibbs energy
H	enthalpy
h	Planck constant, equal to $3.9903126821(\pm57) \times 10^{-10}$ J·s·mol^{-1}; parameter in excess enthalpy
K	equilibrium constant; isothermal bulk modulus
K_S	adiabatic bulk modulus
k	Boltzmann constant, equal to $1.3806504(\pm24) \times 10^{-23}$ J·K^{-1}
LN	defined as $LN(X) = (1-X)\ln(1-X) + X \ln X$
ln	natural logarithm
M	set of variables, and number of elements in it; molar mass
m	mass; mismatch parameter
N	set of equilibrium conditions, and number of elements in it
N_{Av}	Avogadro's number, equal to $6.02214179(\pm30) \times 10^{23}$ mol^{-1}
n	amount of substance; quantum number; number of atoms in formula unit
P	pressure
p	number of phases
R	gas constant, equal to $8.314472(\pm15)$ J·K^{-1}·mol^{-1}
S	entropy
s	parameter in excess entropy
T	thermodynamic temperature
t	Celsius temperature
U	energy
V	volume
v	sound velocity
X	mole fraction; mole fraction of second component
x	variable in general
Y	mole fraction of third component
Z	partition function; number of molecules in primitive cell

x

Greek letters

α	symbol for form / phase
α	cubic expansion coefficient
β	symbol for form / phase
γ	symbol for form / phase
γ	Grüneisen parameter
Δ	operator for difference / change
δ	Kronecker delta function
Θ	characteristic temperature in $AB\Theta$ model
κ	isothermal compressibility
μ	chemical potential
v	frequency
ρ	density
Ω	interaction parameter in magic formula
ω	parameter in excess Gibbs energy
∂ / ∂	partial differential coefficient

Superscripts

E	refers to excess quantity
I,II	for two phases in equilibrium, having the same form
id	refers to ideal-mixing behavior
liq	for liquid
sol	for solid
vap	for vapour
st	for static lattice
vib	related to lattice vibrations
α,β,γ	to refer to forms / phases
o,o	for standard state; for transition temperature, and equilibrium pressure of a pure substance
$*$	for a pure-substance quantity, like entropy

Subscripts

A,B,C	for property of substance / component A, B, C in a given system
BIN	to refer to binodal
c	for critical point
e	to refer to equilibrium
EGC	to refer to equal-G curve
f	for formation from the elements, when attached to Δ
fox	for formation from the oxides, when attached to Δ
m	to refer to melting
o	for a property at a standard state
$SPIN$	to refer to spinodal

§ 301 CENTRAL ROLE OF GIBBS ENERGY

In this section, which is meant to serve as a link between the first and the second volume of Equilibrium between Phases of Matter, the central role of the Gibbs energy is highlighted, the emphasis being on binary systems.

===

the form in which the results of experiment may be expressed

On January 10th 1881, Josiah Willard Gibbs (1839-1903) wrote a letter of acceptance of the Rumford Medal to the American Academy of Arts and Sciences. The following is a citation from that letter (Wheeler 1952).

"A distinguished German physicist has said, - if my memory serves me aright, - that it is the office of theoretical investigation to give the form in which the results of experiment may be expressed. In the present case we are led to certain functions which play the principal part in determining the behavior of matter in respect to chemical equilibrium. The forms of these functions, however, remain to be determined by experiment, and here we meet the greatest difficulties, and find an inexhaustible field of labor. In most cases, probably, we must content ourselves at first with finding out what we can about these functions without expecting to arrive immediately at complete expressions of them. Only in the simplest case, that of gases, have I been able to write the equation expressing such a function for a body of variable composition, and here the equation only holds with a degree of approximation corresponding to the approach of the gas to the state which we call perfect".

Gibbs energy

Gibbs's functions, which are the *free enthalpies* of the bodies, are composed of their *enthalpy H* and *entropy S* as

$$G = H - TS, \tag{1}$$

in which T represents the *thermodynamic temperature*. The enthalpy, which itself is an *auxiliary function*, is composed of *energy U* and *volume V*,

$$H = U + PV, \tag{2}$$

P being the pressure exerted on the body.

Today, the free enthalpy is named *Gibbs energy*, which is short for *Gibbs free energy*.

about the Gibbs function

The *natural variables* of the Gibbs energy are temperature and pressure. Whenever, under reversible conditions, the temperature and pressure of/on a *closed system* are changed, there will be a change in Gibbs energy which is given by

$$dG = -SdT + VdP .\tag{3}$$

If the system is composed of the substances A and B, and the substances can be added to, or withdrawn from the system - the *open system* - the change of its Gibbs energy is

$$dG = -SdT + VdP + \mu_A dn_A + \mu_B dn_B .\tag{4}$$

In this expression the symbol n is for *amount of substance* and the symbol μ for *chemical potential.*
Owing to the fact that the Gibbs energy is a first degree homogeneous function of the amounts of A and B, the *total Gibbs energy* of the system satisfies the relationship

$$G = n_A \mu_A + n_B \mu_B .\tag{5}$$

The mathematical consequence of the two facts expressed by Equations (4) and (5) is the existence of the *Gibbs-Duhem equation*

$$-SdT + VdP - n_A d\mu_A - n_B d\mu_B = 0 .\tag{6}$$

For a system defined on a molar base - system = $\{(1\text{-}X) \text{ mol A} + X \text{ mol B}\}$, X being the *mole fraction* variable - the last three equations change into

$$dG_m = -S_m dT + V_m dP + (\mu_B - \mu_A)dX\tag{7}$$

$$G_m = (1{-}X)\mu_A + X\mu_B\tag{8}$$

$$-S_m dT + V_m dP - (1{-}X)d\mu_A - Xd\mu_B = 0 .\tag{9}$$

From now on, we will define our systems on a molar base, and drop the subscript *m*.

The Gibbs energy is *characteristic* for its natural variables T and P; and it means that all of the system's thermodynamic properties can be derived from its Gibbs energy as a function of T, P, and X.

Another, and for our purpose extremely important property of the Gibbs energy in terms of T and P is the fact that spontaneous/irreversible changes, which proceed at constant temperature and pressure, invariably go together with a lowering of the system's Gibbs energy. *Internal equilibrium* at given T and P is reached when the system has reached its *lowest possible Gibbs energy* (*principle of minimal Gibbs energy*).

binary systems, the magic formula

The most elementary formula for the molar Gibbs energy of a real, non-ideal mixture is the expression

$$G(T,P,X) = (1-X)\, G^{*}_{A}\,(T,P) + XG^{*}_{B}(T,P) + RT\mathrm{LN}(X) + \Omega X(1-X)\,,$$
(10)

with $\mathrm{LN}(X) = (1-X)\ln(1-X) + X\ln X.$ (11)

The quantities G^{*}_{A} and G^{*}_{B} are the Gibbs energies of the pure substances A and B, R is the *gas constant*. The last term in Equation (10) is the *excess Gibbs energy*; in this equation the constant Ω is an *interaction parameter* which depends on the combination of A and B. In Vol.I of this work (Oonk and Calvet 2008) Equation (10) has been referred to as the *magic formula*. The magic of the formula is twofold. First of all, it explains much of the phenomenology of equilibrium between phases – in particular the existence of *critical points of mixing*. And secondly, the thermodynamic properties it generates for very *dilute solutions* of B in A (or A in B) are in full harmony with the laws named after Henry, Raoult, and Van 't Hoff.

binary systems, from Gibbs energy to phase behaviour

The route from a system's Gibbs energy function(s) to the system's phase behaviour is straightforward and univocal – univocal for given T and P and chemical composition. What one has to do is to find - for the given system and experimental circumstances - *the lowest possible Gibbs energy*. For example, for the case that A and B are liquids and their combination at 1 bar pressure does comply with Equation (10) such that Ω is a positive constant, it can easily be shown that below $T=\Omega/2R$ - which is the *critical temperature* T_{c} - the two substances do not mix in all proportions.
The route from Gibbs energy to phase behaviour can be traversed in different manners: geometrically ($\leftarrow$211), analytically ($\leftarrow$212), and numerically ($\leftarrow$213).

The geometric manner is particularly useful for systems composed of two substances, which, besides, play a principal part in the theory of equilibrium between phases. In Figure 1, which is a *GX-diagram* - an isothermal, isobaric section of the Gibbs energy - the *G-curves* are shown for three different forms of matter, α, β, and γ. From pure A to pure B, the system first shows a region where the material takes the form α, next a region where α and β coexist, then a region where the material takes the β form, followed by a region where β and γ coexist, and finally a region where the γ form is taken.

With reference to Figure 1, and at isobaric conditions, the equilibrium between the phases that have the forms α and β is defined by the following *system formulation* (Oonk and Calvet 2008):

$$f = \mathrm{M}\,[T, X^{\alpha}, X^{\beta}] - \mathrm{N}\,[\mu^{\alpha}_{A} = \mu^{\beta}_{A};\ \mu^{\alpha}_{B} = \mu^{\beta}_{B}] = 3 - 2 = 1.$$
(12)

The two conditions of the *set N of equilibrium conditions* act upon the three variables of the *set M of variables.* There is one *degree of freedom,* and it means that the values of X^α and X^β are fixed by the (experimentalist's) choice of T. Or, in other terms, in the *TX*-plane, the *TX phase diagram* there is a curve for the compositions of the phases that have the form α and another curve for the phases that have the form β.

For a limited number of (ideal or idealized) cases the two conditions, the two equations of the set N can be solved in an analytical manner. For most of the cases, however, the equations must be solved numerically.

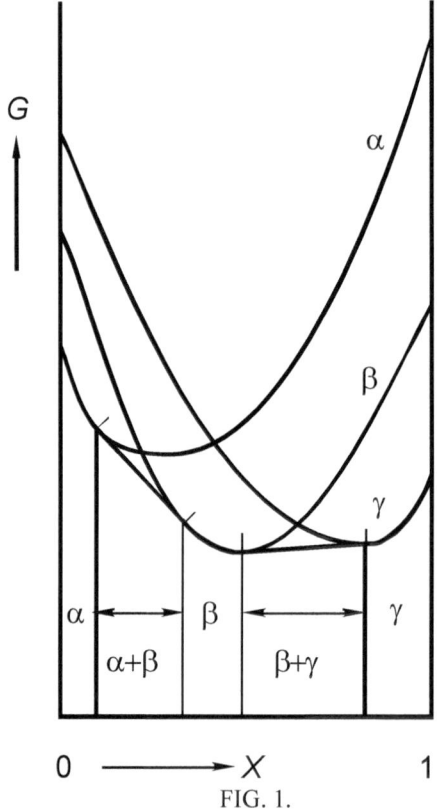

FIG. 1.

The *phase composition* of a system is readily found by stretching a cord along the underside of the *G*-curves. Along the cord the Gibbs energy has its lowest possible value

tools

The *spinodal* and the *equal-G* curve are auxiliary tools, with the help of which the numerical method can be (partly) circumvented ($\leftarrow$212).

The spinodal can be put into action when the two phases in equilibrium have the same form. In that case the problem of two equations with three unknowns reduces to a problem of one equation with two unknowns:

§(301)

$$f_{spin} = \text{M} [T, X] - \text{N} [\partial^2 G/\partial X^2 = 0] = 2 - 1 = 1. \tag{13}$$

For the G-function defined by Equation (10), the spinodal is given by

$$T_{spin} (X) = 2(\Omega/R)X(1-X) . \tag{14}$$

Although the spinodal is not the locus of the equilibrium compositions of the phases - which is the *binodal* - it is inside the region bounded by the binodal, and it shares with the binodal the critical point. These characteristics make that, from the spinodal, a fair estimate can be made of the binodal – the boundary of the *region of demixing.*

In the case of Figure 1, the two components of the system mix in three different forms α, β, and γ. Concentrating on the first two of them, α and β, or rather the equilibrium between α-phases and β-phases, it clearly follows from Figure 1, that the *point of intersection* of the two G-curves necessarily is between the *points of contact*, which define the compositions of the coexisting phases. Or, in other terms, in the *TX*- or *PX*-phase diagram the locus of the points of intersection is a curve - the *equal-G curve* - whose position is inside the two-phase region bounded by the two curves that represent the compositions of the coexisting phases. The equal-G curve analogue of Equation (13) for isobaric equilibria is the formulation

$$f_{egc} = \text{M} [T,X] - \text{N} [\Delta G = 0] = 2 - 1 = 1 . \tag{15}$$

For two G-functions defined by Equation (10), using the Δ operator for β minus α, the equal-G curve is the solution of the equation

$$(1-X)\Delta G^*_A(T) + X\Delta G^*_B(T) + \Delta\Omega X(1-X) = 0 . \tag{16}$$

Any *extremum* in an *equal-G curve* coincides with an *extremum in* the *phase diagram*, unless, of course, the extremum is inside a region of demixing.

An important tool of a different kind is the *linear contribution,* whose usefulness is based on the general fact that nothing in the world of equilibrium between phases will change when a term which is linear in mole fraction(s) is added to or withdrawn from all of the Gibbs functions of the system (←205). By adding, for example, the same linear term AX to each of the three functions shown in Figure 1, the function values at the axis $X=1$ will change by the amount of A; the slopes of the two *common tangent lines* will change, but not the abscissae of their points of contact. And every individual common tangent can be given a zero slope, as a result of which the points of contact become (also) minima – a property which is of great value for drawing sketches of (parts of) GX diagrams; and also for calculating the compositions of coexisting phases (←213).

from phase diagram to Gibbs energy; the isobaric binary region of demixing

Unlike the route from Gibbs function(s) to phase behaviour, the route back, the route from phase diagram to Gibbs energy, is not univocal. The following reasoning should make this clear.

Generally, for real systems at isobaric circumstances, the system-dependent interaction parameter Ω is a function of temperature and composition. For every point in the TX plane there is a corresponding value for Ω. In the case of the route back, there are a limited number of data points – not scattered over the whole TX plane, but only along a *trajectory* in that plane. The trajectory itself is a relation between T and X, and it means that, along the trajectory, Ω can be represented by an expression in temperature, as well as by an expression in mole fraction – and also by a family of mixed functions between the two extreme expressions.

As an illustration, we take the case of the analysis of a set data on the *region of demixing* in the solid state of $\{(1-X) \text{ AgCl} + X \text{ NaCl}\}$. The set of experimental data consists of nine isothermal sections, for which the mole fractions of the coexisting phases, X' and Y', were determined by *X-ray diffraction,* after a 10-20 days period of *equilibration,* see Table 1 (Sinistri et al. 1972). The experimental information has been analyzed by the *computer routine* EXTXD/SIVAMIN (see Oonk 1981) in terms of the *excess Gibbs energy model*

$$G^E(T,X) = X(1-X) \left[(h_1 - Ts_1) + (h_2 - Ts_2)(1-2X)\right]. \tag{17}$$

In this model, the excess Gibbs energy is taken as a linear function in temperature; the more general equivalent of Equation (17) is

$$G^E(T,X) = X(1-X) \left[g_1(T) + g_2(T)(1-2X)\right] . \tag{18}$$

Three different types of calculation have been carried out, and their results are summarized in Table 2. Each time the input data for the computations consisted of a number of *data triplets $T'X'Y'$*. In a first series of computations, the experimental data triplets, Table 1, were taken; and the number of *parameters to be adjusted* was varied from one (h_1 only) to four. Another series of computations were made with a *dummy set of data*, the triplets of which were read from the binodal curve which was drawn by hand through the experimental data. In the case of the third series, computations were made with fixed values for h_1 and h_2, for which the experimentally determined values (Kleppa and Meschel 1965) were taken. For all of the computational results the values are given of the *temperature index Δ_T*, which is the mean difference, all differences being taken as a positive number, between the experimental temperature for a given composition and the temperature for the same composition read from the computed binodal.

The results of the computations, which are displayed in Table 2, give rise to a number of observations that have a general character. In the first place, and in spite of the scatter in the values of h and s, the values obtained for the g-coefficients, for $T = 448$ K, are virtually the same – showing that the Gibbs energy is the key function in phase equilibrium matters.

Table 1: Experimental data pertaining to the region of demixing in the solid state of AgCl+NaCl

T'/K	337	375	416	424	440	447	460	468	470
X'	0.076	0.129	0.219	0.266	0.289	0.349	0.407	0.458	0.428
Y'		0.903	0.903	0.815	0.793	0.727	0.680	0.638	0.583	0.559

Secondly and clearly, the division of Gibbs energy into the real enthalpy and entropy contributions is a matter of great delicacy. The calculated coefficients of enthalpy and entropy are different for the original data set and the dummy set; and, for both sets, the results are far away from the experimental values. A speaking characteristic of the mathematical procedure is the fact that the computed statistical uncertainties (not included in Table 2) in the computed s-coefficients stand in a virtually constant ratio to the uncertainties computed for the partner h-coefficients. For the example at hand, that ratio is 448 K.

As a third observation: the results obtained for the dummy set are in better agreement with the original data than the results obtained with the original data themselves. In a sense this is not surprising, since, after all, in the computational procedure the minimization is not between experimental and computed mole fractions or temperatures, but between 'experimental' and computed values for the two quantities that are generated by the two conditions in the set N of conditions of the system formulation, Equation (12).

Table 2: Results of phase-diagram analysis for the region of demixing in the solid state of AgCl+NaCl. Calculated values of the h and s parameters in Equation (17), in $\mathrm{kJ\,mol^{-1}}$ and $\mathrm{J\,mol^{-1}\,K^{-1}}$, respectively; mean temperature difference in K; values of $g_i = h_i - Ts_i$, $i = 1; 2$, for $T = 448$ K, and expressed in $\mathrm{kJ\,mol^{-1}}$

h_1	s_1	s_2	h_2	Δ_T	g_1	g_2
original data set in Table 1						
7.81				7.2	7.81	
8.13	0.70			7.2	7.82	
8.26	0.98	− 0.19		6.6	7.82	− 0.19
8.31	1.09	0.52	1.62	6.2	7.82	− 0.21
dummy data set						
7.83				7.4	7.83	
9.07	2.77			6.2	7.83	
9.06	2.74	− 0.14		5.4	7.83	− 0.14
9.20	3.06	0.90	2.49	5.0	7.83	− 0.22
dummy set with fixed, experimental h_1 and h_2						
11.32	8.48	− 0.86		13.2	7.52	− 0.86
11.32	7.89	− 0.86	− 1.78	6.6	7.79	− 0.06

the Gibbs energy in computational thermodynamics

From the foregoing example of *thermodynamic analysis* it follows that thermodynamic *excess properties* can be derived from region-of-demixing data, without the need of using *pure-component data.* Pure-component data are needed, on the other hand, if one wants to derive excess properties from data that pertain to a transition, such as the change from solid to liquid.

In practice the (molar) *Gibbs energy* of a material, say a pure component, is obtained by integration over temperature followed by integration over pressure – starting from a reference temperature T_0 (say 298.15 K) and a reference pressure P_0 (say 1 bar):

$$G(T,P) = \Delta_f H^\circ + \int_{T_0}^{T} C_p^\circ \mathrm{d}T - T \left\{ S^\circ + \int_{T_0}^{T} \frac{C_p^\circ}{T} \mathrm{d}T \right\} + \int_{P_0}^{P} V \mathrm{d}P \tag{19}$$

The $\Delta_f H^\circ$ in Equation (19) is the *enthalpy of formation* and S° the *(absolute) entropy,* both at T_0 and P_0. The effects of phase transitions are not included in the expression; these effects are needed when the form in question does not exist at the reference circumstances.

First integrating over temperature has the advantage that *heat capacities* (and heat effects of transitions) have to be known only for the reference pressure (the *atmospheric pressure* in most of the cases). As a rule, experimental heat-capacity data are framed into *polynomial representations,* such as

$$C_p(P_0, T) = C_1 + C_2 \cdot T + C_3 \cdot T^{-1} + C_4 \cdot T^{-2} + C_5 \cdot T^2 + C_6 \cdot \ln T . \tag{20}$$

For the last integration implied in Equation (19), consequently and necessarily, the *volume* of the material has to be known as a function of temperature *and* pressure.

The dependence of volume on temperature is expressed by the *cubic expansion coefficient,* or *thermal expansivity*:

$$\alpha = \frac{1}{V}\left(\frac{\partial V}{\partial T}\right)_P = \frac{1}{V}\left(\frac{\partial^2 G}{\partial T \partial P}\right) . \tag{21}$$

It follows that the volume of the material at P_0 and arbitrary temperature T is given by

$$V(P_0, T) = V(P_0, T_0) \cdot \exp\left(\int_{T_0}^{T} \alpha(P_0, T)\mathrm{d}T \right) . \tag{22}$$

Experimental thermal-expansivity data are represented by an analogue of Equation (20); like

$$\alpha(P_0, T) = \alpha_1 + \alpha_2 \cdot T + \alpha_3 \cdot T^{-1} + \alpha_4 \cdot T^2 . \tag{23}$$

The dependence of volume on pressure is expressed by the *isothermal compressibility* (κ); or its inverse which is the *isothermal bulk modulus* (K):

$$\kappa = -\,(1/V)\,(\partial V/\partial P)_T = -\,(1/V)\,(\partial^2 G/\partial P^2)\;; \tag{24}$$

$$K = -\,V\,(\partial P/\partial V)_T\,. \tag{25}$$

Experimental data on compressibilities as a function of pressure, and at a variety of temperatures of interest, are scarce. It means that pressure analogues of Equation (23) are rare. The consequence is that, for high-pressure work, most of the times use has to be made of *equations of state* ($\rightarrow$306).

remark

As a matter of fact, the real pure-component data that are needed - for the thermodynamic analysis of *TX* data that pertain to a change of form, chemical reactions being absent - are the *difference quantities* for that change. For example for the change from solid to liquid, these quantities are the heat of melting, ΔH, and the change in heat capacity on melting, ΔC_P. The two quantities ΔH and ΔC_P are experimental realities, and for that reason independent of the choice of zero points. In other words, and with reference to Equation (19), knowledge about the heats of formation of the pure components is not required. And the same holds true for their absolute entropies.

The function which plays the principal part in matters of equilibrium between phases is the Gibbs energy – the Gibbs energies as functions of temperature, pressure, and chemical composition of all of the forms in which a system can manifest itself. The route from a set of defined Gibbs functions to the phase behaviour of a (hypothetical) system is univocal. The inverse route, from a (real) system's phase behaviour to the Gibbs functions of its forms, is hampered by the fact that the equilibrium states correspond to non-independent relationships between the variables.

For additional reading we recommend the books by Hillert (1998), Predel et al. (2004), Zhao et al. (2009).

EXERCISES

1. *characteristic*

    For a metal M the *Gibbs energy change*, in $J\,mol^{-1}$ for the transition from solid to liquid at 1 bar pressure as a function of temperature is given by the expression

    $\Delta G(T) = 7692 + 142.97(T/K) + 7.766 \times 10^{-3}(T/K)^2 - 8.3492 \times 10^5 (T/K)^{-1} - 21.260(T/K)\ln(T/K)$.

    - Calculate M's 1 bar *melting point*.
    - Derive the expressions for $\Delta S(T)$; $\Delta H(T)$; and ΔC_P (T). Give the numerical values of these quantities for the melting-point temperature.

2. *a coexistence problem*

    Is there either *i)* a range of temperatures; or *ii)* one temperature; or *iii)* no temperature at which the five substances MgO, CaO, CO_2, $MgCO_3$, and $CaCO_3$ can exist in each others' presence?
    Find the correct answer in two steps:
    1) from $f = M - N$;
    2) using the 1 bar *enthalpies* and *entropies of formation from the oxides* of $MgCO_3$ and $CaCO_3$ (calcite); these are, expressed in *SI units*,
    $MgCO_3$ enthalpy   -116000   entropy -170
    $CaCO_3$ -176000   -154

3. *trigonometric excess Gibbs energy – spinodal and binodal*

    For the excess Gibbs energy function $G^E = (1000 \sin 2\pi X)J\,mol^{-1}$

    - derive the formula for the *spinodal,* and make a graphical representation of the function;
    - determine the limits of the *region of demixing* at zero K (this can be done in a graphical manner; note, however, that these limits follow from $\tan 2\pi X = 2\pi X$);
    - from the spinodal and the limits at zero K, estimate the course of the *binodal* and read from it the mole-fraction values for $T = 0.75$ T_c.

    NB. This exercise is a copy of part of Exc 212:1

4.   *from experimental equal-G curve to phase diagram*

Mixed crystals of {(1-X)1,4-dichlorobenzene + X 1,4-dibromobenzene}, having the α-form of the first component, and prepared by *zone leveling,* were found to melt quasi-isothermally at TX conditions of the *equal-G curve* (Van Genderen et al. 1977). The data pairs (X; T/K) are (0.20; 329.1) (0.21; 329.6) (0.37; 333.2) (0.53; 338.5) (0.66; 344.4) (0.75; 348.7) (0.84; 353.3) (0.86; 354.8) (0.93; 357.7). For the *pure components,* A=ClCl and B=BrBr, the changes in Gibbs energy on melting are, in $J\,mol^{-1}$, $\Delta G^*_A = (18027 - 55.29T/K)$ and $\Delta G^*_B = (20387 - 56.56T/K)$.

- For the data pair (0.53; 338.5), calculate the value of $\Delta\Omega$, Equation (16).
- With the help of the two equations in N, Equation (12) with β=liquid, calculate the equilibrium compositions of the solid phase and the liquid one for $T = 338.5$ K, putting $\Omega^{liq} = 0$.

Clue.   The two equations can be written as
$1-X^{liq}$ = expression (A) in T and $X^{sol;}$
X^{liq} = expression (B) in T and X^{sol}.
By adding the two expressions, an equation in T and X^{sol} is obtained from which the latter can be solved. Subsequently, X^{liq} is obtained by substituting T and X^{sol} in one of the expressions (A) and (B).

- Use the data and the results to construct the solid-liquid phase diagram.

5.   *an elliptical stability field*

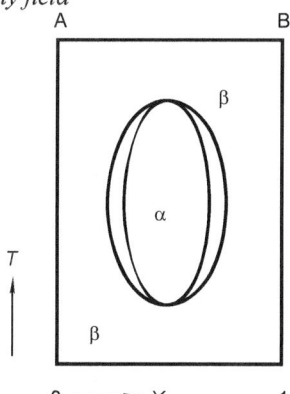

The phase diagram which is shown, is for a system whose forms α and β comply with the magic formula, Equation (10), and such that $G^*_A = G^*_B$ (in α as well as in β) and Ω^α and Ω^β are constants.

- With the help of two or three isothermal GX-sections, find out what the properties are of ΔG^*_A and $\Delta\Omega$ that account for the depicted behaviour. The operator Δ stands for β minus α.

§(301)

12

6. *a metastable form stabilized by mixing*

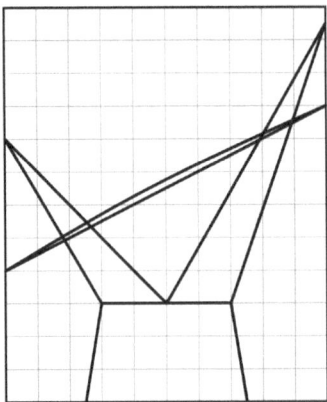

For each of two substances A and B two solid forms, α and β, are found, such that, for A as well as for B, the form β is metastable under all circumstances. In the (*TX*) figure the (calculated) (α + liquid) phase diagram, showing a eutectic *three-phase equilibrium*, and (β + liquid) phase diagram, the narrow loop, are superimposed.

- Construct the stable phase diagram, and observe that there is a stability field for mixed crystals of type β.
  Clue. Draw a sketch of the *GX* section for the *eutectic temperature* of the ($α_I$ + liquid + $α_{II}$) equilibrium.

7. *the equal-G curve and the region of demixing*

For each of the two situations, draw a sketch of the phase diagram that goes together with the given *binodal* and *equal-G curve.*

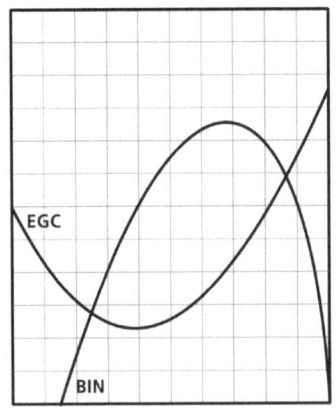
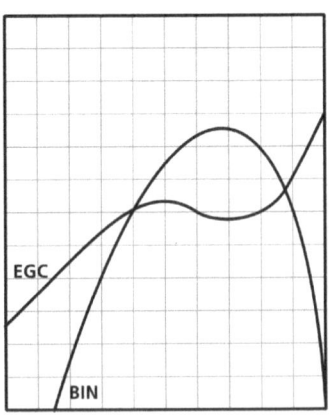

§(301)

8.   *crossed isodimorphism*

The two *TX* diagrams that are shown have two, each other crossing, ideal solid-liquid loops; the two (α + liquid) loops are the same; the two (β + liquid) loops are each other's mirror image.

- For each of the two situations, make a sketch drawing of the stable phase diagram.

NB. The loops were calculated with $\Delta S = 40 \ \mathrm{J\,K^{-1}mol^{-1}}$ for the α to liquid transition and $\Delta S = 80 \ \mathrm{J\,K^{-1}mol^{-1}}$ for the β to liquid transition.

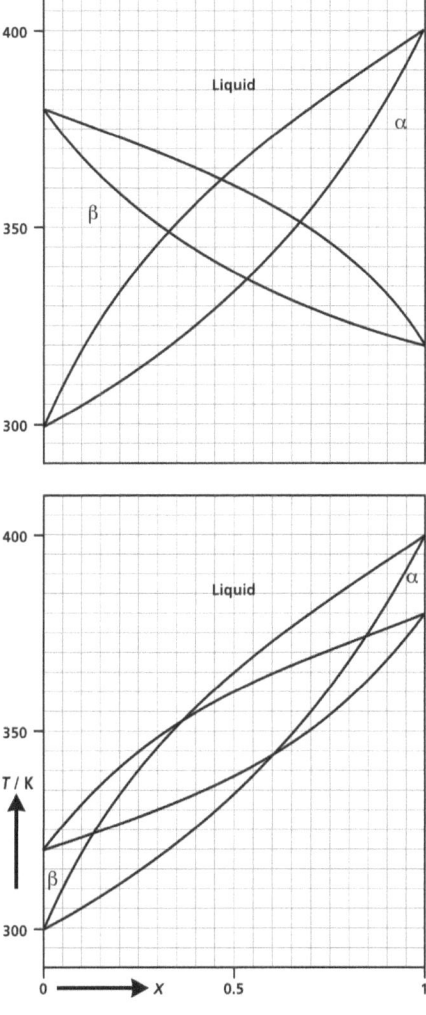

9. *activities and Gibbs-Duhem*

In terms of activities the molar Gibbs energy of a binary system is given by

$$G(T, P, X) = (1-X)G^*_A(T,P) + XG^*_B(T,P) + RT[(1-X)\ln a_A + X\ln a_B].$$

The activities a_A and a_B, shown in the diagram, which was found in a publication, originate from two independent series of experiments at constant T and P.

- Are the two activity curves mutually consistent? If not, draw a pair of improved curves; and also the corresponding GX curve. As a first step, on the vertical axes of the diagram printed, position the values 0 and 1, and in the figure indicate which curve is for a_A and which one for a_B.

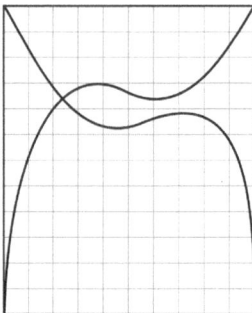

10. *statements about solubilities*

Is it true that "the solubility of a solid substance B in a liquid solvent S, with which is forms an ideal liquid mixture, is independent of the choice of S"?
For two solid forms α and β of a substance B that are sparingly soluble in a number of solvents, is it true that "the solubility of the *metastable form* is higher than that of the *stable form*, and that both solubilities are dependent on the choice of S, whereas the ratio of the two solubilities is not"?

11. *a different analysis of the region of demixing in AgCl + NaCl*

For the *thermodynamic analysis* of a (quasi) *symmetrical region of demixing* in terms of the model $G^E(T,X) = X(1-X)g_l(T) = X(1-X)(h_l - Ts_l)$, an alternative procedure is as follows. For a given data triplet $T'X'Y'$, the value of $g_l(T = T')$ can be calculated from

$$g_l(T=T') = [RT'/2(Y'-X')] \ln[Y'(1-X')/X'(1-Y')]. \tag{A}$$

Next, the various g_i values are plotted against temperature, in order to determine the values of h_i and s_i.

- Show that Equation (A) follows from the special condition (for regions of demixing that are symmetrical with respect to the equimolar composition) $\partial G'/\partial X = 0$, where $G' = RT\ \mathrm{LN}(X) + G^E(T,X)$.
- Carry out the analysis for AgCl + NaCl, using the data triplets in Table 1.

12.   *partial pressure versus concentration*

In the study of *homogeneous chemical equilibria* an important role is played by the *equilibrium constant* ($\leftarrow$007). Here we will first consider the *ammonia equilibrium*, the gas equilibrium between nitrogen, hydrogen, and ammonia

$$N_2 + 3H_2 = 2NH_3 \ . \tag{1}$$

In § 007 we applied the approximation of the ideal gas, and formulated the equilibrium constant (K) in terms of the *partial pressures* of the three substances:

$$K = (X_{NH3}\ P)^2 / (X_{N2}\ P)\ (X_{H2}\ P)^3\ , \tag{2}$$

in which the symbol X is for mole fraction, and the symbol P for the pressure indicated by the manometer connected to the equilibrium vessel. The equilibrium constant is related to the change in Gibbs energy, at standard pressure, of the chemical reaction involved in the equilibrium. The relation is

$$\ln K = -\frac{\Delta G^\circ(T)}{RT}\ . \tag{3}$$

The change of equilibrium constant with temperature, is given by the *Van 't Hoff equation*:

$$\frac{d\ln K}{dT} = \frac{\Delta H^\circ}{R\,T^2}\ , \tag{4}$$

in which ΔH° is the heat effect of the reaction.

The partial pressure of a substance is proportional to its number of moles per unit *volume*, its *concentration*. In the case of homogeneous equilibria in solution, such as the case of acid-base equilibria, the concept of partial pressure loses its significance, and the concept of concentration makes its appearance. The other way round, the concept of concentration is well applicable to equilibria in the gaseous state.

For the ammonia equilibrium the equilibrium constant in terms of concentration, K_c, is given by

$$K_c = [NH_3]^2 / [N_2] \cdot [H_2]^3 .\qquad(5)$$

Most of the times, the concentration is expressed in $mol\,dm^{-3}$.

- For of a pressure unit of 1 bar, write down the relationship between concentration and partial pressure.
- For the ammonia equilibrium, find out how the relationships, Equations (3) and (4), change when K is replaced by K_c.

13. *reminiscence of James Boswell* ($\leftarrow$Exc 12)

The gas equilibrium between sulphur dioxide, oxygen and sulphur trioxide

$$2SO_2 + O_2 = 2SO_3 ,$$

is the subject of this exercise. The equilibrium system is contained in a closed vessel that can be thermostated at any desired temperature between 800 K and 1200 K. The pressure exerted by the gas can be read from a manometer attached to the vessel. The chemical composition of the system and its thermodynamic properties are such that there is an equilibrium state for which the amounts of the three substances are equal and for which the manometer indicates a pressure of 3 bar.

For the three substances, the table gives, for the reference pressure of 1 bar, and for $T = 1000$ K, the *heat (enthalpy) of formation* and the *absolute entropy*, expressed in $J\,mol^{-1}$ and $J\,K^{-1}\,mol^{-1}$, respectively (Robie et al. 1978). In answering the questions, the change of the numerical values with temperature may be ignored.

Oxygen		243.58
Sulphur dioxide	− 362111	305.78
Sulphur trioxide	− 459748	334.83

- What is the temperature of the equilibrium state defined above? And for this state calculate the equilibrium constant K_c ($\leftarrow$Exc 12).

Starting from the defined equilibrium state, the temperature of the vessel and its contents are increased by 50 degrees, after which the system acquires a new equilibrium state.

- For the new equilibrium state calculate the relative amounts of the three substances and also the pressure indicated by the manometer. Apply the partial-pressure description.

We now change from *relative* to *absolute amounts*, and state that in the first equilibrium state the amounts of the three substances all are 1 mol.

- Calculate the net volume of the vessel.

One extra mole of oxygen is supplied to the system and, next, the temperature of the vessel and its contents are set to 1000 K.

- Applying the partial-pressure description, calculate the amounts of the three substances in the equilibrium state at 1000 K. To verify the result, carry out a second calculation using concentrations instead of partial pressures.

14. *vapour pressures over liquid mixtures of methanol and methyl acetate*

This exercise is about the derivation of the excess Gibbs energy function for a binary liquid mixture from a set of data pertaining to the isothermal equilibrium between liquid and vapour. The data which are shown in the table are for the system $\{(1-X)$ methanol $+ X$ methyl acetate$\}$ at $T = 318.15$ K (Figurski 1974; see also Figurski et al. 1997).

X_e^{liq}	X_e^{vap}	P/kPa	X_e^{liq}	X_e^{vap}	P/kPa	X_e^{liq}	X_e^{vap}	P/kPa
0.0000	0.0000	44.503	0.1715	0.374	61.326	0.6200	0.650	73.117
0.0392	0.138	49.708	0.2330	0.435	64.676	0.7060	0.695	73.210
0.0805	0.236	54.262	0.3810	0.534	69.761	0.9240	0.8799	69.570
0.1184	0.304	57.669	0.4830	0.585	71.845	1.0000	1.0000	65.639

The *idealized description* is adopted: the vapour phase is an ideal gas, and the excess Gibbs energy of the liquid mixtures is not dependent of the pressure exerted by the gas. The model for the excess function is

$$G^E(X) = X(1-X)\{g_1 + g_2(1-2X)\} \tag{1}$$

Taking the composition of the liquid phase as the independent variable (X), the vapour pressure and the composition of the vapour phase are given by the following equations ($\leftarrow$212):

§(301)

$$P(X) = (1 - X) P_A^o \exp(G_A^E / RT) + X P_B^o \exp(G_B^E / RT) \qquad (2)$$

$$X^{vap}(X) = [P(X)]^{-1} X P_B^o \exp(G_B^E / RT). \qquad (3)$$

The equal-G curve (EGC) is given by the equation

$$\ln P_{EGC}(X) = (1 - X)\ln P_A^o + X \ln P_B^o + \frac{G^{E\,liq}(X)}{RT}. \qquad (4)$$

In Equations (2)-(4) the P^o are for the equilibrium pressures of the pure components A and B. In Equations (2) and (3) the excess properties are the *partial excess Gibbs energies* of the two components.

In the following three tasks the significance of the result can be assessed by calculating the *pressure index* Δ_P (mean absolute difference between experimental pressure and pressure calculated by Equation (2)), and the *mole-fraction index* Δ_X (mean absolute difference between vapour composition and composition calculated with Equation (3)).

- Taking $g_2 = 0$, calculate the value of g_1 just by using the coordinates of the maximum in the phase diagram.

- From the experimental data, calculate the partial excess energies of the components by means of Equation (3), and its counterpart for component A. Combine the two partial properties to the integral excess Gibbs energy, and so for each of the ten data triplets. From the ten G^E values, derive the values of g_1 and g_2 (e.g. from a plot of G^E divided by $X(1-X)$ against X).

The *EGC fraction method*. From Equation (4) it follows that the complete excess Gibbs energy function (read: more than just one value from the maximum) can be derived from the full course of the EGC. Generally, X_{EGC}, the mole fraction value of the point of intersection of the G-curves, is not exactly half-way between X^{liq} and X^{vap}, but at the side of the phase whose G-curve has the smallest radius of curvature, in other words, the highest second derivative. In order to have a relation between EGC position and curvatures, we define the two properties f and Q. The property f gives the position of the EGC as a fraction:

$$f = (X_{EGC} - X^{liq}) / (X^{vap} - X^{liq}). \qquad (5)$$

The property Q is the quotient of the second derivatives of the G-functions of the phases at their equilibrium compositions:

$$Q = (\partial^2 G^{liq}/\partial X^2)_{X=Xliq} / (\partial^2 G^{vap}/\partial X^2)_{X=Xvap} \qquad (6)$$

§(301)

Experience has shown that good results are obtained by the relation (Oonk et al. 1971):

$$f = 0.5 - 0.2 \log_{10} Q. \tag{7}$$

- For each data triplet calculate the values of Q and f and from them the position of the EGC and next the value of G^E. From the set of excess values, calculate g_1 and g_2. NB. For the calculation of the second derivative of G^{liq}, include the g_1 value found in the first task.

15.   *a consistency test*

The circumstance that the two partial excess Gibbs energies, figuring in the second task of the preceding exercise, can be derived from the experimental data offers the opportunity to perform an indirect check on the internal consistency of the data set. The two partial quantities G^E_A and G^E_B originate from the same integral quantity and should therefore be mutually consistent. As a simple consequence the area under the curve of $(G^E_B - G^E_A)$ versus X, when taken over the whole mole-fraction range, should sum up to zero, because

$$\int_0^1 (G^E_B - G^E_A)\, dX = \int_0^1 (\partial G^E / \partial X)\, dX = G^E(X{=}1) - G^E(X{=}0) = 0 - 0 = 0.$$

- For the outcome of the second task in the preceding exercise (see the solutions part of this book), make a pot of $(G^E_B - G^E_A)$ versus X, and find out to what extent the area rule is satisfied. See also McGlashan (1963).

16.   *assessment of methodology*

The table printed below pertains to the idealized isothermal equilibrium, at $T = 318.5$ K, between liquid and vapour in the system $\{(1{-}X)\,A + X\,B\}$. The mole fraction of the liquid phase is exact (independent variable); the mole fraction of the vapour phase and the equilibrium vapour pressure (in hectopascal) are given in three significant figures; the equilibrium pressures of the pure components are exact.

X_e^{liq}	X_e^{vap}	P/hPa	X_e^{liq}	X_e^{vap}	P/hPa	X_e^{liq}	X_e^{vap}	P/hPa
0.000	0.000	450	0.400	0.538	697	0.875	0.824	700
0.050	0.166	514	0.500	0.586	712	0.950	0.917	675
0.125	0.313	584	0.600	0.634	721	1.000	1.000	650
0.200	0.403	631	0.700	0.688	722			

- To assess the precision of the three different methods, carry out the three tasks formulated in Exc 14.

- Is the origin of the lack of agreement between experimental and calculated equilibrium states in Exc 14, a matter of methodology, or rather a matter of incompatibility between the (idealized) thermodynamic description and the nature of the experimental data?

17. *dimolybdenum nitride*

The solid substance dimolybdenum nitride (Mo_2N) is used as a catalyst. Thermochemical data at standard pressure (1 bar) are available for temperatures between 298.15 and 1400 K (Barin 1989). For the two temperature limits, the *enthalpy of formation* and the *Gibbs energy of formation* are

T / K	$\Delta_f H^o / J\,mol^{-1}$	$\Delta_f G^o / J\,mol^{-1}$
298.15	-81588	-54810
1400	-61865	31143

From the data it follows that the substance, when heated under ambient pressure, will decompose, in the vicinity of 1000K, into molybdenum and nitrogen.

- Why is it that the substance will decompose in the vicinity of 1000 K?

- For the two temperatures determine the numerical values of $\Delta_f S^o$, the *entropy of formation*.

- From the changes in $\Delta_f H^o$ and in $\Delta_f S^o$, determine the values of the parameters a and b in in $\Delta_f C_p{}^o = a + bT$.

- Calculate the temperature at which the substance, in a cylinder-with-piston experiment ($\leftarrow$003) under 1 bar pressure, will decompose into molybdenum and nitrogen.

- To what pressure nitrogen should be added to the vessel at room temperature, in order to prevent Mo_2N from decomposing before 1400 K, when heated in a vessel-with-manometer ($\leftarrow$003).

18.   *the Helmholtz energy in its capacity of equilibrium arbiter*

This exercise is a follow-up of Exc 202:1, in which "$(1-\alpha)$ mole of ammonia
(NH$_3$), 0.5 α mole of nitrogen (N$_2$) and 1.5 α mole of hydrogen (H$_2$) are
present in a cylinder-with-piston under 1 bar external pressure and kept at a
temperature of 400 K. Under these conditions the Gibbs energy of formation
of NH$_3$ is equal to $-$ 5984 J·mol^{-1} (Robie 1978)". From a plot of the system's
Gibbs energy as a function of the *degree of dissociation* α, it could be derived
that the lowest Gibbs energy is obtained for $\alpha = 0.33$. In that state the system
has 1.33 mol of substance; accordingly its (ideal-gas) volume is 44.233 dm^3.
This time the system is kept in a vessel, provided with a manometer, and
having a constant *volume* of 44.436 dm^3.

- For values of α from 0.20 to 0.50 in steps of 0.05 calculate the systems
  Gibbs and Helmholtz energies and make plots of these quantities as a
  function of α. What are your observations?

19.   *the Wilson equation*

The *'Wilson equation'* (Wilson 1964) is an expression for the excess Gibbs
energy of a liquid mixture, $\{X_A$ mol A $+ X_B$ mol B$\}$, of two molecular
substances A and B (see Prausnitz et al. 1986):

$$G^E / RT = - X_A \ln (X_A + \Lambda_{AB} X_B) - X_B \ln (X_B + \Lambda_{BA} X_A) . \qquad (1)$$

In Wilson's theory the constants Λ_{AB} and Λ_{BA} are given by

$$\Lambda_{AB} = (V^*_B / V^*_A) \exp \{- (\lambda_{AB} - \lambda_{AA}) / RT \}; \text{ and}$$
$$\Lambda_{BA} = (V^*_A / V^*_B) \exp \{- (\lambda_{AB} - \lambda_{BB}) / RT \},$$

where V^* is for *molar volume* and λ for *interaction energy.*

- Derive the expressions for the *excess chemical potentials* (*excess
  partial Gibbs energies*) of A and B.

- Demonstrate that the condition $(\partial \mu^E_A / \partial X_B)_{X_B \to 0} = 0$ is satisfied
  ($\leftarrow$207).

- Is the Wilson equation applicable to systems showing limited liquid
  miscibility? NB. This is typically a second-derivative problem.

In Ohe's compilation of vapour-liquid equilibrium data (Ohe 1989)
experimental phase diagrams are shown along with tables of smoothed data

(pressure or temperature and compositions of the two phases); the *Antoine constants* (←110 Exc 5) of the components; and the two '*Wilson parameters*' (given as $\Lambda 12$ and $\Lambda 21$). The Wilson equation, for that matter, is one of the tools of the optimization procedure. Realizing that the aim of the computations is not to assess the excess Gibbs energy, the question arises to what extent the computed $\Lambda 12$ and $\Lambda 21$, when introduced in equation (1) will yield a reliable set of G^E values.

The isothermal system $\{(1- X)$ mol water (1) $+ X$ mol methanol (2)$\}$ at $T =$ 308.15 K, measured by McGlashan and Williamson (1976), may serve as a representative example. For the three compositions $X = 0.25$; $X = 0.50$; and $X = 0.75$, McGlashan's G^E values, expressed in J mol^{-1}, are: 333; 350; and 225, respectively. Ohe's Wilson parameters for the liquid state of the system are $\Lambda 12 = 0.92588$ and $\Lambda 21 = 0.60703$.

- Observe that the agreement between McGlashan's values and those calculated with Ohe's parameters is qualitative. Next, by trial and error, fit the values of Λ_{AB} and Λ_{BA} to McGlashan's data.

20. *Marius Ramirez's view from the arc*

The equation
$$\ln (P / P_c) = (a\tau + b\tau^{1.5} + c\tau^{2.5} + d\tau^{5}) T / T_c,$$
which has been proposed by Wagner (Wagner 1973), is a powerful expression for the representation of *vapour-pressure curves* (liquid + vapour equilibrium). The P_c and T_c are the critical pressure and temperature; the variable τ is defined as $\tau = 1 - (T/T_c)$; and a, b, c, and d are substance-dependent constants. Ambrose and Walton (Ambrose and Walton 1989) applied the equation to the vapour pressures of n-alkanes and 1-alkanols – up to n-eicosane ($C_{20}H_{42}$) and 1-eicosanol. As an example, for hexadecane ($P_c =$ 1.435 MPa; and $T_c = 722$ K) the values of the constants were calculated as $a = -10.03664$; $b = 3.41426$; $c = -6.6827$; and $d = -4.863$.

The equation has been used by M. Ramirez (Ramirez 2002) to construct artificial data sets for 1-alkanols; valid for temperatures close to their melting points. The artificial data were combined with available experimental data, with the aim of calculating the *enthalpy of vaporization* at the melting point. Subsequently, the enthalpies of vaporization were combined with the enthalpies of melting to yield the enthalpies of sublimation - the latter serving as a key to know the *lattice energies* of the substances (see e.g. Ewig et al. 1999).

For the assessment of the vapour-pressure data Ramirez made use of the *arc representation* or '$\ln f$ representation' (←110). Instead of making a plot of $\ln P$ against $1/T$, a plot is made of $\ln f$ against $1/T$. The $\ln f$ property is given by $\ln f = \ln P - \alpha + \beta/T$; and its parameters α and β are taken such that $\ln f$ is zero for each of the two extreme data pairs of a given set of data. The $\ln f$ representation of a high quality set of data has the appearance of a rainbow. See also p.30.

§(301)

The concave nature reflects the general fact that for liquid+vapour equilibria the heat-capacity difference is negative: heats of vaporization decrease with increasing temperature. The equation for the slope of the $\ln f$ function is

$$d\,(\ln f)\,/\,d(1/\,T) = -\Delta H_B^o\,/\,R + \beta \quad,$$

where the subscript B refers to a substance in general. For the temperature at the top of the *arc* the heat of vaporization is equal to the product of gas constant R and parameter β. The difference in heat capacity between vapour and liquid can be calculated from the characteristics of the arc: its height h, base length b, and the reciprocal temperature at the top $1/T_{max}$:

$$\Delta C_{P_B}^o = -8R\{T_{max}\cdot b\}^{-2}\cdot h \quad.$$

- For hexadecane, create a set of T,P data pairs, for $295 \leq T/K \leq 340$ and in steps of 5 K; i.e. for a temperature range just above the melting point of the substance. Next, construct the arc representation of the data.
- From the characteristics of the arc calculate the difference in heat capacity between vapour and liquid. Calculate the heat of vaporization for $T = 298.15$ K.

21. *the Rackett equation*

In Lide and Kehiaian's handbook (Lide and Kehiaian (1994) a table is given for liquid *molar volume* and *saturated density* at 298.15 K, along with the values of the substance-dependent constants A_1 and A_2 of the so-called Rackett equation (Rackett 1970). "Saturated densities are important input parameters to process design calculations and in particular the saturated liquid density" (Spencer and Adler 1978). The empirical Rackett equation allows the calculation of molar volume as a function of temperature. The equation is $V(T) = A_2\,A_1^u$, where $u = [1 + (1 - T/T_c)^{2/7}]$, and T_c the *critical temperature*.
The constant A_2 has a value which is close to $RT_c\,/\,P_c$, where P_c is the *critical pressure*.
According to Lide and Kehiaian, the equation, moreover, permits one to calculate an 'approximate value' of the *cubic expansion coefficient α*.

- Derive the expression for the calculation of the cubic expansion coefficient meant above. NB. $da^u\,/\,dX = a^u\,(\ln a)\,du\,/\,dX$. Why is it that the expression yields an *approximate* value?
- For benzene at 298.15 K, calculate the numerical values of molar volume and cubic expansion coefficient. What is about the value of benzene's critical pressure?

For benzene, C_6H_6, the data are $T_c = 562.15$ K; $A_1 = 0.26967$; $A_2 = 0.95325$ x 10 m^3·mol^{-1}, the range of validity being 273 K – 558 K.

22. *the Simon equation*

The Simon equation (Simon and Glatzel 1929) is an empirical relationship, just like the Rackett equation (Exc 21) and the relationship proposed by Wagner (Exc 20). This time the relationship is between the pressure (P) exerted on a substance and the substance's melting point (T_m):

$$P - P_o = A \left[(T_m / T_o)^b - 1 \right] .$$

where A and b are system-dependent constants, adjustable parameters. The P_o and T_o are the coordinates of a reference point on the *melting curve*, such as the *normal melting point*, or a *triple point* (solid+liquid+vapour, or solid1+liquid+solid2; see e.g. Richter and Pistorius 1973, for acetone with its two triple points).

The data in the table below, which pertain to the substance potassium, were measured by Bridgman (1914). The data are taken from a table in a publication by Babb (1963), according to whom "Bridgman always smoothed his values before publication". The constants of the Simon equation given by Babb are $A = 4266$ bar, and $b = 4.44$.

T / K	P / bar	T / K	P / bar	T / K	P / bar	T / K	P / bar
335.7	0	365.6	1982	408.6	5948	440.2	9913
351.9	991	377.9	2974	425.7	7930	452.8	11985

- To appreciate the power of the Simon equation, calculate, for the temperatures in the table, the pressures generated by the equation and the constants given. Calculate the *pressure index* Δ_P, which is the mean of the absolute difference between experimental and calculated pressures. You could also think of making an arc representation (cf. Exc 20), in terms of the property $\Pi = P + \alpha + \beta / T$, plotted against temperature.

- In the footsteps of Simon and Glatzel, solve the constants A and b from the two equations generated by the two data pairs (408.6 K; 5948 bar) and (452.8 K; 11985 bar) in combination with ($T_o = 335.7$ K; $P_o = 0$ bar). With the two constants calculated, all other things being the same, repeat the first task.

- For sodium molybdate, Na_2MoO_4, which was studied by Pistorius (1966), the constants of the Simon equation are $A = 11.4$ kbar and $b = 3.52$. The normal melting point of the substance is 961 K; at this temperature the heat of melting is 24.5 kJ mol^{-1}. Calculate the *change in molar volume* on melting at 961 K.

23. *an exploratory calculation*

The data in the table, *entropy, volume,* and *Gibbs energy of formation from the elements,* pertain to the solid forms of aluminium silicate (Al_2SiO_5), and are valid for $T = 298.15$ K and $P = 0$ GPa (Barin 1989). These data, which, most certainly, fail to make an accurate predicion of the *PT* phase *diagram,* lend themselves, however, to making an exploratory calculation (cf. Exc 004:7).

- For each of the three *forms,* determine the numerical values of the constants *a, b,* and *c* of the Gibbs energy equation $G(T,P) = a + b(T - 298.15$ K$) + cP$.

form	entropy	volume	Gibbs energy
	$J \cdot K^{-1} \cdot mol^{-1}$	$J \cdot GPa^{-1} \cdot mol^{-1}$	$J \cdot mol^{-1}$
sillimanite (I)	96.090	49490	− 2441070
andalusite (II)	93.776	51530	− 2442890
kyanite (III)	84.467	44090	− 2443881

- Set up the equations for the three *two-phase equilibrium lines,* and calculate the coordinates of the *triple point.*
- Construct the phase diagram, for $300 \leq (T / K - 273.15) \leq 800$ and $0 \leq (P / GPa) \leq 1.0$; and compare your result with the diagram in Tonkov's (1992) compilation, read, the outcome of Exc 004:7.

NB. See also Althaus (1969) ; Richardson et al. (1969)

24. *fluid carbon dioxide*

The table (Angus et al. 1976) pertains to fluid carbon dioxide, for which it gives, for a number of selected pressures, the values of *molar volume* at $T = 500$ K .

P / bar	V/cm^3·mol^{-1}	P / bar	V/cm^3·mol^{-1}	P / bar	V/cm^3·mol^{-1}
1	41536	10	4127.0	100	389.14
2	20753	20	2048.9	200	187.16
5	8283.4	50	802.79		

- By *graphical integration* of volume against pressure, or rather $P \times V$ against $\log P$, determine the numerical value of the change in Gibbs energy $\Delta G = G$ (500 K, 200 bar) $- G$ (500 K, 1 bar). What is the value of f (for *fugacity* ←Exc 108:6), if one wants to represent the result by $\Delta G = RT \ln f$?

25. *compounds in the role of component*

The substances considered here are the *ternary compounds* A_2BC and AB_2C; the binary compound A_3B; and two different binary compounds AB. The ternary compounds appear in the first, and the binary compounds in the second and third tasks. In each of the tasks the compounds are made the *components* of a binary system; in these systems they do not mix when solid; and on melting they dissociate fully, to yield ideal liquid mixtures of the entities A, and B, (and C, in the case of the ternary ones).

In each case, the task comes down to setting up the equations for the two liquidi. The most convenient approach is to make use of equilibrium constants, substituted in the integrated form of Equation (4) in Exc 12:

$$R \ln (K / K_o) = - \Delta H^\circ \left[(1 / T) - (1 / T_o) \right] ,$$

where ΔH° is the heat of melting of the pure component of the system and T_o its melting point.

The system of the first task is defined as $\{(1 - X)$ mole of $A_2BC + X$ mole of $AB_2C\}$ and such that one mole is the amount of substance which yields, on melting, the sum of four mole of particles A, B, and C. The components have the same melting point, which is 1500 K; and also the same heat of melting, which is 150 kJ·mol^{-1}.

- For the equilibrium between solid A_2BC and liquid mixture, derive the formula for the relation between temperature and composition variable X, i.e. the *liquidus equation*. Give the coordinates of the *eutectic point*; and make a sketch of the *TX* solid-liquid phase diagram.

The system of the second task is $\{(1 - X)$ mole of $A_3B + X$ mole of AB$\}$. The component A_3B melts at 1500 K with a heat effect of 150 kJ·mol^{-1}; for component AB these values are 1475 K and 75 kJ·mol^{-1}, respectively.

- Set up the equations for the two liquidi; calculate the coordinates of the eutectic point; and make a sketch of the phase diagram.

The system for the third task is the same as the one for the second task, with the distinction, however, that AB's melting point is changed to 1400 K. This time a liquid mixture composed of 0.5 mol A and 0.5 mol B is cooled down from 1600 K to 1300 K. The melting points of the substances A and B are below 1300 K.

§(301)

    •   What is the temperature at which crystals first form; and at which temperature does the last liquid disappear?

26.   *Professor Geus's nickel coin*

In 1869 A. Horstmann published his influential paper describing the *heterogeneous chemical equilibrium* between solid NH_4Cl and the equimolar gaseous mixture of NH_3 and HCl produced by the solid (←007). Owing to the fact that the composition of the gaseous phase is fixed, the equilibrium is monovariant. The *system formulation* (←003), accordingly, is given by

$$f = M\,[T, P] - N\,[\,\mu_{NH4Cl} = \mu_{NH3} + \mu_{HCl}\,] = 2 - 1 = 1\,.$$

The table gives a set of equilibrium pressures for Celsius temperatures at intervals of 20 degrees (←007).

$t\,/\,°C$	$P\,/\,$Torr	$t\,/\,°C$	$P\,/\,$Torr	$t\,/\,°C$	$P\,/\,$Torr
240	25	280	122	320	435
260	58	300	235	340	736

Let us now consider the following case. In a vessel provided with a manometer an amount of salammoniac (NH_4Cl) is introduced, along with an amount of nickel – a catalyzer for the decomposition of ammonia. In this situation the 'double equilibrium'

$$NH_4Cl\ (solid) = \{(NH_3 = 0.5\ N_2 + 1.5\ H_2) + HCl\}\ (gas)$$

is realized, in which the fraction α (*degree of dissociation*) of NH_3 has dissociated into N_2 and H_2. The temperature of the vessel and its contents is set at 600 K.

At the temperature of 600 K the Gibbs energies of formation (from Barin 1989) are: for NH_4Cl −92193; for NH_3 +15834; and for HCl −97975; all expressed in $J\,mol^{-1}$ and for the standard pressure of 1 bar. In what follows, ideal-gas behaviour may be assumed.

    •   First demonstrate that Horstmann's data are in line with the expectation that NH_3 did not dissociate.

    •   By writing down its system formulation, demonstrate that the double equilibrium is monovariant (i.e. everything is fixed by the choice of temperature).

    •   Calculate, for $T = 600$ K the numerical values of *i)* the mole fractions of the components in the gas phase; and *ii)* the pressure indicated by the manometer.

§(301)

27. *montroydite and dephlogisticated air*

When heated, the mineral montroydite, or red oxide of mercury = mercuric oxide (HgO), decomposes into mercury and oxygen:

$$HgO \rightarrow Hg + 0.5\,O_2 \,.$$

And because of this fact the substance played an important part in the discovery of the element of oxygen (see e.g. Elmsley 2001). Historical heating experiments with mercuric oxide were carried out between 1771 and 1773 by Carl Wilhelm Scheele (1742 - 1786) and in 1774 by Joseph Priestley (1733 - 1804). Priestley, who spoke of '*dephlogisticated air*', wrote: "But what surprised me more than I can well express was that a candle burned in this air with a remarkable brilliant flame."

$T\,/\,K$	$\Delta_f H^\circ$	$\Delta_f G^\circ$	$T\,/\,K$	$\Delta_f H^\circ$	$\Delta_f G^\circ$
298.15	−90798	−58528	700	−146260	−9550
400	−90388	−47558	800	−144338	9853
500	−89633	−36928	900	−142301	29009
600	−88647	−26478	1000	−140170	47926

The table with HgO's formation properties (from Robie et al. 1978; for the standard pressure of 1 bar, and expressed in $J\,mol^{-1}$) reveals that these properties undergo changes between 600 and 700 K. These changes are related to the transition in mercury, from liquid to vapour.

- From the data, derive mercury's 1 bar *boiling point*; and also its *heat of vaporization*.

The decomposition of mercuric oxide can be beautifully demonstrated in a test tube: the gaseous mercury which is formed condenses to a cylindrical mirror in the cold upper part of the tube; and the released oxygen can be detected with a burning candle.

- From the data in the table, estimate the temperature at which the decomposition takes place.

In order to study the equilibrium between solid mercuric oxide, liquid mercury, and a gaseous mixture of mercury and oxygen, an amount of mercuric oxide and an amount of mercury are introduced in a vessel. The vessel, which is provided with a manometer, can be set at any temperature between room temperature and 800 K.

- By writing down the $f = M - N$ *system formulation* ($\leftarrow$Exc 26) demonstrate that the system is monovariant.
- Ignoring the difference in heat capacity between gaseous and liquid mercury, and assuming ideal-gas behaviour, calculate for $T = 700$ K the pressure indicated by the manometer, and also oxygen's mole fraction in the vapour phase.

§(301)

28.   *the adiabatic bulk modulus*

The *adiabatic bulk modulus* is defined as

$$K_S = - V \, (\partial P / \partial V)_S:$$

it is related to the change of volume under adiabatic conditions; read at constant entropy.

- Making use of the difference between C_p and C_V ($\leftarrow$Exc 107:10), which is given by

$$C_p - C_V = \alpha^2 K \, V \, T,$$

  show that the adiabatic bulk modulus is related to the *isothermal bulk modulus* as

$$K_S = K \, (C_P / C_V).$$

from §(110)

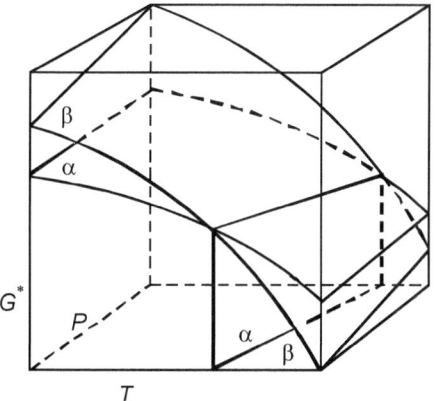

G^*PT surfaces for two forms, α and β, of a pure substance

§(301)

Vapour pressures over liquid water in the range from 10 to 40 °C

from §(110)

T/K	$P/Torr$
283.15	9.209
288.15	12.788
293.15	17.535
298.15	23.756
303.15	31.824
308.15	42.175
313.15	55.324

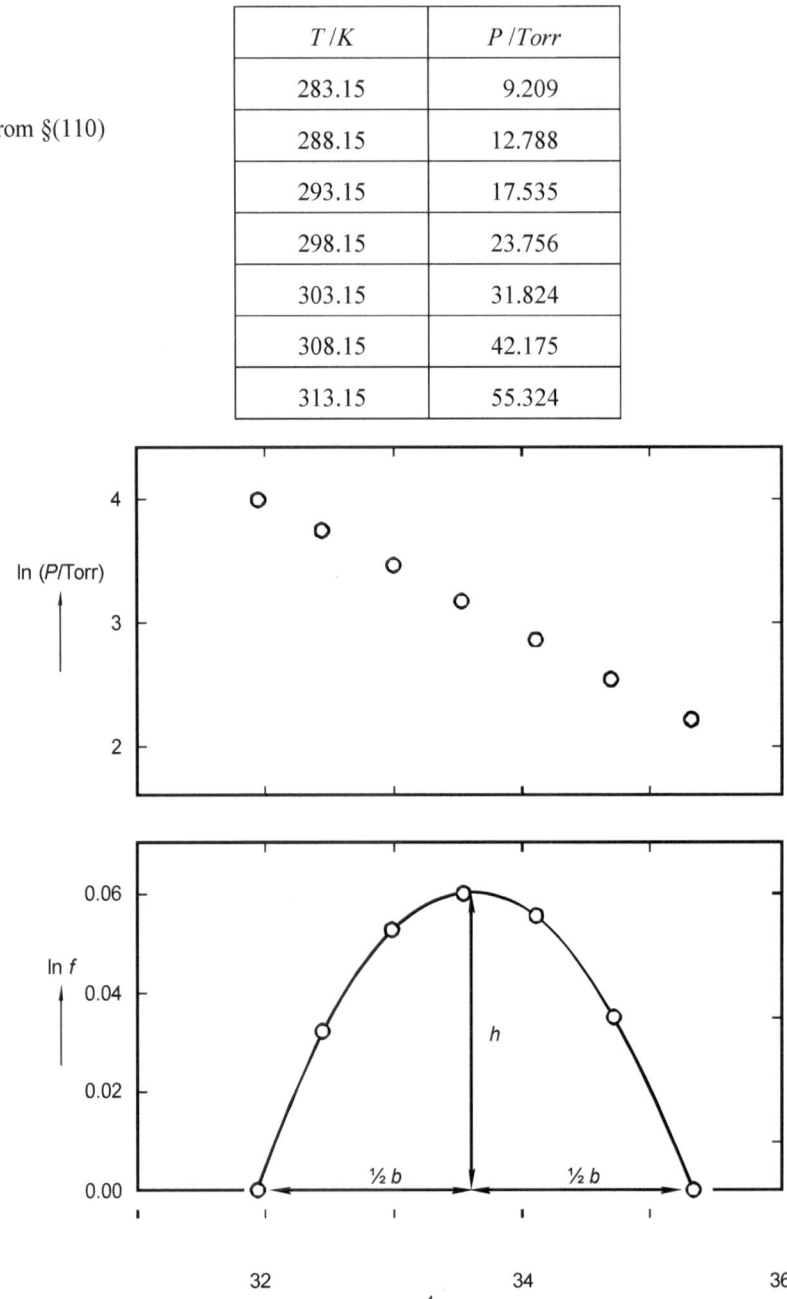

Two graphical representations of the vapour-pressure data in Table 1. Top: the traditional Clausius-Clapeyron plot. Bottom: the arc representation

§(301)

§ 302 SLOPES OF CURVES IN PHASE DIAGRAMS

The derivation of the Clapeyron equation, for the slope of an equilibrium curve in a pure substance's PT phase diagram, serves as a model for the derivation of general equations for the slopes of curves in TX and PX phase diagrams of binary systems.

---

a derived system formulation

The equilibrium between two forms, α and β, of a pure substance B satisfies the *equilibrium condition* $\Delta_\alpha^\beta G_B^* = 0$. It implies that the *equilibrium states* are the solution of the equation $\Delta_\alpha^\beta G_B^* = 0$. The solution is represented by a curve in the *PT* plane (see figure on p 29).

And owing to the fact that along the curve the condition remains satisfied, it is obvious that the change of $\Delta_\alpha^\beta G_B^*$ along the curve is also zero: $d(\Delta_\alpha^\beta G_B^*) = 0$. This observation implies that the effects of the changes in *T* and *P* compensate one another:

$$d(\Delta_\alpha^\beta G_B^*) = \left(\partial \Delta_\alpha^\beta G_B^* / \partial T\right)_P dT + \left(\partial \Delta_\alpha^\beta G_B^* / \partial P\right)_T dP$$
$$= -\Delta_\alpha^\beta S_B^* dT + \Delta_\alpha^\beta V_B^* dP = 0 \tag{1}$$

As a result, there is a relation between d*T* and d*P*:

$$dP / dT = \Delta_\alpha^\beta S_B^* / \Delta_\alpha^\beta V_B^*, \tag{2}$$

and this relation is the equation named after Clapeyron (←110).

The first part of the foregoing remarks finds expression in the *system formulation* (Oonk and Calvet 2008)

$$f = M[T,P] - N\left[\Delta_\alpha^\beta G_B^* = 0\right] = 2 - 1 = 1 . \tag{3}$$

In Equation (3) the symbol M is for the *set of variables,* and the symbol N for the *set of conditions.*

For the second part of the foregoing remarks, and for the purpose of this section, we introduce a *'derived system formulation'*, which is

$$f' = M'[dP / dT] - N'\left[d\left(\Delta_\alpha^\beta G_B^*\right) = 0\right] = 1 - 1 = 0 . \tag{4}$$

The first formulation, Equation (3), is a case of one equation and two unknowns. In the language of phase equilibria there is one *degree of freedom*: the investigator has the freedom to set the pressure (temperature) and, after that, the equilibrium temperature (pressure) is adjusted by the system.

Mathematically speaking, the equilibrium condition makes that one of the variables becomes a dependent one: the necessity to introduce a value for one of the variables, in order to find the other. The derived formulation, Equation (4), on the other hand, is just the case of one equation and one unknown.

For the case of isobaric equilibrium between two mixed forms α and β in the binary {(1 − X) mole of A + X mole of B} the two formulations are

$$f = M\left[T, X^{\alpha}, X^{\beta}\right] - N\left[\Delta_e \mu_A = 0; \Delta_e \mu_B = 0\right] = 3 - 2 = 1, \tag{5}$$

and

$$f' = M'\left[dX^{\alpha}/dT, dX^{\beta}/dT\right] - N'\left[d(\Delta_e \mu_A) = 0; d(\Delta_e \mu_B) = 0\right] = 2 - 2 = 0. \tag{6}$$

The operator Δ_e has the function of indicating that a difference is taken between the value of a quantity in β and the value of the same quantity in α, and such that α and β have taken their equilibrium compositions. In the following the operator Δ without the subscript 'e' is used for a difference property such that α and β have the same composition.

The solution of the derived set of equations consists of two expressions for the *slopes of the two equilibrium curves* in the *TX* phase diagram.

Of special practical interest is the limiting situation where $X^{\alpha} \rightarrow 1$, and $X^{\beta} \rightarrow 0$, and the substances A and B get the status of *solvent* and *solute*, respectively. In that situation, the chemical potentials of A lose their excess parts, and $RT \ln (1 - X)$ simply becomes $- RT X$ ($\leftarrow$ 207). And because the difference in A's chemical potentials is taken, an expression is obtained for the difference between the slopes of the equilibrium curves at $X = 0$. For this limiting case the derived formulation reads

$$f' = M'\left[\left(dX^{\beta}/dT - dX^{\alpha}/dT\right)_{X \rightarrow 0}\right] - N'\left[d(\Delta_e \mu_A) = 0\right] = 1 - 1 = 0. \tag{7}$$

derivation of the van der Waals equations

Figure 1 is a part of a *TX* phase diagram representing the equilibrium between the two mixed states α and β, corresponding to the formulations represented by Equations (5) and (6).

The two equilibrium conditions in the set of conditions N in Equation (5) can also be formulated as

$$N\left[\mu_A^{\alpha} = \mu_A^{\beta}; \ \mu_B^{\alpha} = \mu_B^{\beta}\right]. \tag{8}$$

Geometrically, and for a given *T*, the two conditions are expressed by the *common tangent* drawn to the Gibbs functions of the two forms α and β, such as is shown in Figure 2.

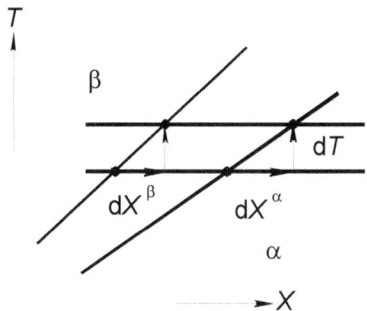

FIG. 1.

Part of TX phase diagram

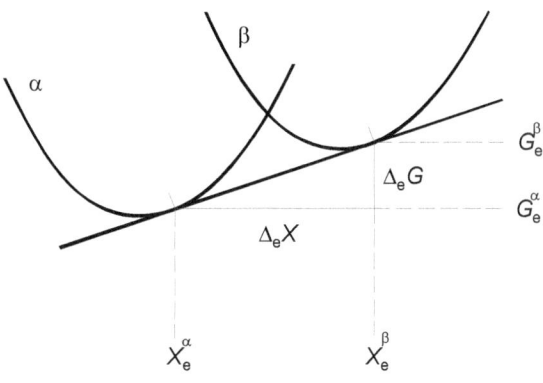

FIG. 2.

Towards a 'more mathematical' formulation of the two conditions for the common tangent

In a purely mathematical way, the two conditions for the common tangent line are

$$N \left[\left(\frac{\partial G^{\alpha}}{\partial X} \right) = \left(\frac{\partial G^{\beta}}{\partial X} \right) = \frac{\Delta_e G}{\Delta_e X} \right] \quad . \tag{9}$$

And they give rise to the following two conditions (two signs of equality) in the *derived set of conditions* N' in Equation (6):

$$d(\partial G^{\alpha}/\partial X) = d(\partial G^{\beta}/\partial X) = d(\Delta_e G/\Delta_e X) \quad . \tag{10}$$

§(302)

In terms of the variables in the set M of variables, Equation (5), the *total differential* of $(\partial G^\alpha/\partial X)$ is

$$d\,(\partial G^\alpha / \partial X) = (\partial / \partial X^\alpha)\,(\partial G^\alpha / \partial X)\,dX^\alpha + (\partial / \partial X^\beta)\,(\partial G^\alpha / \partial X)\,dX^\beta \\ + (\partial / \partial T)\,(\partial G^\alpha / \partial X)\,dT, \tag{11}$$

in which

i) $(\partial / \partial X^\alpha)\,(\partial G^\alpha / \partial X) = \partial^2 G^\alpha / \partial X^2$;

ii) $(\partial / \partial X^\beta)\,(\partial G^\alpha / \partial X)$ vanishes, because the Gibbs energy of form β is not a function of the composition of any other form;

iii) $(\partial / \partial T)\,(\partial G^\alpha / \partial X) = (\partial / \partial X)\,(\partial G^\alpha / \partial T) = -\,(\partial S^\alpha / \partial X)$,

so that

$$d\,(\partial G^\alpha / \partial X) = (\partial^2 G^\alpha / \partial X^2)\,dX^\alpha - (\partial S^\alpha / \partial X)\,dT\;. \tag{12}$$

By analogy, for the total differential of $\partial G^\beta / \partial X$,

$$d\,(\partial G^\beta / \partial X) = (\partial^2 G^\beta / \partial X^2)\,dX^\beta - (\partial S^\beta / \partial X)\,dT\;. \tag{13}$$

As regards $\Delta_e G/\Delta_e X$, both its derivatives with respect to X^α and X^β are zero, because of the mathematical nature of the common tangent:
$(\Delta_e G / \Delta_e X)_{X^\beta + dX^\beta} = (\Delta_e G / \Delta_e X)_{X^\beta}$, see Figure 3.

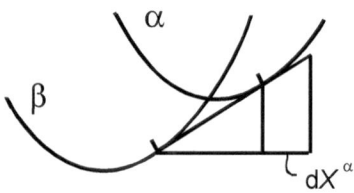

FIG. 3.

As a result, for the (total) differential of $\Delta_e G/\Delta_e X$:

$$d\,(\Delta_e G/\Delta_e X) = (\partial / \partial T)\,(\Delta_e G/\Delta_e X)\,dT = -(\Delta_e S/\Delta_e X)\,dT\;. \tag{14}$$

§(302)

By combination of Equations (13) and (12) with Equation (14) the so-called *Van der Waals relations*, Equations (15 a) and (15 b), are obtained:

$$\frac{dX^\alpha}{dT} = \frac{\partial S^\alpha / \partial X - \Delta_e S / \Delta_e X}{\partial^2 G^\alpha / \partial X^2} \; ; \tag{15a}$$

$$\frac{dX^\beta}{dT} = \frac{\partial S^\beta / \partial X - \Delta_e S / \Delta_e X}{\partial^2 G^\beta / \partial X^2} \; . \tag{15b}$$

The entropy properties in Equations (15) are *integral molar quantities*, as are the Gibbs energies in Equations (9) and (10).

Starting the derivation from the equilibrium conditions in terms of *chemical potentials*, which are *partial molar Gibbs energies*, a result is obtained having partial entropies (←201-203):

$$\frac{dX^\alpha}{dT} = -\frac{\left(1 - X_e^\beta\right)\Delta_e S_A + X_e^\beta \Delta_e S_B}{\Delta_e X \left(\partial^2 G^\alpha / \partial X^2\right)} \; ; \tag{16a}$$

$$\frac{dX^\beta}{dT} = -\frac{\left(1 - X_e^\alpha\right)\Delta_e S_A + X_e^\beta \Delta_e S_B}{\Delta_e X \left(\partial^2 G^\beta / \partial X^2\right)} \; ; \tag{16b}$$

Note that the equilibrium mole fractions in the equation for dX^α/dT are for the β phase, and that, inversely, X_e^α appears in the equation for dX_e^β / dT.

The isothermal counterparts of the Equations (15) and (16) follow from the latter in a straightforward manner (as always in such cases: when T is replaced by P, minus S has to be replaced by V):

$$\frac{dX^\alpha}{dP} = -\frac{\partial V^\alpha / \partial X - \Delta_e V / \Delta_e X}{\partial^2 G^\alpha / \partial X^2} = -\frac{\left(1 - X_e^\beta\right)\Delta_e V_A + X_e^\beta \Delta_e V_B}{\Delta_e X \cdot \partial^2 G^\alpha / \partial X^2} \tag{17a}$$

$$\frac{dX^\beta}{dP} = -\frac{\partial V^\beta / \partial X - \Delta_e V / \Delta_e X}{\partial^2 G^\beta / \partial X^2} = -\frac{\left(1 - X_e^\alpha\right)\Delta_e V_A + X_e^\alpha \Delta_e V_B}{\Delta_e X \cdot \partial^2 G^\beta / \partial X^2} \tag{17b}$$

The quantities operating in the Equations (15) and (16) are shown in Figure 4. In the graphical representation $\partial S^\alpha / \partial X$ and $\Delta_e S/\Delta_e X$ are given by DC/AD and BD/AD; their difference, which is minus BC/AD, is equal to minus $\left\{ (1 - X_e^\beta)\, \Delta_e S_A + X_e^\beta \Delta_e S_B \right\}/\Delta_e X.$

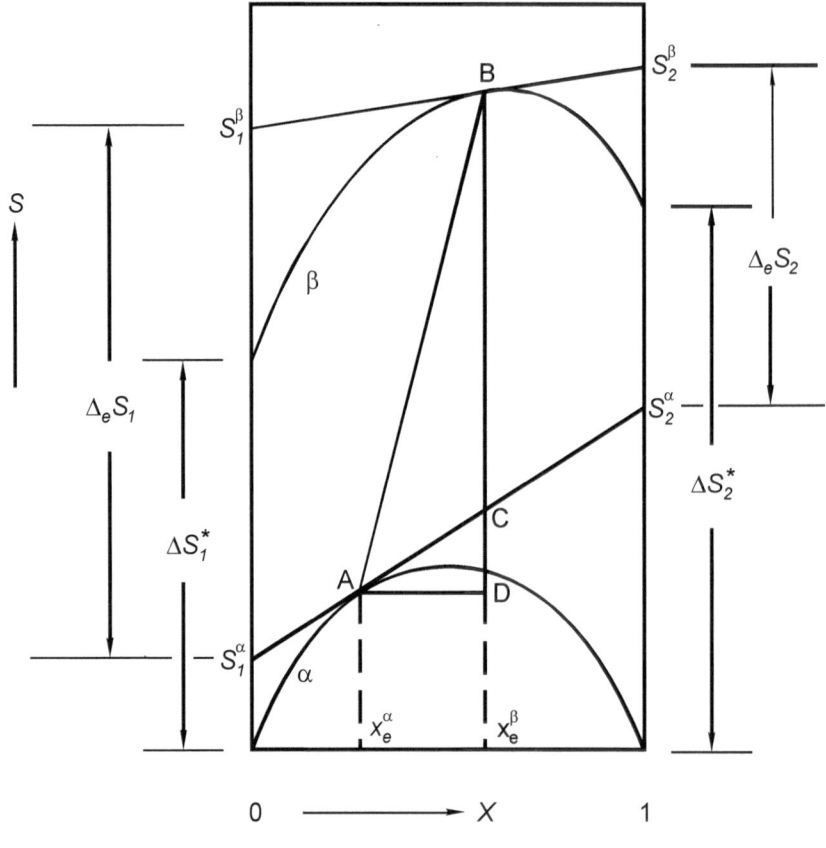

FIG. 4.

The quantities operating in the isobaric Van der Waals equations are shown in this entropy versus mole fraction diagram. NB. In this figure (from Oonk 1981) the subscripts 1 and 2 refer to the components A and B, respectively

corollary

The Van der Waals equations, isobaric and isothermal, are free from approximations, and for that reason they have a general validity. In a simple and unambiguous manner the equations clarify a series of principal phase diagram characteristics.

One of the consequences of the principle of minimal Gibbs energy is that, at the equilibrium compositions, the Gibbs functions must be convex. In other words, the second derivative with respect to mole fraction has to satisfy the condition.

$$\partial^2 G / \partial X^2 \geq 0. \tag{18}$$

The sign of equality in this condition corresponds to *critical mixing.*

In the vicinity of a *critical point* the second derivative has a small value. As a consequence dT/dX is small: the equilibrium curve will have a rather flat part with a *point of inflexion* near the centre of it (→Exc 6, for an arithmetical example). The other way round, a point of inflexion is an indication for the proximity of a *(metastable) region of demixing* (not always, however →Exc 7).

A speaking example of a system with a metastable region of demixing is the combination of copper and cobalt, see Figure 5. The top of the ROD in *supercooled liquid mixtures* was determined by Nakagawa (1958). Supercooled liquid mixtures were kept, for a few minutes, at a fixed temperature and then quenched to room temperature. The quenched samples were analyzed, and were either homogeneous (T, X conditions outside ROD) or consisted of two layers of different compositions (T, X conditions inside ROD).

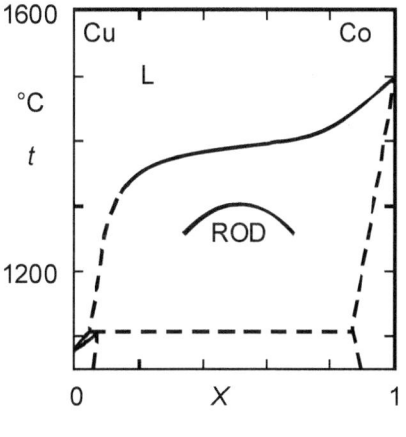

FIG. 5.

Region of demixing (ROD) in the supercooled liquid state of the system copper + cobalt

From the Equations (16a and b) it follows that whenever the phases α and β have equal equilibrium compositions ($\Delta_e X = 0$), both dT^α/dX and dT^β/dX will be zero: the phase diagram will show a (so-called) *stationary point*. That point is either a maximum or a minimum. Theoretically, there is the possibility of a third type of stationary point, which is the *horizontal point of inflexion*. Examples of phase diagrams with a maximum or a minimum are numerous. No example is known of a phase diagram with a point of horizontal inflexion (→Exc 7), i.e. a diagram as shown by Figure 6.

38

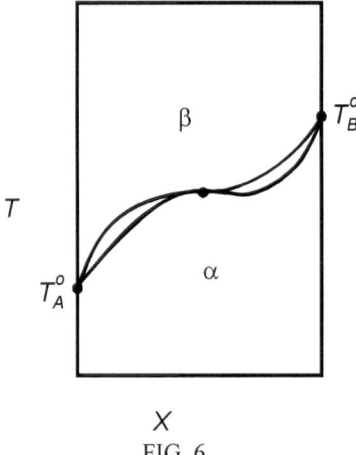

X

FIG. 6.

Phase diagram with a horizontal point of inflexion as stationary point. Examples of such a diagram are not known: the occurrence would correspond to an exceptional coincidence of system properties (note that the figure was drawn by hand: at the right-hand side the equilibrium curve for the composition of the α phase incorrectly shows a minimum)

In the majority of cases, the thermodynamic properties of the system are such that in Equations (15a and b) $\partial S^{\alpha}/\partial X$ and $\partial S^{\beta}/\partial X$ are dominated by $\Delta_e S/\Delta_e X$. In terms of Figure 4: AB generally is steeper than the tangent lines at A and B. The dominating role of $\Delta_e S/\Delta_e X$ implies that in TX phase diagrams the slopes of the two equilibrium curves have the same sign – determined by $\Delta_e S/\Delta_e X$ ($\rightarrow$Exc 3).

Obviously, similar conclusions can be drawn from Equations (17a and b) for equilibria between two mixed states at constant temperature.

The observations implied in the foregoing paragraphs were formulated by Konowalow (1881) in the rules named after him. For the isobaric (TX) equilibrium between liquid and gaseous mixtures the *Konowalow rules* read:

I. The boiling point as a function of composition shows a stationary point only if liquid and vapour have the same composition.

II. The boiling point is raised by the addition of the component whose mole fraction in the vapour phase is lower than its mole fraction in the coexisting liquid phase.

III. The compositions of the coexisting vapour and liquid phases change in the same sense with temperature.

§(302)

the Van 't Hoff equations

In *dilute solutions* for $X \rightarrow 0$, the chemical potential of A, the solvent, simply is ($\leftarrow$207)

$$\mu_A = G_A^* - RTX \ . \tag{19}$$

The difference property $\Delta_e \mu_A$, accordingly, is given by

$$\Delta_e \mu_A = \Delta G_A^* - RTX^\beta + RTX^\alpha, \tag{20}$$

and the total differential of $\Delta_e \mu_A$,

$$d(\Delta_e \mu_A) = -\Delta S_A^* \ dT - RT \ dX^\beta + RT \ dX^\alpha \ . \tag{21}$$

Herewith, and realizing that $T \rightarrow T_A^o$, the *difference in initial slopes* - meant in the derived formulation, Equation (7) - is given by

$$dX^\beta / dT - dX^\alpha / dT = (\Delta_e \mu_A) = -\Delta S_A^* / RT_A^o \qquad (X \rightarrow 0) \ . \tag{22}$$

The important observation to be made is that the difference in initial slopes is fully and only given by the transition properties of the main component, substance A. These properties are the transition temperature, $T^o{}_A$, and the *entropy of transition*, $\Delta S^*{}_A$; the latter being the quotient of $\Delta H^*{}_A$; the heat effect of the transition, and $T^o{}_A$.

Obviously, the isothermal counterpart of Equation (21) is

$$d(\Delta_e \mu_A) = \Delta V_A^* \ dP - RT \ dX^\beta + RT \ dX^\alpha; \tag{23}$$

and it gives rise to

$$dX^\beta / dP - dX^\alpha / dP = \Delta V_A^* / RT \qquad (X \rightarrow 0) \ . \tag{24}$$

For β = vap, and α = liq, and in the approximation $\Delta V_A^* = RT/P$:

$$dX^{vap} / d\ln P - dX^{liq} / d\ln P = 1 \qquad (X \rightarrow 0) \ . \tag{25}$$

This time, the result, the difference in initial slopes in $\ln P$ vs. X diagrams depicting the equilibrium between liquid and vapour, is independent of anything in the physical world!

$$\S(302)$$

the equal-G curve

The formulation for the *equal-G curve* (EGC) in *TX* phase diagrams is Equation (301:15); accordingly, the derived formulation for the EGC is

$$f' = M'[dT / dX] - N'[d(\Delta G) = 0] = 0 \ , \tag{26}$$

in which

$$\Delta G = \Delta G(T, X) = (1 - X) \, \Delta G_A^*(T) + X \, \Delta G_B^*(T) + \Delta G^E (T, X) \ . \tag{27}$$

The *slope of the EGC* is the quotient of the partial derivatives of $\Delta G(T, X)$:

$$\frac{dT_{EGC}}{dX} = - \frac{\partial \Delta G / \partial X}{\partial \Delta G / \partial T} = \frac{\Delta G_B^* - \Delta G_A^* + \partial \Delta G^E / \partial X}{(1 - X)\Delta S_A^* + X \Delta S_B^* + \Delta S^E} \ . \tag{28}$$

For $X \to 0$, and $T \to T_A^o$ the last equation changes into

$$\left(\frac{dT_{EGC}}{dX} \right)_{X \to 0} = \frac{\Delta G_B^*(T = T_A^o) + \left(\partial \Delta G^E / \partial X \right)_{T = T_A^o, X \to 0}}{\Delta S_A^*} \ ; \tag{29}$$

and so because of the fact that the $\Delta G_A^* (= 0$, for $T = T_A^o)$ and ΔS^E ($= 0$, all excess properties being zero for $X = 0$) vanish. The importance of Equation (29) is that it allows for a simple relationship between the *initial slope of the EGC* and k_0^o, which is the *equilibrium distribution coefficient* for $X \to 0$, defined as

$$k_0^o = (k_0)_{X \to 0} = \left(\frac{X_e^\alpha}{X_e^\beta} \right)_{X \to 0} \ . \tag{30}$$

The derivation of that relationship starts from the second equilibrium condition in the general formulation, Equation (5):

$$\Delta_e \mu_B = \Delta G_B^* + RT \ln (X^\beta / X^\alpha) + \Delta_e G_B^E = 0 \ . \tag{31}$$

Remembering ($\leftarrow$203) that component B's partial Gibbs energy is defined as

$$G_B = G + (1 - X)(\partial G / \partial X), \tag{32}$$

it becomes clear that for $X \to 0$

$$\left(\Delta_e G_B^E \right)_{X \to 0} = (\partial \Delta G^E / \partial X)_{X \to 0}, \tag{33}$$

and with that:

$$\ln k_0^o = \left\{\ln\left(\frac{X_e^\alpha}{X_e^\beta}\right)\right\}_{X\to 0} = \frac{\Delta G_B^*(T=T_A^o)+(\partial \Delta G^E/\partial X)_{X\to 0, T=T_A^o}}{RT_A^o} \quad . \quad (34)$$

And after substitution of Equation (29):

$$\ln k_0^o = (\mathrm{d}T_{EGC}/\mathrm{d}X)_{X\to 0}\cdot\Delta S_A^*/RT_A^o \quad . \tag{35}$$

The second of these two expressions for k_0^o has the advantage, over the first, that it can be applied to the phase diagram data themselves, i.e. without the need to subject the data to a thermodynamic analysis and, next, use the outcome for Equation (34).

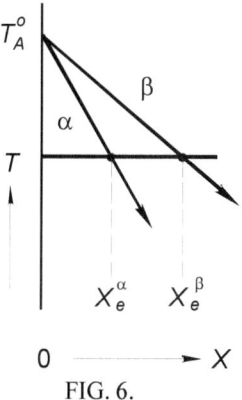

FIG. 6.

Part of TX phase diagram for $X\to 0$, with its opening angle

From Figure 6 it follows that for $X\to 0$, the quotient of X_e^α and X_e^β is equal to the quotient of $\mathrm{d}X^\alpha/\mathrm{d}T$ and $\mathrm{d}X^\beta/\mathrm{d}T$. It means that $\mathrm{d}X^\alpha/\mathrm{d}T$ and $\mathrm{d}X^\beta/\mathrm{d}T$, for $X\to 0$, can be solved, individually, from one of the two Equations (34) and (35) for their quotient, and Equation (22) for their difference.

===

On the lines of derivation of the Clapeyron equation, general expressions have been derived for i) the slopes of equilibrium curves in TX and PX phase diagrams (Van der Waals equations), ii) the difference between the initial slopes of two equilibrium curves in TX and PX phase diagrams (Van't Hoff equations); iii) the slope and initial slope of the equal-G curve in a TX phase diagram, and the relation of the initial slope to the equilibrium distribution coefficient.

===

EXERCISES

1. *the isothermal counterparts*

   Starting from the isothermal counterpart of Equation (6), write down the derivation of the *isothermal van der Waals equations*, Equations (17a and b).

2. *a variant of the linear contribution property*

   As a variant form of the linear contribution property (←205): for the application of the isobaric Van der Waals equations, Equations (15a and b), it is not necessary to know all four absolute entropies ($S_A^{*\alpha}$, $S_A^{*\beta}$, $S_B^{*\alpha}$, $S_B^{*\beta}$), the entropy changes ΔS_A^* and ΔS_B^* being sufficient.
   - Prove the validity of this statement.

3. *exceptional circumstances*

   Find out, guided by Figure 4, what the circumstances are that favour the (exceptional) change from $\left|\Delta_e S / \Delta_e X\right| > \left|\partial S^\alpha / \partial X\right|$ to $\left|\partial S^\alpha / \partial X\right| > \left|\Delta_e S / \Delta_e X\right|$.

4. *negligible solid solubility*

   For the isobaric equilibrium between pure solid A and liquid mixture of A and B, i.e. for α = sol and β = liq, and X^{sol} invariably being equal to zero, Equation (16a) loses its significance and Equation (16b) reduces to

   $$\frac{dX^{liq}}{dT} = \frac{-\Delta_e S_A}{X_e^{liq}\,(\partial^2 G^{liq} / \partial X^2)}$$

   - Derive this equation starting from $d(\Delta_e \mu_A) = 0$.
   - Calculate dX^{liq}/dT for the hypothetical system in Exc 5 for $X_e^{liq} = 0.5$ and $T = 980.1$ K.

5. *a simple eutectic system*

   A hypothetical system composed of the substances A and B has the following characteristics. Solid A and B do not mix. Liquid mixtures have $G^E(T,X) = -2000\,R \cdot K \cdot X\,(1-X)(1-T/500K)$.

   Melting points: $T_A^o = 1000$ K; $T_B^o = 900$ K.

   Entropies of melting: $\Delta S_A^* = \Delta S_B^* = 10 \cdot R$.

- Calculate the TX phase diagram for $850 \leq T/K \leq 1100$ (first determine the nature - upper or lower - of the critical point, and calculate its position).

6.   *the metastable region of demixing in copper + chromium*

From the slope of the liquidus at $X = 0.5$ and the composition of the coexisting solid phase in the system Cu + Co (Figure 5), calculate, with the help of one of the Van der Waals equations, the critical temperature of the metastable region of demixing.

Take $G^{E\,liq}(T, X) = G^{E\,liq}(X) = \omega\,X\,(1-X)$; calculate ω and from that T_c. The entropies of melting of the pure components are 1.15 R for Cu, and 1.04 R for Co.

7.   *'statistics' of stationary points*

A TX (or PX) two-phase region with one extremum is quite common; a two-phase region with two extrema is exceptional; and one with a horizontal point of inflexion has still to be found. To appreciate this, we take a hypothetical family of binary systems, characterized by the following set of properties.

The components of the member systems are miscible in two distinct forms: the high-temperature form β in which the components mix ideally; and the low-temperature form α which is defined by the excess function

$G^{E\,\alpha}$ (only X) $= X(1-X)\,\{1000 - 1000(1-2X)\}$ $\mathrm{J\,mol^{-1}}$.

The components of the systems all have the same entropy of transition, whose value is 10 $\mathrm{J\,K^{-1}\,mol^{-1}}$. The transition point of component A invariably is $T^o_A = 400$ K. All heat capacities are zero. The only distinction between the member systems of the family is in the transition temperature T^o_B of component B.

- Calculate the one and only T^o_B for which the $(\alpha+\beta)$ two-phase region will show a horizontal point of inflexion.
- Define the range of T^o_B temperatures for which the member systems will have a phase diagram with a minimum; and also the range for which the phase diagram will display a minimum plus a maximum.

8.   *from van der Waals to van 't Hoff*

Owing to the fact that the equation for the limiting slopes $dX^\beta/dT - dX^\alpha/dT = -\Delta S^*_A/RT^o_A$, Equation (22), owes its validity to vanishing excess chemical potentials, it is logical that the equation can be derived from Equations (16 a and b), by inserting ideal behaviour and taking $X \to 0$.

- Write down the derivation referred to.

9. *the idealized liquid + vapour equilibrium*

Starting from the isothermal Van der Waals equations, Equations (17a and b), demonstrate that for the isothermal equilibrium between ideal liquid mixtures of negligible molar volume and ideal gas mixtures the following relation is valid – a relation for the difference in slope of liquidus and vaporus at an isobaric section in the ln$P$ vs. X phase diagram

$$\frac{dX_e^{liq}}{d\ln P} - \frac{dX_e^{vap}}{d\ln P} = -1 + X_e^{vap} + X_e^{liq}.$$

10. *idealized equilibrium and initial slopes*

For the (idealized case of) isothermal equilibrium between a liquid phase, whose volume is virtually negligible, and a vapour phase, which is an ideal gas mixture, the initial slopes in ln$P$ vs. X phase diagrams are given by

$$(dX^{vap}/d\ln P)_{X\to 0} = \left[1 - (P_A^o/P_B^o)\, \exp\left\{ -(dG^{Eliq}/dX)_{X\to 0}/RT \right\} \right]^{-1};$$

$$(dX^{liq}/d\ln P)_{X\to 0} = (dX^{vap}/d\ln P)_{X\to 0} - 1.$$

The second of the two expressions - which is Equation (25) for the difference beween the initial slopes - is the consequence of the equilibrium condition d $(\Delta_e\mu_A) = 0$; and it is the case of two unknowns and one equation. And obviously a second equation, a second condition is needed to solve the two unknowns individually.

- Demonstrate that the role of second equation is fulfilled by $\Delta_e\mu_B = 0$, taken for $X \to 0$. In other words, formulate the derivation of the first of the two expressions.

11. *the azeotropic equilibrium*

The azeotropic equilibrium between binary mixtures of liquid and vapour can be represented by the formulation

$$f = M\left[T, P, X_{az} \right] - N\left[F_1 = 0;\ F_2 = 0 \right] = 1,$$

where $F_1 = G^{vap} - G^{liq}$, and $F_2 = (\partial G^{vap}/\partial X)_{T,P} - (\partial G^{liq}/\partial X)_{T,P}$. The solution of the equations is a trajectory in PTX space.

- Derive expressions for the change of azeotropic temperature with pressure $(dT_{az}/d\ln P)$, and the change of azeotropic composition with pressure $(dX_{az}/d\ln P)$, and so for the idealized case where i) the vapour is an ideal gas; ii) the liquid's molar volume is negligible; and iii) the liquid's excess Gibbs energy is given by $G^{E\,liq}(X) = \omega X(1-X)$.

§(302)

12.   *the eutectic point under pressure*

Under isobaric conditions, the equilibrium between pure solid A, pure solid B, and liquid mixture of A and B is invariant. The equilibrium temperature is referred to as the eutectic temperature (T_{eut}), and the composition of the liquid phase as the eutectic composition (X_{eut}).

- Derive expressions for the change of T_{eut} and X_{eut} with pressure. Start from $d(\mu_A^{liq} - G_A^{*sol}) = 0$ and $d(\mu_B^{liq} - G_B^{*sol}) = 0$.

- Calculate the values of dT_{eut}/dP and dX_{eut}/dP for an ideal liquid mixture, and for $\Delta S_A^* = \Delta S_B^* = 5\,R$, and $\Delta V_A^* = 1\,cm^3 \cdot mol^{-1}$ and $\Delta V_B^* = 10\,cm^3 \cdot mol^{-1}$, X_{eut} and T_{eut} being 0.5 and 298.15 K, respectively.

13.   *initial slopes of liquidus and solidus*

Silver chloride and sodium chloride are isomorphous substances: above the critical temperature of the region of demixing and at subsolidus conditions the two mix in all proportions. Their melting points and heats of melting are (Barin 1989):

AgCl   730 K   12322 J mol^{-1}
NaCl   1073.95 K   28158 J mol^{-1}

The solid state excess Gibbs energy of the combination, as given by the last row in Table 301:2, is

$$G^E = X(1-X)\{g_1 + g_2(1-2X)\}, \text{ with}$$
$$g_1 = (11320 - 7.89\,T/K)\,J\,mol^{-1}$$
$$g_2 = (-860 + 1.78\,T/K)\,J\,mol^{-1}, \text{ and } X \text{ representing AgCl's mole fraction.}$$

- Using the information given, ignoring heat capacities and also the fact that the liquid mixtures might deviate from ideal mixing behaviour, calculate the initial slopes of equal-*G* curve, liquidus, and solidus at either side of the composition diagram.

- Next, from the initial slopes, estimate the full course of the liquidus and solidus curves.

- If you are curious to know how well your estimate goes with the phase diagram dictated by the thermodynamics of the system, you could calculate, for two or three temperatures, the mole fractions of the coexisting liquid and solid phases (←Exc 301:4).

14. *the 'opening angle' in TX phase diagrams*

The following statement pertains to the situation sketched by Figure 6, and, in it, the opening angle, i.e. the angle at T^o_A between the two equilibrium curves.

"The greater the value of $(\Delta S^*_A / T^o_A)$, which is a property of component A, and the steeper the equal-G curve, which is a property of the combination of A and B, the greater the opening angle in the TX phase diagram".

- Demonstrate the validity of this statement, just by drawing, for given $\Delta T = T^o_A - T$, two modifications of Figure 6, indicating the EGC point at $T = T$ by an open circle.

15. *a rule of thumb for heteroazeotropes*

The case is considered where, in an isobaric binary system, the equal-G curve for the change from liquid to vapour has a minimum, and such that the minimum is inside a region of demixing with an upper critical solution temperature ($\leftarrow$211). In this situation the liquid+vapour phase diagram shows three two-phase regions: $L_I + V$ and $V + L_{II}$ above, and $L_I + L_{II}$ below the three-phase equilibrium line.

A heterogeneous liquid mixture, when heated, will start to produce vapour at the three-phase equilibrium temperature. The temperature of the system and the composition (X_{AZ}) of the vapour that is formed remain unchanged, until one of the liquids is exhausted. The system is said to have a *heteroazeotrope*.

In the following, the *slopes* (dT/dX) of the two vaporus curves at the three-phase equilibrium are denoted by V_L and V_R, and the slopes of the two liquidi by L_L and L_R.

As a rule of thumb, the ratio between V_L and V_R, is approximated by

$$V_L : V_R = (X_L - X_{AZ}) : (X_R - X_{AZ}), \qquad (A)$$

where X_L and X_R represent the compositions of the liquid phases at the three-phase equilibrium.

- With reference to Equation (16b): under what assumption the ratio $V_L : V_R$ is exactly given by $(X_L - X_{AZ}) : (X_R - X_{AZ})$?

- What are the circumstances that make that, for liquid+vapour equilibria, expression (A) is respected in reasonable approximation?

To get an idea of the strength of the relationship, Equation (A), we take a set of three hypothetical binary systems $\{(1-X) \text{ mol A} + X \text{ mol B}\}$, that have a number of common characteristics: gaseous mixtures are ideal; all heat capacities are zero; component A's boiling point and entropy of vaporization are 320 K and $10R$, respectively; component B's boiling point is 340 K.

The systems (1) and (2) have $G^{Eliq} = 6000\,X\,(1-X)$ J mol^{-1}; system (3) has $G^{Eiq} = 6000$ $X(1-X)[1+B(1-2X)]$ J mol^{-1}, with $B = 0.2$. Component B's entropy of vaporization is $12R$ for system (1); and $10R$ for systems (2) and (3).

The table gives, for the three systems: the temperature of the three-phase equilibrium and the compositions of the phases; the values of the slopes of liquidi and vapori, expressed in K; the two quotients of the slopes; and the ratio defined in the right-hand member of expression (A).

	T_3 / K	X^{liql}	X^{vap}	X^{liqII}	L_L	V_L	V_R	L_R	V_L/V_R	L_L/L_R	(A)
(1)	312.64	.2008	.3069	.7992					-.25	-.20	-.22
(2)	310.88	.1961	.3445	.8039	-8	-20	61	26	-.33	-.31	-.32
(3)	310.42	.0974	.3145	.7435	-31	-31	59	27	-.52	-1.15	-.51

- Calculate the numerical values for the four open places in the table.

- From the numbers in the table, it follows that for the systems (1) and (2), and unlike system (3), the ratio of the slopes of the two liquidi, just like the ratio for the vapori, is closely approximated by the value of (A).

- Why is it that for system (3) the quotient of L_L and L_R does deviate substantially from the value of (A), whereas for systems (1) and (2) such is not the case?

48

from §(004)

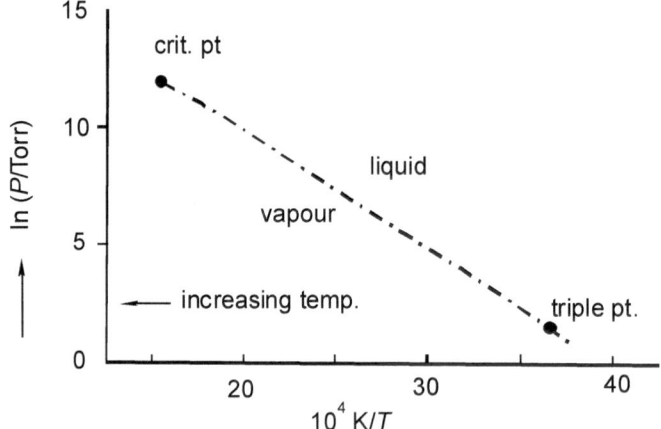

Clausius-Clapeyron plot of the boiling curve for water; from below the triple point up to its end point, the *critical point* where the distinction between liquid and vapour comes to an end (←206)

§ 303 RETROGRADE EQUILIBRIUM CURVES

===

Normally, the course of equilibrium curves in TX phase diagrams is such that along the curve, on increasing temperature, dX/dT does not go through zero, changing its sign. Occasionally, equilibrium curves display a retrograde behaviour, in that, on increasing temperature, the value of the composition variable, X, passes through a maximum or a minimum.

===

closed region of demixing

As an introduction to this chapter we recall the *magic formula*, Equation (301:10):

$$G(T,P,X) = (1\text{-}X)\, G^*_A\,(T,P) + XG^*_B(T,P) + RT\text{LN}(X) + \Omega X(1\text{-}X)\,. \tag{1}$$

And this time we change the parameter Ω from a constant to a non-linear function of temperature:

$$\Omega = (a + bT + cT^2)\, RT\,. \tag{2}$$

In Figure 1, the property $\Omega/2R$ is plotted versus temperature for $a = -1$; $b = 0.016$ K^{-1}; and $c = 0.00002$ K^{-2}. There are two points of intersection with the line $\Omega/2R = 1$; between the points of intersection $\Omega/2R$ is greater than T. These characteristics ($\leftarrow$206 $\leftarrow$212) are related to a *closed region of demixing*, the two points of intersection corresponding to a *lower* and an *upper critical point* – see also Exc 212:4.

In the canted position, Figure 1 bottom, the phase diagram has a maximum, point a, and a minimum, point b. The points a and b are points on equilibrium curves where the composition variable reaches its lowest (a), or highest value (b). Equilibrium curves showing a point like a or b are said to be *retrograde*; they have a retrograde course.

Clearly, as follows from a comparison with Figure 212:2, which is copied here, Figure 2, the retrograde behaviour displayed by the phase diagram in Figure 1, is caused by the typical, non-linear course of the excess Gibbs energy. The G^E function shown in Figure 1 goes together with a strong influence of the *excess heat capacity*, which makes that the *excess enthalpy* is negative at the lower critical point (LCP), and positive at the upper critical point (UCP) ($\leftarrow$Exc 212:3). The *excess entropy*, like the excess enthalpy, undergoes a great change; this can be read from the entropy (-of-mixing) cross-sections shown in Figure 3.

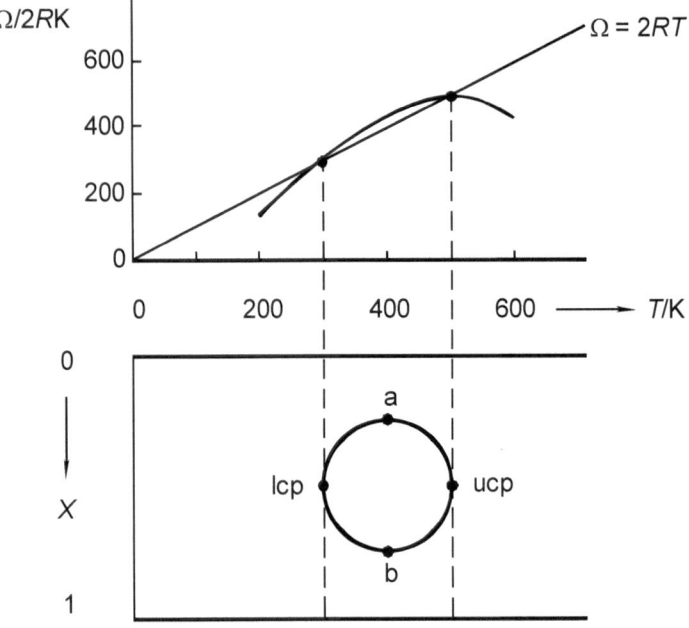

FIG. 1.

In terms of the model $G^E = \Omega X (1 - X)$, demixing will take place, for a given temperature T, when Ω's value exceeds $2RT$

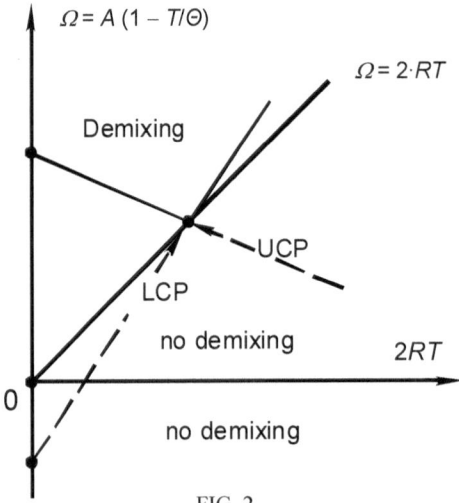

FIG. 2.

In terms of the model $G^E = \Omega X (1 - X)$, with $\Omega = A (1 - T/\Theta)$, there will be demixing whenever $\Omega > 2 \cdot RT$. The model accounts for the existence of regions of demixing with upper critical points (UCP), as well as for regions of demixing with lower critical points (LCP)

§(303)

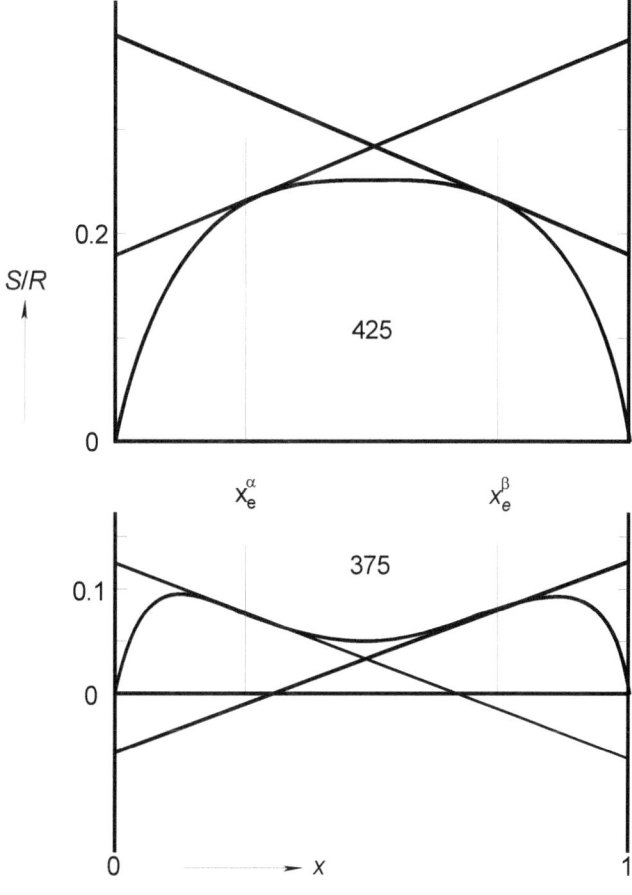

FIG. 3.

Entropy versus mole fraction diagrams for the case of the closed region of demixing shown in Figure 1. The two isothermal sections show that $\partial S^{\alpha}/\partial X$ and $\partial S^{\beta}/\partial X$ change sign

With reference to the *Van der Waals equations*, Equations (302:15a and b), the entropy diagrams reveal that, for the case at hand, the retrograde behaviour is fully related to the change of signs of $\partial S^{\alpha}/\partial X$ and $\partial S^{\beta}/\partial X$, $\Delta_e S$ being zero for this symmetrical case.

An example of a closed region of demixing is provided by the system 2,4-dimethylpyridine + water (Andon and Cox 1952). When measured in sealed glass tubes, the liquid-state region of demixing is found to extend from 22°C to about 180°C. An extensive study into the thermodynamic mixing properties of the system was made by Kortüm and Haug (1956). They found, among other things, that the *excess enthalpy*, the *enthalpy of mixing* (indeed) is negative at the lower side and positive at the upper side of the two-phase region.

§(303)

52

For the theoretical example in Exc 212:4, referred to as *a re-entrant region of demixing*, the Ω function, Equation (2), is given by $\Omega = [5 - 0.016(T/K) + 0.00002(T/K)^2]RT$. For that case, there is an UCP at 300K and an LCP at 500K; between these two temperatures there is miscibility in all proportions. Note that this time retrograde behaviour is absent.

An example is found in the combination of acetone and the *polymer* polystyrene at elevated pressure. Experimental data on the system are given by Zeman and Patterson (1972); for a thermodynamic analysis, see Koningsveld et al. (2001).

Whatever the case may be, regions of demixing like those in 2,4-dimethylpyridine + water and in acetone + polystyrene are evidence of an unusual, substantial change with temperature and/or pressure of the interaction between the two *elementary entities* of the system. Systems whose components belong to a *chemically coherent group of substances* (such as alkali halides, or n-alkanes) never display closed or re-entrant regions of demixing.

re-entrant behaviour

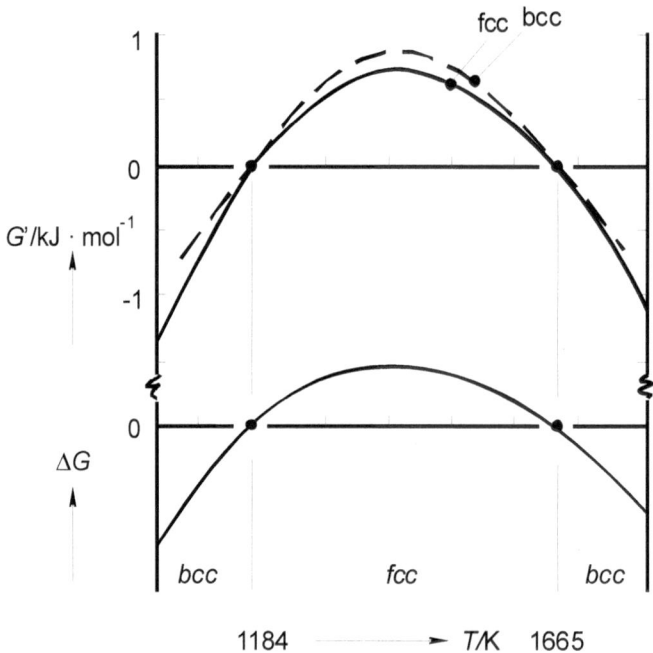

FIG. 4.

The Gibbs energies of the bcc and fcc forms of Fe between Curie temperature (1042 K) and melting point (1809 K). Top: modified Gibbs energies (G'); dashed lines are for metastable parts/extensions. Bottom: sketch of $\Delta G = G_{bcc} - G_{fcc}$

§(303)

Figure 4 pertains to the substance iron (Fe); in particular its bcc and fcc forms. At 1184 K bcc changes into fcc, the change being accompanied by an increase in entropy of 0.09 R, and a decrease in heat capacity of 0.9 R. These values are such that the entropy difference between the two forms is going to change – with the effect that the Gibbs energy difference is becoming zero again. This occurs at 1665 K where fcc changes back, into bcc, with an increase in both entropy and heat capacity of 0.06 R and about 0.4 R, respectively.

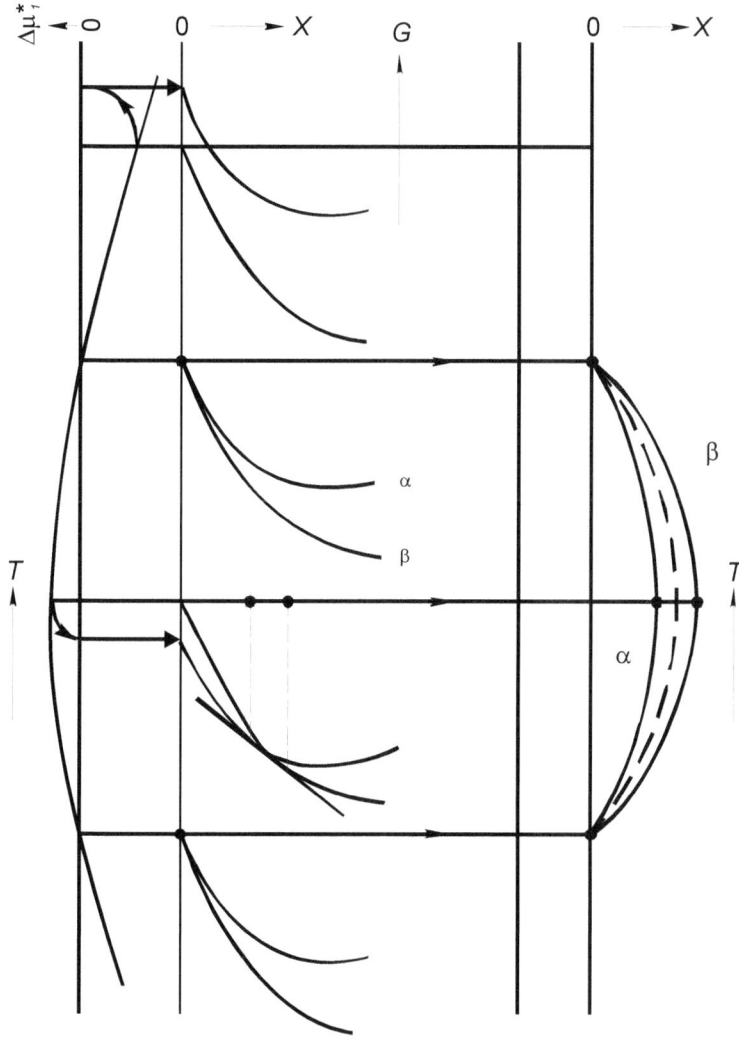

FIG. 5.

Relation between uncommon Gibbs-energy-of-transition function of the pure first component and the TX phase diagram showing a gamma- loop (from Oonk 1981, with $\Delta\mu^*$ for ΔG^*)

54

In Figure 5 it is shown what can happen when a substance like iron is alloyed with another substance. In the *TX* phase diagram there is a two-phase region of a retrograde nature, in metallurgy referred to as a *gamma-loop* (the fcc form of iron is known as γ-Fe).

As an example, Figure 6 shows the experimentally determined γ-loop for the system iron + silicon. Owing to the great difference in *magnetic susceptibility*, the α-γ transition was studied with a magnetic balance.

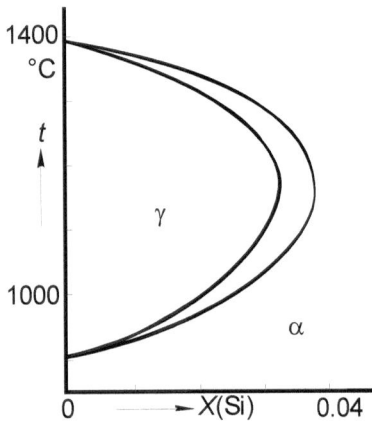

FIG. 6.

Gamma-loop in the *TX* phase diagram of the system iron + silicon (Fischer et al. 1966)

As an opening to the thermodynamic treatment – of the kind of retrograde behaviour shown by Figure 6 – we apply the EGC approach (←301), assuming that the two substances mix ideally in each of the two forms. The equal-*G* curve in that case is the solution of the equation

$$\Delta G = (1 - X)\, \Delta G_A^* + X \Delta G_B^* = 0 \ . \tag{3}$$

And the solution is given by

$$X_{EGC}(T) = \frac{\Delta G_A^*(T)}{\Delta G_A^*(T) - \Delta G_B^*(T)} \ . \tag{4}$$

From Equation (4) it follows that in the range of physically real *X* values, i.e. for $0 \leq X \leq 1$, EGC points are found if

§(303)

i) ΔG_A^* and $(\Delta G_A^* - \Delta G_B^*)$ have the same sign, and

ii) $\left|(\Delta G_A^*)\right| \leq \left|(\Delta G_A^* - \Delta G_B^*)\right|$.

In Figure 7 it is demonstrated how this turns out for the case in which both ΔG_A^* and ΔG_B^* have unusual courses, and intersect below the zero level

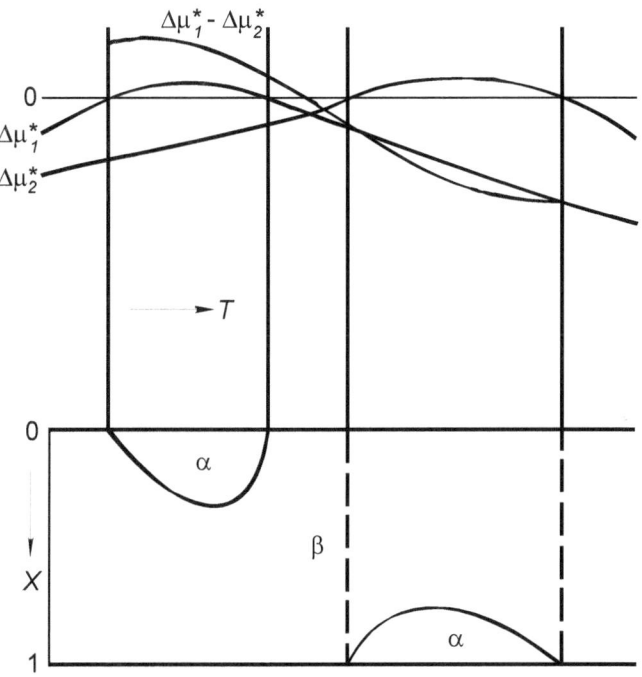

FIG. 7.

Isobaric equilibrium between two ideal mixed states. The pure second component and the first have unusual Gibbs-energy-of-transition functions. These functions are such that in the TX diagram there are two gamma-loops (from Oonk 1981, with $\Delta\mu^*$ for ΔG^* and 1=A, and 2=B)

§(303)

56

In the case of the system iron + chromium the gamma-loop has a minimum, Figure 8; it is an expression of non-ideal mixing (→Exc 3).

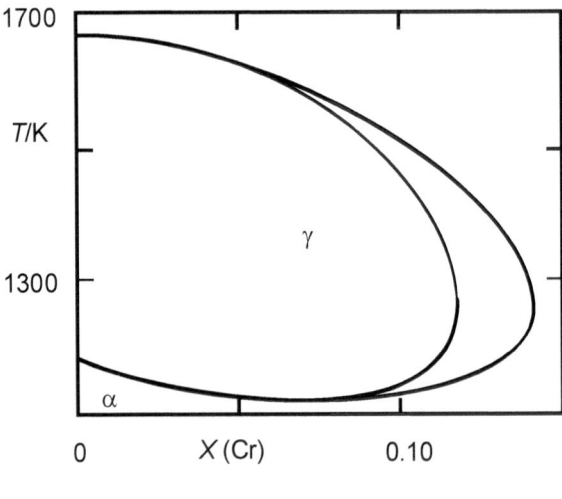

FIG. 8.

Gamma-loop with minimum in the *TX* phase diagram of the system Fe + Cr (Kirchner et al. 1973)

The type of behaviour displayed by Figure 8 is also found in binary *liquid-crystalline* systems. An example is the system 4-(n-octyloxy)-4'-cyanobiphenyl + 4-n-heptyl-4'-cyanobiphenyl, in which the first component is re-entrant (Gobl-Wunsch et al.1981; see alsoVan Hecke 1985 for a detailed thermodynamic treatment).

retrograde solubility

Generally, for the equilibrium between two mixed phases in two different forms, α and β, the two equilibrium curves have either both a positive slope or both a negative slope. This is related to the fact that $\partial S^\alpha/\partial X$ and $\partial S^\beta/\partial X$ both are dominated by $\Delta_e S/\Delta_e X$ (←302; Figure 302:4 in particular). It is, however, quite conceivable that there are circumstances that make that, say, $|\partial S^\alpha/\partial X|$ is going to become greater than $|\Delta_e S/\Delta_e X|$ - and that dX^α/dT is going to change sign, and, at the same time, dX^β/dT does not.

Guided by Figure 302:4, and knowing that for $X \rightarrow 0$ the partial differential coefficient $\partial S/\partial X \rightarrow \infty$, one can imagine that a retrograde course of the α equilibrium curve is favoured by the combination of

§(303)

i)   a small difference in entropy between the forms α and β;

ii)   a very small X^{α}; and

iii)   a large X^{β}. See Figure 9.

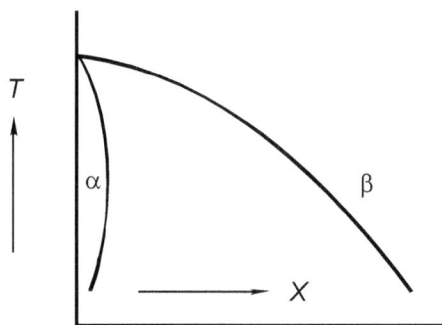

FIG. 9.

The kind of retrograde behaviour shown in this figure is referred to as retrograde solubility – of component B in the α-form of practically pure A

In contrast to closed regions of demixing and gamma-loops, retrograde solubility does not require any particular C_P influence to account for its existence. The following elementary treatment should make this clear – for the case where β is an ideal liquid mixture (superscript liq) and α is a solid (sol) for which $G^{Esol} = \Omega X (1 - X)$, and such that Ω is a constant.

The way to deal with the problem is as follows. First, out of the two equilibrium conditions, $N\left[\mu_A^{liq} = \mu_A^{sol}; \mu_B^{liq} = \mu_B^{sol} \right]$, the one in μ_A is used to yield the composition of the liquid as a function of temperature. Subsequently the result from μ_A is substituted in the relationship between X_e^{sol} and X_e^{liq}, provided by the equilibrium condition in terms of μ_B.

Out of the *excess chemical potentials*, generated by $\Omega X (1 - X)$, the one for A vanishes for $X \to 0$; and the one for B, which is $\Omega (1 - X)^2$, becomes $\mu_B^{E\ sol} = \Omega$ ($\leftarrow$207). The fact that $\mu_A^{E\ sol} \to 0$ means that the equilibrium composition of the liquid is given by the *ideal liquidus equation* ($\leftarrow$210):

§(303)

58

$$\ln\left(1 - X_e^{liq}\right) = -\frac{\Delta H_A^*}{R}\left(\frac{1}{T} - \frac{1}{T_A^o}\right) \quad . \tag{5}$$

From this equation, along with $(1 - X)\, d\ln\,(1 - X) = -X\, d\ln X,$

$$\frac{d\ln X_e^{liq}}{dT} = -\frac{\left(1 - X_e^{liq}\right)}{X_e^{liq}}\cdot\frac{\Delta H_A^*}{RT^2} \quad . \tag{6}$$

From the condition in μ_B:

$$\ln X_e^{sol} = \ln X_e^{liq} + \Delta G_B^* / RT - \Omega / RT \,, \text{ and} \tag{7}$$

$$\frac{d\ln X_e^{sol}}{dT} = \frac{d\ln X_e^{liq}}{dT} - \frac{\Delta H_B^*}{RT^2} + \frac{\Omega}{RT^2} \quad . \tag{8}$$

As a result,

$$\frac{d\ln X_e^{sol}}{dT} = -\frac{\left(1 - X_e^{liq}\right)}{X_e^{liq}}\cdot\frac{\Delta H_A^*}{RT^2} - \frac{\Delta H_B^*}{RT^2} + \frac{\Omega}{RT^2} \quad . \tag{9}$$

This derivative is equal to zero – the condition for X_e^{sol} having a maximum value – for

$$\Omega = \frac{\left(1 - X_e^{liq}\right)}{X_e^{liq}}\cdot\Delta H_A^* + \Delta H_B^* \quad . \tag{10}$$

As an example, let $T_A^o = T_B^o = \Delta H_A^* / R = \Delta H_B^* / R = 1000\,\text{K}$, $\Delta S_A^* = \Delta S_B^* = R$, and let the requirement be that X_e^{sol} reaches its maximum value when $X_e^{liq} = 0.25$. Then, from Equation (10), it is read that Ω should have the value of 4000 RK.

The analysis given above and leading to Equation (10) goes back to Van Laar (1908). In Meijering's (1948) paper one can read that Van Laar did not particularly stress the point of a retrograde solidus, and it would explain the fact that the discovery of the first real example (Zn-rich solidus in Zn + Cd) came as a surprise (which "was considered so abnormal that a tentative interpretation was given, which made the maximum a discontinuous one, requiring an allotropic transformation in Zn").

See Thurmond and Struthers (1953) for the retrograde solid solubilities of Sb in Ge, Cu in Ge, and Cu in Si.

==

Three different types of retrograde behaviour in isobaric binary systems have been examined: the closed region of demixing, the γ-loop (or re-entrant behaviour), and retrograde solubility. In the first two cases, where each of the two conjugate equilibrium curves is retrograde, the cause of the behaviour is in the relative strong change of (excess) enthalpy/entropy with temperature. In the case of retrograde solubility, where only one of the two curves is retrograde, the phenomenon is the result of a more or less coincidental concurrence of thermodynamic properties – not requiring any heat capacity.

==

EXERCISES

1.   *two similar arcs, the arc representation*

The *heat capacity* of the material is fully responsible for the arc-like appearance of the modified Gibbs energy function, Figure 3, top.

- Assuming a constant Cp, derive an expression for the relation between Cp and the dimensions of the arc; the latter are the width of the arc (b) at the zero level and its height (h) above that level. Apply Taylor series expansion, and truncate after the second derivative's term.

- Calculate fcc iron's Cp from the arc in Figure 3, i.e. for $b = (1665 - 1184)$ K, and $h = 0.73$ kJ $\cdot$ mol^{-1}.

  NB. Appreciate the similarity between the arc and the one in § 110

2.   *types of phase diagram for re-entrant behaviour*

Inspired and guided by Figure 7, invent a number of *types of phase diagram* for re-entrant behaviour in combination with ideal mixing.

§(303)

3. *the gamma-loop in iron + chromium and its EGC analysis*

Taking into account the aspect of deviation from ideal mixing, the most elementary extension of Equation (3), to define the *EGC*, comes down to

$$(1-X)\Delta G_A^* + X\Delta G_B^* + \Delta\Omega\,X(1-X) = 0,$$

which is a quadratic equation in X, from which, for given T, the unknown X can be solved in an easy manner.

- Calculate the position of the EGC in the TX plane for the arithmetical example in which

  i) $T_A^o(\alpha \to \beta) = 1665\,K$;   ii) $\Delta S_A^*(\alpha \to \beta,\ at\ 1665\,K) = 0.060\,R$; and

  iii) $\Delta C_{PA}^*(\alpha \to \beta) = 0.374\,R$.

- Show, that for $X \to 0$, the slope of the EGC is given by $dX/dT = \Delta S_A^* / (\Delta G_B^* + \Delta\Omega)$.

- Give a semi-quantitative interpretation $(\Delta G_B^*\,?;\Delta\Omega?;$ or rather$\Delta\Omega(T)\,?)$ of the phase diagram, Figure 8.

NB. The consequences of the equation given above, which is $(1-X)\Delta G_A^* + X\Delta G_B^* + \Delta\Omega\,X(1-X) = 0,$ have been explored by Wada (1961), who used the term *allotropic phase boundary* (APB = our EGC), making reference to Roesler U, Sato H, and Zener C, "Theory of Alloy Phases" ASM (1955), p.255.

4. *retrograde solubility*

For the example of system defined below Equation (10), where ideal liquid mixtures are in equilibrium with solid solutions showing retrograde solubility, make a table in which for $600 \le T/K \le 950$, in steps of 50 K, the following information is presented:

  i)   the equilibrium compositions of the coexisting phases;

  ii)   the values of $\partial S^{sol}/\partial X$; $\partial S^{liq}/\partial X$; $\Delta_e S/\Delta_e X$, all three divided by R ($\leftarrow$ Exc 302:3);

  iii) the sign of the slopes of liquidus and solidus.

5. *Breusov's retrograde liquidi*

Starting from the expression $\Delta H_i^* = a_i + b_i T$ (with i = component A, component B) for the enthalpy of melting, the following equation can be derived for the equilibrium between pure solid component i and ideal liquid mixture

$$X_i(T) = \left(T_i^\circ\right)^{-b_i/R} \cdot e^{a_i/RT_i^\circ} \cdot T^{b_i/R} \cdot e^{-a_i/RT}$$

A mathematical analysis of this equation, for the various possible combinations of signs of a_i and b_i, has been given by Breusov (1966). The only constraint on a_i and b_i is $\left(a_i + b_i T_i^\circ\right) > 0$: the enthalpy of melting has to be positive at T_i°

- Give the derivation of the equation.
- Demonstrate that retrograde liquidi are obtained for the combination of $a_i < 0$ and $b_i > 0$.
- Calculate the phase diagram for

	T_i° / K	a_i / RK	b_i / R
component A	100	- 37.5	0.75
component B	100	30	0

6. *the elliptical stability field once more*

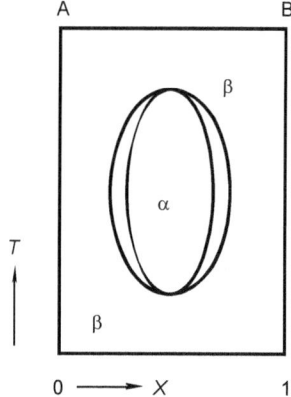

This time, as a follow-up of Exc 301:5, the symmetric, hypothetical phase diagram has a minimum at 300 K and a maximum at 500 K. The diagram pertains to a system which is fully symmetrical, and whose properties are such that the heat capacities of the two forms α and β are constants, and the excess Gibbs energies are given by Equation (1) with constant Ω's.

§(303)

With $\Delta G^{*}{}_{A} = \Delta G^{*}{}_{B} = \Delta G^{*}$, the EGC equation shown in Exc 3 reduces to

$$\Delta G^{*} + \Delta\Omega\, X(1\text{-}X) = 0.$$

- Demonstrate that the position of the EGC is fully determined by the numerical values of $\Delta\Omega$ and $\Delta C_{P}{}^{*}$. Demonstrate also that the temperature, at which the elliptic region bordered by the EGC is at its maximum width, is fixed by the choice of the temperatures of the minimum and the maximum.

- For $\Omega^{\alpha} = 4500$ J mol^{-1}; $\Omega^{\beta} = 5500$ J mol^{-1}; and $\Delta C_{P}{}^{*} = 5$ J mol^{-1} calculate the position of the EGC and draw a sketch of the phase diagram you expect.

Four isobaric binary sample systems act as vehicles to describe the delicate route from experimental observations to the diagram that depicts the states of equilibrium between coexisting phases.

introduction

In the index of Volume I of *Equilibrium between Phases of Matter* the terms *time* and *speed* are absent. Quite understandable, because, whenever there is equilibrium between phases, a situation has been created where time stands still, so to say. In experimental work on the phase behaviour of a chemically defined system, on the other hand, time and speed are factors that have a determinative influence on the signals given off by the system – and, as a consequence, on the observations made by the investigator.

For the purpose of this chapter on phase-equilibrium research, carried out on four isobaric binary systems, we will make a distinction between the *true phase-equilibrium diagram* and what we will call the '*observation diagram*'. The latter is a graphical representation with a collection of symbols that reflect the observations made by the investigator. Ideally, the symbols in the observation diagram are situated on the equilibrium curves and the three-phase equilibrium lines of the true phase diagram. And whatever the case may be, thermodynamic calculations are the best way to assess the status of an observation diagram – and to turn the observation diagram into the true phase-equilibrium diagram.

To illustrate these matters, we have selected four systems – systems with which we have been involved ourselves, and which all in their own way reflect the influence of time and speed and other aspects of experimentation. The four are sodium chloride + potassium chloride; *n*-nonadecane + *n*-heneicosane; silica + magnesium oxide; and limonene + carvone.

The three first mentioned systems are members of a *family of similar systems.* We demonstrate that, in addition to thermodynamic computations, correlations by means of *system-dependent parameters* are a powerful tool to raise the scientific status of both the members of a family and the family as a whole.

The measure of agreement between a calculated curve in a TX phase diagram and a set of experimental data points will be represented by the *temperature index* Δ_T and *mole-fraction index* Δ_X. The index Δ_T ($\leftarrow$301) gives the mean of the absolute difference between experimental and calculated temperature. It will be used when composition is the independent experimental variable. The index Δ_X which gives the mean of the absolute difference between experimental and calculated mole fraction, will be used when temperature is the independent variable.

64

the system sodium chloride + potassium chloride

The isobaric NaCl + KCl system is one of the best investigated binary systems and a key system in inorganic mixed-crystal research. It has a solid-liquid loop with a minimum and well separated from the minimum a region of demixing, of which the top is at about 500 °C. At room temperature the two components are virtually insoluble in one another.

the observation diagram and the true phase diagram

Making reference to the earlier investigations by Le Chatelier (1894) and Ruff and Plato (1903), Kurnakow and Żemczużny (1907) were among the first who studied the phase behaviour of the combination of sodium chloride and potassium chloride. They allowed liquid mixtures to cool, measuring the temperature of the samples as a function of time. A typical *cooling curve* is shown in Figure 1. In the figure, two significant features are marked: the circle corresponds to the *onset of crystallization*; the triangle marks a slow-motion region, which Kurnakov and his co-worker, correctly, considered as evidence for *solid-state demixing*. The complete set of observations is shown in the *observation diagram* Figure 2, left-hand side. In Figure 2, right-hand side the observations are superimposed on the true phase-equilibrium diagram. The two diagrams give rise to a number of remarks.

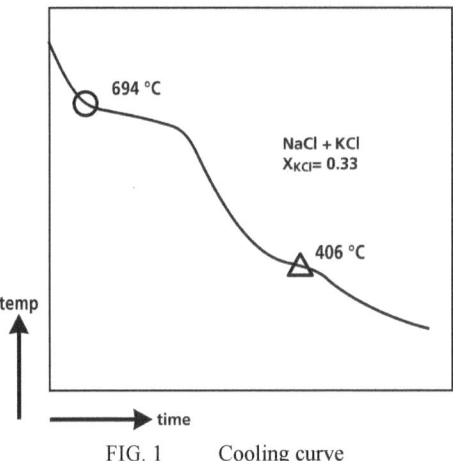

FIG. 1   Cooling curve

First of all, and in spite of the limitations of their methodology, the conclusions drawn by the investigators, as regards the true nature of the system's phase behaviour, is correct (a fortunate circumstance being that the two chlorides have the same crystal structure, the *NaCl-type of structure*).

§(304)

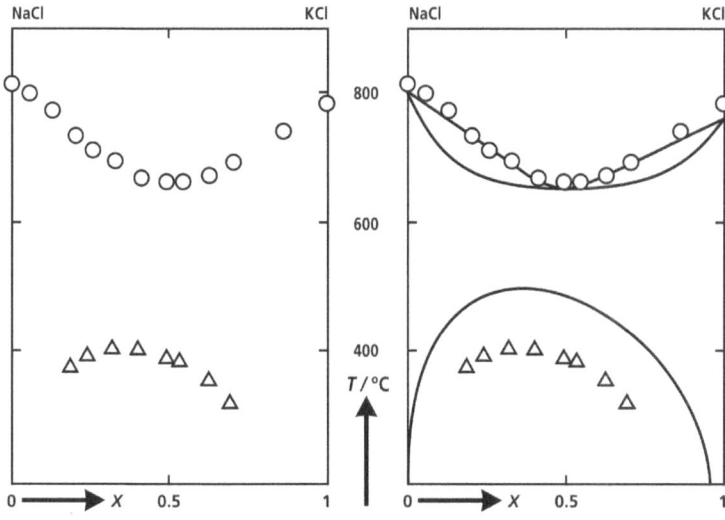

FIG. 2.
Left: observation diagram. Right: Superposition of observation
diagram and true phase diagram

Table 1:
A collection of transition and melting points, revealing systematicdeviations of the
thermometer system (*pyrometer*) used in St. Petersburg (Kurnakow and Żemčużny
1907; Żemčużny and Rambach 1909)

substance	change	St. Peter	modern	Δt	authors	year
Zn	melting	419	**419.50**			
CsCl	transition	451	470	− 19	Z + R	1909
LiCl	melting	614	610	4	Z + R	1909
Sb	melting	631	**630.85**			
CsCl	melting	646	645	1	Z + R	1909
NaBr	melting	768	747	21	K + Z	1907
KCl	melting	790	771	19	K + Z	1907
NaCl	melting	819	801	18	K + Z	1907
KF	melting	837	857	− 20	K + Z	1907
Ag	melting	962	**960.80**			
NaF	melting	997	996	1	K + Z	1907

§(304)

As a striking fact, the observed onset temperatures of crystallization are above the true *liquidus* – only in part, however, and most pronounced for the pure components. The differences in 'observed' and true melting temperatures are evidence of a *systematic deviation*. A close examination of the melting temperatures reported by Kurnakow, for a variety of susbstances, reveals that the deviations are a function of temperature and such that there are negative and positive ones; see Table 1. Making allowances for the deviations, one can draw the conclusion that Kurnakow's liquidus data are in full agreement with the true liquidus.

The triangles in Figure 2 are far below the boundary curve of the *region of demixing*; it is a phenomenon related to kinetics rather than thermometry. The solid-state separation of phases requires much time; more than the time taken for the registration of cooling curves.

The establishment of the equilibrium between solid phases requires isothermal equilibration experiments of long duration (*annealing*); in combination with *X-ray diffraction* to determine the compositions of the phases in equilibrium. Such experiments were carried out by Bunk and Tichelaar (1953), see Table 2, by Barrett and Wallace (1954), Nguyen-Ba-Chanh (1964), and Luova and Tanilla (1966). The *binodal* curve in Figure 2, right-hand side, is the outcome of those experiments. Special attention should be given to Nacken (1918), who measured *refractive indices* to determine composition.

Table 2:
Experimental data pertaining to the equilibrium between two solid phases, I and II, both having the NaCl-type of structure, in the system $\{(1-X)\ NaCl + X\ KCl\}$; Bunk and Tichelaar (1953). The order of the columns is Celsius temperature; X-value phase I; X-value phase II

309	0.021	0.889
391	0.061	0.771
447	0.137	0.634
466	0.195	0.542
472	0.191	0.521
488		0.335
490	decomposed, but no sharp	diffraction pattern

In Bunk and Tichelaar's investigation homogeneous mixed crystalline samples were obtained by keeping a finely powdered mixture of NaCl and KCl at 550 °C, for at least 20 hours. After *quenching* to room temperature, the samples were used to determine, by X-ray powder diffraction, the value of the lattice constant as a function of composition. The authors reported that the atmosphere had to be carefully excluded: as even traces of water give rise to decomposition of the mixed crystals into the pure components. The data in Table 2 were obtained by heating a homogeneous mixed crystal for 24 hours at the chosen temperature, and by measuring, after quenching, the two lattice constants involved in the diffraction pattern. The uncertainty in composition, as estimated by the authors, is 1.5 mole %.

§(304)

solidus versus liquidus

The classical *cooling-curve method* and the modern methods of *differential microcalorimetry* all have the advantage of being fast and user-friendly. The other side of being fast is the difficulty of realizing – in a continuous and overall manner – a state of thermodynamic equilibrium, read, the state of minimal Gibbs energy. In experimental practice, and concentrating on the change from liquid to mixed-crystalline solid or *vice versa*, this difficulty comes down to the observation that reported *solidus temperatures* often are inaccurate, and sometimes even against the rules of phase equilibrium. *Liquidus temperatures*, on the other hand, are less sensitive to the side effects of fast experimentation. Reported liquidus temperatures, as a rule, are accurate.

The best thing to do, is to derive the course of the solidus from the experimental liquidus, applying the rules of thermodynamics. Even the simplest thermodynamic notions are capable of giving results that are superior to reported experimental data (→Exc 1, 2).

A computer program for deriving the solidus from the liquidus is LIQFIT (Bouwstra et al. 1980). The procedure is elaborated below for the system C19+C21.

thermodynamic analysis of the region of demixing

The compositions of the coexisting phases X^I and X^{II} are the solution of the set N of equations

$$N = N\left[\mu_A^I = \mu_A^{II}; \ \mu_B^I = \mu_B^{II} \right], \tag{1}$$

where the *chemical potentials* follow from the Gibbs energy function. The other way round, with the help of the two equations in N, Equation (1), information on the Gibbs function can be derived from the compositions of the coexisting phases as a function of temperature.

A *thermodynamic analysis* of phase-diagram data starts by the choice of a *model* for the Gibbs function – in the case at hand, the choice of a *model for the excess Gibbs energy*. For our purpose, a convenient model is the excess Gibbs energy as given by the expression

$$G^E(T,X) = X(1-X)\{g_1(T) + g_2(T)(1-2X)\}, \tag{2}$$

and such that the coefficients g_1 and g_2 are linear functions of temperature:

$$g_i(T) = h_i - Ts_i; \quad i = 1, 2 \tag{3}$$

§(304)

The corresponding expressions for the chemical potentials of the components A(=NaCl), and B(=KCl) are

$$\mu_A = G_A^* + RT \ln(1-X) + X^2 \left\{ g_1 + g_2 (3-4X) \right\} ; \qquad (4a)$$

$$\mu_B = G_B^* + RT \ln X + (1-X)^2 \left\{ g_1 + g_2 (1-4X) \right\} . \qquad (4b)$$

Let us now suppose that the compositions, X^I and X^{II}, of the coexisting phases are known for a given temperature T. Then it follows from the substitution of two times the Equations (4) in Equation (1) that, for that temperature, the values of g_1 and g_2 are the solution of the two equations

$$\left\{ (X^{II})^2 - (X^I)^2 \right\} g_1 + \left\{ (X^{II})^2 (3-4X^{II}) - (X^I)^2 (3-4X^I) \right\} g_2 = RT \ln(1-X^I) - RT \ln(1-X^{II}) ;$$

$$(5a)$$

$$\left\{ (1-X^{II})^2 - (1-X^I)^2 \right\} g_1 + \left\{ (1-X^{II})^2 (1-4X^{II}) - (1-X^I)^2 (1-4X^I) \right\} g_2 = RT \ln X^I - RT \ln X^{II}$$

$$(5b)$$

A concrete example of the application of these two equations is given in Exc 3 – which corresponds to the 2/2 solution of EXTXD. EXTXD (EXcess properties from TX phase Diagrams) is a computer program whose *regression equations*, like Equations (5), are based on the *equality of chemical potentials* (Brouwer and Oonk 1979; see Oonk et al. 1986, for a listing in PASCAL).

EXTXD operates with the model contained in Equations (2) and (3) extended by a term $g_3(T) (1-2X)^2$; and such that the number of *parameters to be adjusted* to the experimental information is increased from one (the parameter h_1) to six (h_1, s_1, h_2, s_2, h_3, and s_3). A typical EXTXD result is displayed in Table 3; obtained for an artificial, smoothed data set, composed of 27 *data triplets* (T, X^I, X^{II}) in the range from 620 K to 750 K at 5 K intervals, based on the experimental information in Table 1. In Table 2 the mole-fraction index Δ_X pertains to the experimental mole fractions given in Table 1 (with the exception of X^{II} for the highest temperature) and the mole fractions calculated from the output excess properties.

With regard to the Δ_X values of about 0.006: it may be observed that the reported uncertainties in the experimental mole fractions are as large as 0.015. The latter figure seems to be realistic; that is to say, when the data are compared with data from the other sources mentioned above.

Table 3:
Result of analysis by EXTXD of artificial data set based on the experimental information in Table 1. Values of h_i and s_i are in kJ·mol^{-1} and J·K^{-1}·mol^{-1}, respectively. Values for g_i are in kJ·mol^{-1}, and valid for $T = 685$ K

soln	h_1	s_1	h_2	s_2	h_3	s_3	Δ_x	g_1	g_2	g_3
1/0	12.93							12.93		
1/1	11.13	−2.56						12.88		
2/1	29.53	24.17	3.22				0.0064	12.97	3.22	
2/2	29.10	23.48	7.37	5.82			0.0053	13.02	3.38	
3/2	25.03	17.21	7.95	7.28	0.52		0.0055	13.24	2.96	0.52
3/3	17.96	6.27	8.36	9.10	4.37	4.01	0.0056	13.67	2.12	1.62

mathematical versus physical significance

The last four solutions in Table 2, in spite of their divergent h_i and s_i values, reproduce the experimental region of demixing with equally high precision. Apparently, an experimental phase diagram with its experimental uncertainties is not the expression of only one, a unique set of thermodynamic properties. In other words, the computed h and s coefficients in the first place have a *mathematical significance*: the power of reproducing the experimental phase diagram with adequate precision. Their *physical significance* is limited; for the computed h to a lesser degree than for the computed s coefficients.

Experimental *heat-of-mixing data*, read excess enthalpy data come down to a value for h_1 of 17.7 kJ·mol^{-1}; the mean for data from different sources, showing a dispersion of about 0.5 kJ·mol^{-1} (see Van der Kemp et al. 1992). The fact that the computed h_1 of the 3/3 solution is close to the experimental value seems to be coincidental.
The full information contained in Table 2 does suggest that the physical significance of the computed result is concentrated in the g coefficients; in that they represent the true excess Gibbs energy. Or more precisely, the excess Gibbs energy computed for the mean temperature of the data, which is in the vicinity of 685 K. Apparently, the difficulty is in dividing the *excess Gibbs energy* into its *excess enthalpy* and *excess entropy* components. This difficulty also finds expression in the circumstance that, for all of the h_i/s_i pairs in Table 3, the quotient of the computed *mathematical uncertainties* (not shown in the table) in h_i and s_i is ≈ 685 K.

the system n-nonadecane + n-heneicosane

The normal alkanes n-nonadecane and n-heneicosane (unlike NaCl and KCl) have a rich polymorphic nature, in combination with the exixtence of a mesostate. At subsolidus conditions the two alkanes are completely miscible (like NaCl and KCl). At low temperature the stable phase diagram is the result of a competition between four forms that have subtle structural differences.

The *TX* diagram shown in Figure 3 is an instructive starting point for the discussion of the system *n*-nonadecane + *n*-heneicosane ($C_{19}H_{40}$ + $C_{21}H_{44}$); C19 + C21 for short. Apart from a 'normal' solid state and the liquid one, the two components and their mixtures can take up a *mesostate*. The latter is referred to as the *rotator state*, owing to the fact that in it the molecules have a certain degree of *rotational freedom* around their long axis. The rotator state has a number of distinct forms, which differ from one another in molecular stacking and degree of rotational freedom (see e.g. Sirota et al. 1993). In the case of C19 + C21 the *rotator form* is the orthorhombic form RI, *space group Fmmm*, with four molecules per unit cell ($Z = 4$).

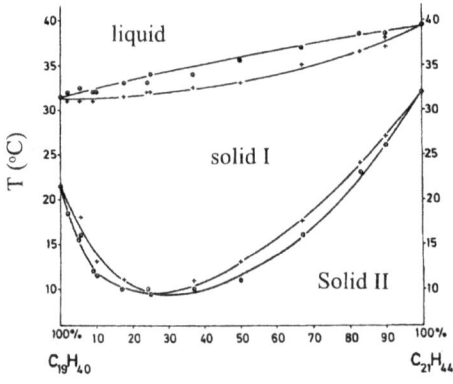

FIG. 3.

The system nonadecane + heneicosane. *TX* diagram, reflecting the existence of a mesostate: the rotator state, which is crystalline and in which the molecules have a certain degree of rotational freedom around their long axis (Würflinger and Schneider 1973)

One of the curious things about the *TX* diagram, Figure 3, is reflected by the *opening angles* (←Exc 302:14) of the two loops. For the upper loop, i.e. for the change from rotator to liquid, the opening angles are too large to correspond to true thermodynamic equilibrium. For the lower loop, for the change from normal solid to rotator, on the other hand, the opening angles are surprisingly small.

§(304)

Clearly, the diagram has the status of an observation diagram rather than a true phase-equilibrium diagram. Besides, diagrams that are similar to Figure 3 have been published among others by Mazee (1957) and Maroncelli et al. (1985).

Extensive investigations - including crystallographic, spectroscopic and thermodynamic methods – carried out after the publication of the diagram, Figure 3, have provided the keys to understanding and appreciating the diagram as it is. Two of these keys are *polymorphism* and *experimental time scales.*

polymorphism

Like the *alkali halides,* the *normal alkanes* represent a *chemically coherent group of substances.* But unlike the former, the latter have a rich polymorphic nature. For the pure *n*-alkanes, the appearance of a distinct form not only is related to temperature and pressure, but also to the value of 'n' in C_nH_{2n+2} (Cn for short) and, in particular n's parity, i.e. odd or even.

The odd alkanes C19 and C21, which at ordinary pressure give rise to a stable appearance of the rotator form RI, are orthorhombic at low temperature. The intermediate, even alkane C20, on the other hand, is triclinic and it just fails to undergo, before melting, the transition to the rotator mesostate – C22 being the first even alkane with a stable rotator phase.

Pairs of *n*-alkanes (Cn' and Cn") easily give rise to *mixed crystals* - in the rotator as well as in the normal solid state – provided that the relative difference between n' and n" is small.

The *syncrystallization* of two alkanes has a large effect on the polymorphic relationships. The system *n*- tetradecane (C14) + *n*-hexadecane(C16), as an example, has a large (intermediate) *TX* stability field for the form RI, whereas for the pure components C14 and C16 the form RI is far from stable (see above). The general observation is that by alloying Cn' with Cn" forms are stabilized that make for the pure alkanes a first appearance for values of n that are higher than n' and n".

For the system C19 + C21 the role of polymorphism is illustrated by the *TX* diagram shown in Figure 4 (Metivaud 1999). Instead of just one, the normal solid state of the system has four distinct forms. These forms are the monoclinic form Mdci and the three orthorhombic forms Oi, Op, and Odci. The distinction between the three intermediate forms, Odci, Mdci, and Op, is rather subtle – both in structural and energetic terms (Rajabalee et al. 1999, 2000).

§(304)

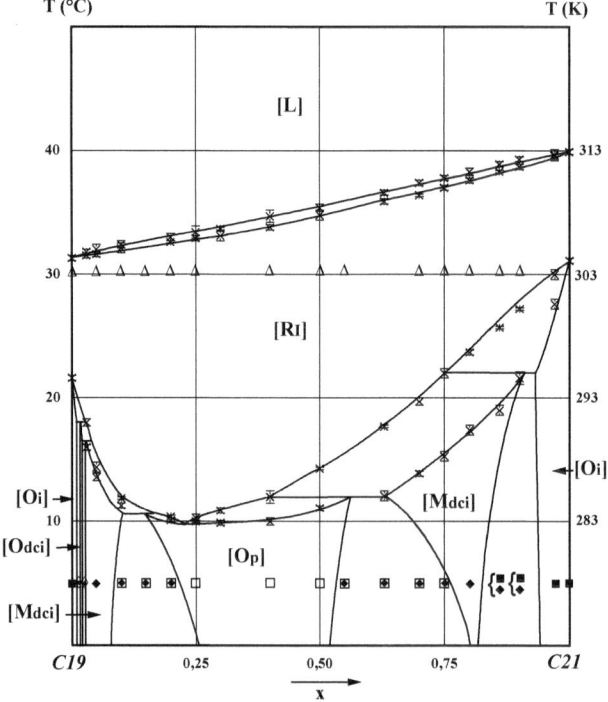

FIG. 4.
The system nonadecane + heneicosane. *TX* diagram related to the existence of
a rotator mesostate and revealing the polymorphic fine structure of the
'normal' solid state (Metivaud 1999)

time scales

Passing over the fact that there are four forms rather than one for the
'normal' solid state, one can observe that the two *TX* diagrams, Figures 3 and 4, show
notable differences as regards the widths of the two transition loops. Especially in the
case of the lower loop, these differences are too significant to be ascribed to different
interpretations of thermo-energetic signals. In other terms, the differences have to
correspond to differences in experimental procedures in combination with absence of
full thermodynamic equilibrium.   The fact is that for a mixed crystal a full-
equilibrium transition does require a continuous redistribution of the component
molecules over each of the two phases. In this respect the transition in a mixed-
crystalline sample differs from the transition in each of its pure components.
Apparently, the redistribution of molecules requires more time than the change from
one state to the other.

§(304)

In certain cases and under certain circumstances it is possible to fully suppress the redistribution of molecules during the change of state. And when such is the case, the transition will proceed in a quasi-isothermal manner – resembling the transition in a pure substance. A speaking example is offered by the system 1,4-dichlorobenzene + 1,4-dibromobenzene, pertaining to the melting of mixed-crystalline material prepared by *zone levelling* (Van Genderen et al. 1977; ←Exc 301:4). When studied by *microcalorimetry* at high speed, a sample prepared by zone leveling will melt isothermally – and so at its *equal-G curve* temperature. When studied by *adiabatic calorimetry*, that is to say at very low speed, the same mixed material will start to melt at its *solidus temperature*, to become completely liquid at the *liquidus temperature* (Van der Linde et al. 2002).

quasi-isothermal transition from normal solid to rotator

In the system C19 + C21 mixed samples can be conditioned such that they change isothermally from the normal solid to the rotator state. This time the recipe is a prolonged stay in the rotator (!) state (Métivaud 1999; Mondieig at al. 2003). The experimental evidence is demonstrated by the collection of *thermograms* shown in Figure 5 – for the sample containing 63 Mol% of C21. Samples were kept at 305 K, for a length of time t, subsequently cooled down to about 275 K, and finally heated in a *Differential Scanning Calorimeter* (DSC).

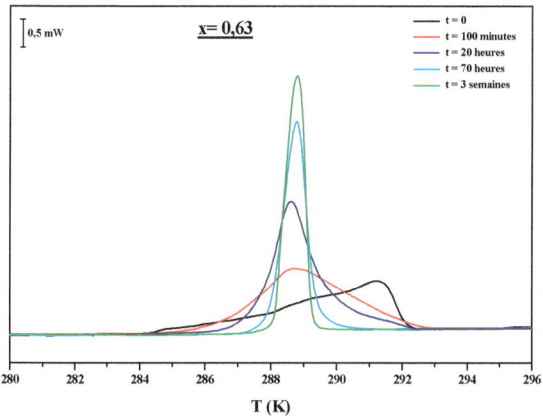

FIG. 5.

The system nonadecane + heneicosane at composition $X = 0.63$. Change from normal solid to rotator. Evolution of DSC signals over time, displayed by samples that have been kept in the rotator state at 305 K

§(304)

Figure 5 clearly and unambiguously reveals that, on increasing the duration of the stay in the rotator state, the signals become more and more sharp. At $t = \infty$, i.e. in practice after three or four weeks, the signals reach the sharpness of the thermograms recorded for the pure components of the system. The experimental data pertaining to the change from Mdci to RI are collected in Table 3. The table also displays the characteristics of the thermograms recorded for freshly prepared samples (quenching of a homogeneous liquid mixture into liquid nitrogen)

Table 3:
The system C19 + C21. The transition from Mdci to RI. First column: mole fraction C21. Second column: temperature of isothermal change (EGC temperature). Third and fourth columns: onset and end temperatures measured by Differential Scanning Calorimetry directly after preparation of a fresh sample

X	T_{EGC}/K	T_{on}/K	T_{end}/K
0.05	286.9	286.8	287.6
0.10	284.7	284.5	285.1
0.63	288.1	285.3	290.9
0.70	290.4	287.1	292.9
0.75	291.4	288.5	295.2
0.80	293.9	290.6	296.9

phase diagram analysis

The lower part of the (C19 +C21) system's phase diagram, Figure 4, ignoring the appearance of the form Odci, can be regarded as the stable end result of three phase diagram loops that cross one another. This is shown in Figure 6, and on its own, the figure is an illustration of *crossed isotrimorphism.* One of the loops is for the combination Mdci + RI; its course in the diagram is from C19's *metastable transition point* (for the change Mdci→RI), via a minimum, to the metastable transition point of C21. Near the edges of the diagram the (Mdci+RI) loop is 'overruled' by the (Oi+RI) loop, and in the central part by the loop for (Op+RI).

Each of the individual loops can be subjected to a thermodynamic analysis on its own, and a convenient method to do so is LIQFIT (Bouwstra et al. 1980; Bouwstra et al. 1986; Jacobs 1990). The LIQFIT computer routine, in essence, is a succession of phase diagram calculations - directed by the EGC – by means of which either the upper or the lower equilibrium curve is made to pass through the respective experimental points. The key formula is the EGC equation ($\leftarrow$Equations 212:36; 37):

$$T_{EGC}(X) = T_{ZERO}(X) - \frac{G_{EGC}^{E\alpha}(X)}{(1-X)\Delta S_A^* + X\Delta S_B^*} \, , \tag{6}$$

§(304)

where

$$T_{ZERO}(X) = \frac{(1-X)\Delta H_A^* + X\Delta H_B^*}{(1-X)\Delta S_A^* + X\Delta S_B^*} .$$

(7)

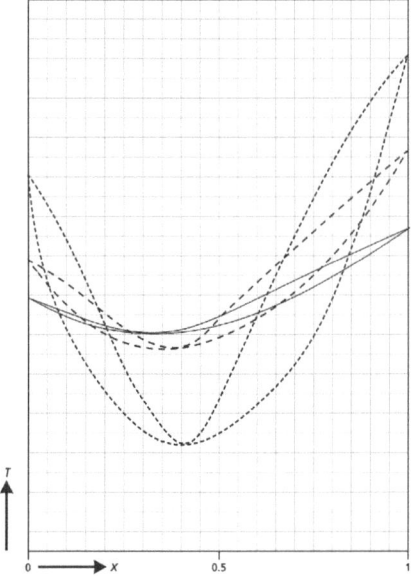

FIG. 6.
The lower part of the C19 + C21 phase diagram is a case of crossed isotrimorphism

For the underlying case, the property minus $G^{E\alpha}$ in Equation (6) has to be replaced by ΔG^E, which is the difference in excess Gibbs energy (RI minus Mdci) along the EGC and for the mean temperature of the transition range.

In normal practice, the *pure-component properties* in Equation (7) are known, and LIQFIT computations start with an estimated EGC position - to yield a first approximated ΔG^E. For the case at hand, the situation is different in two ways. First, the pure components do not undergo the transition (Mdci→RI): the pure-component properties have to be found in an indirect manner. Secondly, the position of the EGC is already known (albeit only in part): the property ΔG^E is directly found (→Exc 5).

§(304)

the system silica + magnesium oxide

In contrast to NaCl + KCl and C19 + C21, the components of the system SiO_2 + MgO belong neither to a family of chemically coherent substances, nor to the same class of solid materials. Magnesium oxide is ionic and not polymorphic; silica on the other hand has a variety of solid forms, which have three-dimensional network structure with atomic bonds. This time, the complexity of the system is in the liquid state – where a unilateral region of demixing makes its appearance.

structural entities

In comparison with the two foregoing systems, the system MgO + SiO_2 is much more inaccessible – both theoretically and experimentally.

In the case of NaCl + KCl one mole of mixture contains *Avogadro's number* of cations and also anions; and in C19 + C21 one mole of mixture has Avogadro's number of molecules. The fact that the number of ions / molecules does not change with composition makes that the *thermodynamic description* of the two systems remains relatively simple.

For the system $\{(1\text{-}X)$ mole $= 40.304$ g of MgO $+ X$ mole $= 60.085$ g of $SiO_2\}$ the situation is completely different – and especially so at the side of silica. Silica has a rich polymorphic nature, cristobalite being the solid form which is stable at the normal melting point. Cristobalite, like diamond, has a three-dimensional network structure – and when it melts, it certainly does not break down in free SiO_2 molecules. Its low heat of melting is evidence of the fact that, on melting, the substance retains much of its network structure. Liquid silica, for that matter, is said to have a *random network structure.*

The addition of MgO to liquid silica gives rise to a mixture of Mg^{2+} ions and a variety of polymeric anions – such that the *degree of polymerization* decreases with increasing fraction of MgO. Generally, and in other terms, by the addition of oxides like MgO the *Si : O ratio* is enhanced – that ratio is a key property in rationalizing the structures of liquid and crystalline silicates (see Griffen 1992).

experimental

The difficulties, which are encountered in experimental work on the system, are related to the high temperatures involved, and, in addition, to the presence of the region of demixing; see Figure 7. The fact that the melting point of MgO is given by the round number of 3100 K ($\pm$ 25 K) speaks for itself (see e.g. Swamy et al. 1994). Not only that the measurement of high temperatures has its difficulties, a fact is also that, in *equilibration-and-quenching* experiments, phases in equilibrium have to be quenched to room temperature for their analysis.

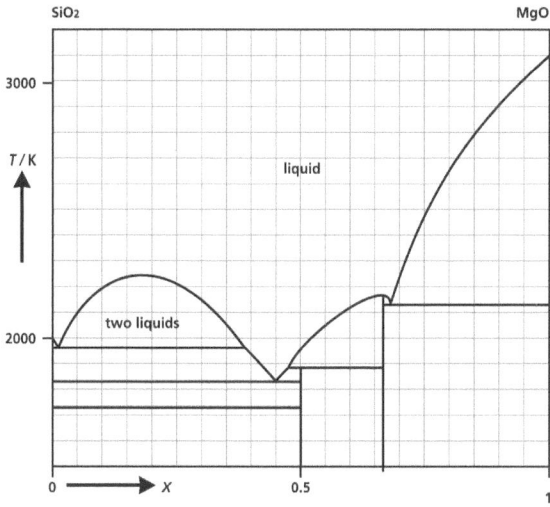

FIG.7.
The system {(1-X) mole of SiO_2 + X mole of MgO}. Phase diagram, showing the presence of two compounds, which are (proto)enstatite ($MgSiO_3$)and forsterite (Mg_2SiO_4); a region of demixing; and the polymorphic change from trydimite to cristobalite

Experiments, in which samples were equilibrated at high temperature and subsequently quenched, were carried out by Bowen and Andersen (1914, including a discussion on the geological significance of the results; →306); Greig (1927); Ol'shanskii (1951); and Hageman and Oonk (1986). In the studies of Bowen and Anderson, and Greig the maximum temperature of equilibration was about 2000 K; Ol'shanskii went to 2473 K; and Hageman to about 2350 K.

In the study by Bowen and Andersen the samples, after equilibration, were quenched in mercury, and subsequently examined under the microscope. In that manner the nature of the crystalline phases was established; and the composition of the glass, the originally liquid phase, was found by measuring the refraction index.

Greig found that silica-rich mixtures of MgO, CaO, SrO, MnO, ZnO, FeO, NiO, or CaO and SiO_2 all melt to give two immiscible liquids. In each case the temperature of the three-phase equilibrium - between cristobalite and the two liquid phases - appeared to be close to 1700 °C.

Greig observed that cristobalite crystallizes so rapidly from the less silicious phase - read, less viscous phase - that quenching to a clear glass is exceedingly difficult (which, besides, is in agreement with Hageman's findings). For the melting point of cristoballite Greig gave (1713 ± 5) °C.

§(304)

Ol'shanskii determined the full extent of the regions of demixing in $MgO + SiO_2$; $CaO + SiO_2$; and $SrO + SiO_2$. The critical temperatures, for the systems in the same order, given by Ol'shanskii are >2200 °C; 2110 °C; and ~1900 °C. The critical temperatures reported by Hageman are systematically about 200 K lower than Ol'shanskii's values: 2240 K ($MgO + SiO_2$); 2159 K ($CaO + SiO_2$); and 2015 K ($SrO + SiO_2$).

quenching through a region of demixing

This time we will take a closer look at the experiments through which, by equilibration and quenching, the extent of the liquid-state region of demixing is determined.

Ideally, and simply speaking, the outcome of a series of equilibration-and-quenching experiments can be represented in an observations diagram by means of two different symbols: one type of symbol for the homogeneous samples and the other for the heterogeneous samples. Next, in the diagram a curve can be drawn such that the two types of symbol are separated from one another. That curve is the binodal, the boundary of the region of demixing.

In experimental reality this simple picture of equilibration and quenching is seriously obscured by two different, and at the same time related causes. First of all, in the vicinity of the critical temperature, the *driving force for phase separation* is at its lowest, and the difference between the two coexistent phases at its smallest. A separation into two clearly separated phases (by gravity due to difference in density) is not so evident. Secondly, some of the samples that are homogeneous at the temperature of equilibration have to traverse the region of demixing when they are quenched. Depending on the rate of quenching, and the viscosity of the liquid, the material may start to separate into two phases. This is especially so when the sample comes into the region enclosed by the spinodal; the region where the Gibbs function is concave and where, as a result, every local fluctuation in composition is irreversible (because of the lowering of Gibbs energy).

The distinction, in quenched glasses, between incomplete separation (in the vicinity of the critical temperature) and onset of separation (during quenching) is not unambiguous – and it means that the investigator is left with uncertainty about the position of the critical temperature. In this context it is meaningful that Hageman and Oonk (1979, 1987) had to reconsider their interpretation of the phenomena in the system $B_2O_3 + BaO$: the critical temperature of 1539 K reported in the 1987 publication is 137 K lower than the one in the 1979 publication!

Materials that have been prepared by quenching through the field enclosed by the spinodal, as a rule, have their own characteristic structure. Examples are found in metallurgy and ceramics (see, Jantzen and Herman 1978).

For *spinodal decomposition* see also Hillert (1998), and Papon et al. (2002).

§(304)

thermodynamic analysis

Papers written on the thermodynamic analysis of the system $MgO + SiO_2$ are quite numerous; we mention the papers signed by Blander and Pelton (1987); Michels and Wesker (1987, 1988); Hillert and Wang (1989); and Swamy et al. (1994). The Investigations by Michels and Swamy have in common that *solid solubility* of SiO_2 in MgO (periclase) is not taken into account – disregarded in the case of Swamy. In the phase diagrams calculated by Hillert and Blander there is a single-phase field for periclase (MgO with little SiO_2), which is based on observations by Schlaudt and Roy (1965).
As regards the thermodynamic description of the liquid mixtures, the investigations by Blander and Michels are alike; and so are, in a different way, the investigations by Hillert and Swamy.
The descriptions by Blander and Michels are based on the model by Toop and Samis (1962), in which three different *oxygen species* are distinguished:

$O°$ = bridging oxygen
O^- = singly bound oxygen
O^{2-} = free oxygen ion

The three species take part in the chemical equilibrium

$(Mg^{2+})O^{2-} + Si-O°-Si = Si-O^- + (Mg^{2+}) + {}^-O-Si$;
or for short $O^{2-} + O° = 2O^-$.

The descriptions by Hillert and Swamy are in terms of the *two-sublattice model* (Hillert et al. 1985). In this model, when applied to the liquid state in $MgO + SiO_2$, there is a cationic sublattice for Mg^{2+}; and an anionic sublattice for which the species O^{2-} ; SiO_4^{4-} ; and SiO_2 are taken. The *site fractions* of the three species on the anionic sublattice as a function of composition are part of the output of the computations.
For each of the four investigations, and notwithstanding the different nature of the models, the calculated phase diagram is in perfect agreement with the experimental data. And from this observation it follows, once more, that agreement between calculated and experimental phase diagram not necessarily means that the adopted model is fully representative of the true physical nature of the system.

the system under high pressure

In the last three sections of this work the combination of SiO_2 and MgO will be the principal system of study.

§(304)

80

the combination of limonene + carvone

Limonene and carvone with their chiral molecules give rise to four binary systems, two of which being thermodynamically independent – laevorotatory limonene + laevorotatory carvone, and dextrorotatory limonene + laevorotatory carvone. The four molecules are rather flat: each of the four is quite superimposable with the three others. From an a priori point of view, and for each of the binary combinations, one can state that the liquid state will have virtually ideal mixing properties; and that, because of the superimposability of the molecules, a certain degree of solid solubility cannot be precluded.

optically active substances

Limonene and carvone have *chiral molecules* and it means that the molecules cannot be superimposed by their mirror images; see Figure 8.

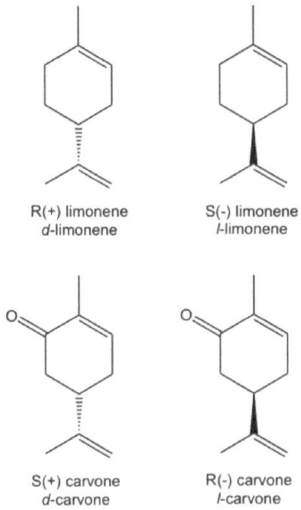

R(+) limonene
d-limonene

S(-) limonene
l-limonene

S(+) carvone
d-carvone

R(-) carvone
l-carvone

FIG. 8.
The molecules of the enantiomers of limonene and carvone

The four different molecules are characteristic of four different *optically active substances*: *dextro-* and *laevorotatory* limonene, and *dextro-* and *laevorotatory* carvone – *d-* and *l*-limonene, and *d-* and *l*-carvone for short. The four substances are components of a variety of essential oils extracted from plants (see e.g. Bauer and Grabe 1985).

§(304)

When cooled in a (micro)calorimeter the four substances, which are liquid at room temperature, fail to crystallize - the *supercooled liquid* enters into the *vitreous state* (←004). On heating from the vitreous state, the material first undergoes the *glass transition*, and subsequently it crystallizes in most of the times. Carvone crystallizes more readily than limonene; in DSC experiments supercooled liquid limonene crystallizes only once every five times (Gallis et al. 2000).

experimental observations by Differential Scanning Calorimetry

DSC experiments have been carried out on the system {(1−X) mole of *laevorotatory* limonene + X mole of *laevorotatory* carvone}. Samples, having a mass of 13-15 mg, were cooled to 153 K; maintained at that temperature for ten minutes; and then heated at a rate of 2 K·min⁻¹ (Calvet et al. 1996). The thermograms that were obtained are shown in Figure 9, and the numerical information they contain in Table 4. From 153 K on, the thermograms first show the *glass transition* - i.e. the change from vitreous to (normal) supercooled liquid - then an *exothermic peak* due to crystallization, and subsequently the *endothermic peak* of melting. Mixtures having X < 0.5 showed neither the glass transition nor the events of crystallization and melting.

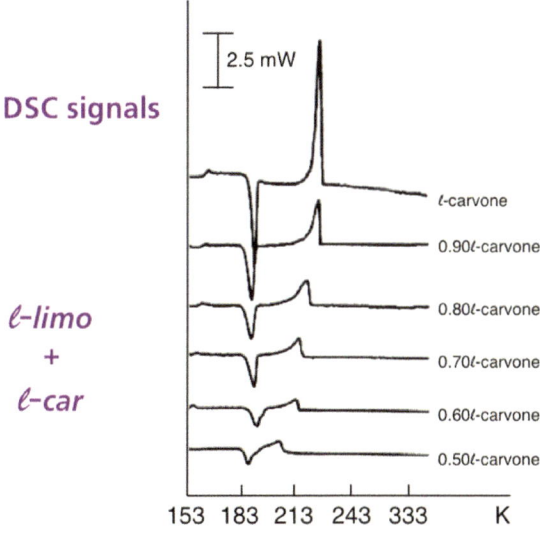

FIG. 9.
The system laevorotatory limonene + laevorotatory carvone. DSC thermograms as a function of composition. See also Table 4 to observe that the figures on the temperature scale show some imperfections

§(304)

Table 4:
Numerical information provided by the DSC experiments as a function of composition: T_g, temperature of glass transition; T_{crys}, temperature of crystallization; T_{on}, onset of melting; T_{end}, end of melting; and Q, heat effect of endothermic change

X	T_g / K	T_{crys} / K	T_{on} / K	T_{end} / K	Q/ kJ·mol^{-1}
0.5		199		226	
0.6	156	201	221	231	4.1
0.7	160	199	225	233	5.2
0.8	162	195	229	237	6.2
0.9	165	193	237	241	7.7
1.0	168	199	244	244	11.3

DSC experiments were also carried out by Gallis (1998) on the system *dextrorotatory* limonene + *laevorotatory* carvone. Gallis applied virtually the same experimental circumstances as did Calvet. It is important to note that Gallis's results are fully in line with the results obtained by Calvet.

experiments by adiabatic calorimetry

Experiments by *adiabatic calorimetry* have been carried out in the laboratory of the (former) Chemical Thermodynamics Group at Utrecht University – using the instruments designed by J. Cees van Miltenburg, and constructed by Gerrit J.K. van den Berg (Van Miltenburg et al. 1987). Adiabatic calorimeters are the instruments *par excellence* to study the phase behaviour and thermodynamic properties of materials at subambient temperatures. The samples used for the measurements typically have a mass of about 5 g.
For the experiments by adiabatic calorimetry, four mixed samples were prepared out of *dextrorotatory* limonene and *laevorotatory* carvone, the carvone mole fraction of the samples being $X = 0.1$; $X = 0.31$; $X = 0.50$; and $X = 0.85$. The samples were subjected to a variety of experimental procedures of heating and cooling; these are detailed in the following paragraphs.
The melting properties of the 'optically pure' substances, as obtained by adiabatic calorimetry, are as follows. *Laevorotatory* limonene melts, just like *dextrorotatory* limonene, at (199.18 ± 0.05) K. The heat of melting of the two *enantiomers* is (11.38 ± 0.02) kJ·mol^{-1} (Gallis et al. 1996). For the optically pure carvones these data are 249.5 K and 11.73 kJ·mol^{-1} (Gallis et al. 1996).

heating from the vitreous state to fully liquid

The so-called *heat capacity diagrams*, Figure 10, clearly display the glass transition, and also the phenomena of crystallization followed by melting. In this respect the curves are similar to the ones in Figure 9.

§(304)

adiabatic calorimetry

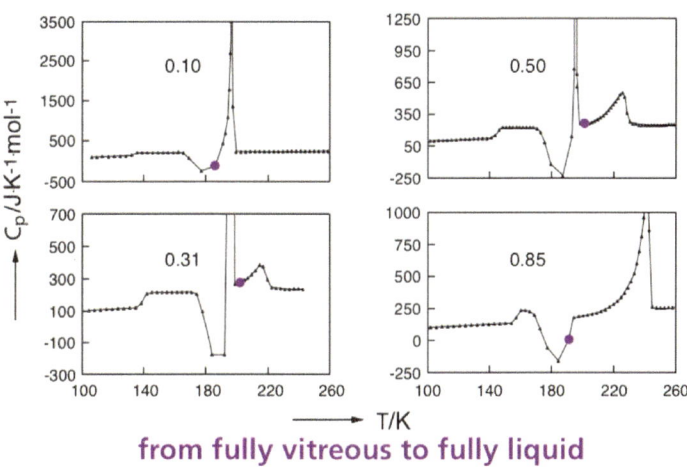

from fully vitreous to fully liquid

FIG. 10.

The system d-limonene + l-carvone. Heat-capacity diagrams from experiments by adiabatic calorimetry, the samples being vitreous at the start

cooling before the end of melting, followed by reheating.

In the following enumeration of observations - pertaining to experiments where, at the start of the second heating run, the samples are partly vitreous and partly crystalline - an important part is played by the glass transition. In the first place, the *temperature* of the glass transition, which changes substantially with composition, is an indicator for the composition of the vitreous part. In the second place, the *jump in heat capacity* at the glass transition is an indicator for the (relative) amount of the vitreous part. For the pure components d-limonene and l-carvone the jumps in heat capacity are 82 and 94 $J K^{-1} mol^{-}$, respectively.

The sample having the composition $X = 0.1$, and at the start being fully vitreous, was first heated to 185 K, then cooled to 120 K, and subsequently heated again to 190 K. In the first heating run the glass transition was at 133 K. In the second heating run the glass transition was at about 150 K, and from the jump in heat capacity it follows that about one-sixth of the sample had been vitreous at the second start, see Figure 11.

The samples having the compositions $X = 0.31$ and $X = 0.50$ were heated, in the first run, to 202 K, which is at the end of the large endothermic effect. The samples were subsequently cooled to 156 K and 179 K, respectively. In the second heating run the samples were heated until they were fully liquid: they did not show the large endothermic effect anymore.

§(304)

The sample having $X = 0.85$ was first heated to 190 K, then cooled to 120 K, and, in the second run, heated to 190 K again. In the first run the glass transition was at 158 K, and in the second run it was close to the glass transition temperature of pure limonene, see Figure 11. In the second run the onset of (further) crystallization was at 170 K. From the jumps in heat capacity, it follows that, at the second start at 120 K, about one-sixth of the sample had been vitreous.

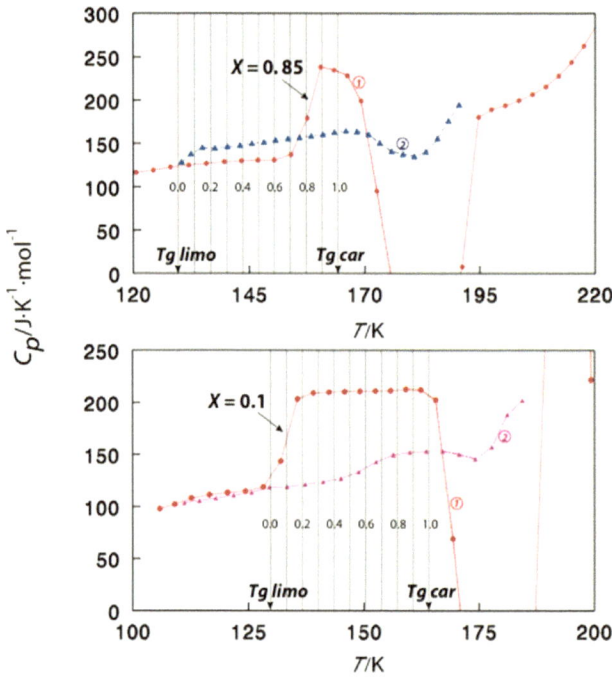

FIG.11.
The system d-limonene + l-carvone. Heat-capacity diagrams from experiments by adiabatic calorimetry, samples at the second start being partly vitreous and partly crystalline. Courtesy of J.C. van Miltenburg

from fully crystalline to fully liquid.

In the last series of experiments, fully crystalline samples - obtained by a series of cooling cycles – were heated from 140 K to 260 K. The heat capacity diagrams are shown in Figure 12.

§(304)

adiabatic calorimetry

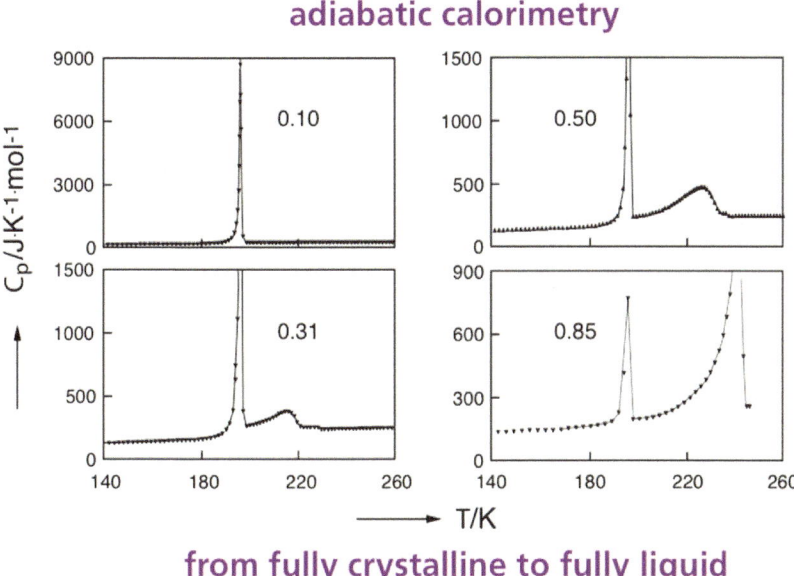

from fully crystalline to fully liquid

FIG.12.

The system *d*-limonene + *l*-carvone. Heat-capacity diagrams from experiments by adiabatic calorimetry, samples at the second start being fully crystalline

interpretation

The ensemble of experimental results, visualized in Figure 13, along with the data obtained for the pure components, give rise to two main conclusions. The first is that the equilibrium change from crystalline solid to liquid corresponds to a phase diagram of the *simple eutectic* type. The second is that the combination of the two optically active substances gives evidence of *dual crystallization*: the phenomenon that, from a supercooled liquid mixture of two similar substances, the two (can be made to) crystallize one after the other.

equilibrium solid-liquid phase diagram

The two *liquidi*, *lc* and *ll* in Figure 13, follow from the properties of the pure components, taking their heat of melting independent of temperature, and introducing the expression of $1500\,X(1-X)\,\mathrm{J\,mol^{-1}}$ for the excess Gibbs energy of the liquid mixtures. The *eutectic point* is calculated at $T = 196.38$ K and $X = 0.103$.

§(304)

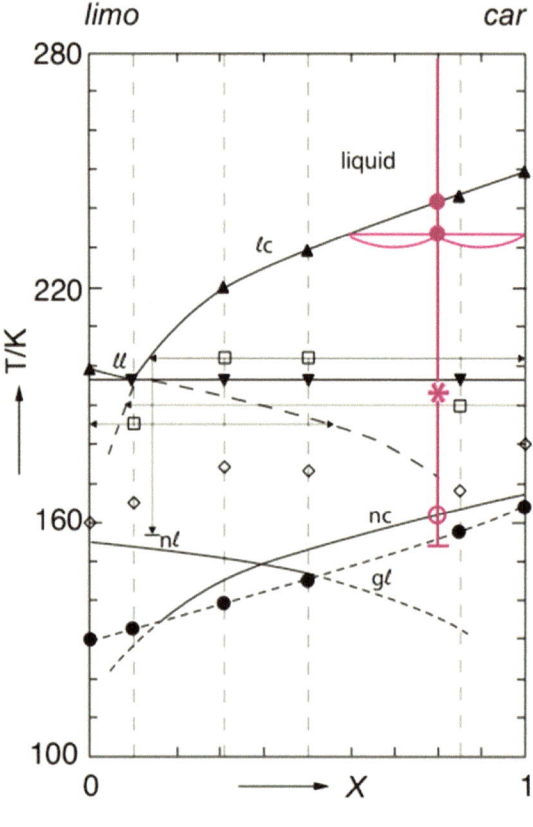

FIG.13.

The system $\{(1-X)$ d-limonene $+ X$ l-carvone$\}$: multifunctional TX diagram. Observations by adiabatic calorimetry: filled circles, glass transition; diamonds, onset of crystallization; filled upside down triangles, eutectic event; filled triangles, end of melting; open squares and arrows pertain to experiments in which the sample is heated from the vitreous state, and arrested, see text. Curves: ll, limonene's liquidus; lc, carvone's liquidus; nl, limonene's nucleation curve; nc, carvone's nucleation curve

dual crystallization

To give a coherent interpretation of the observations on partial crystallization, three assumptions are made; *i*) crystallization is preceded by *nucleation*, such that nucleation will not take place below the glass transition temperature; *ii*) the two components nucleate independently; and *iii*) the driving force for the nucleation of a component is a certain constant value for the difference between its liquid and solid *chemical potentials*.

§(304)

These assumptions find expression in the two *nucleation curves, nl* and *nc* in Figure 13. The nucleation curves were calculated with temperature-independent heats of melting (using the values with which the liquidi were calculated); with 110 $X(1-X)$ $Jmol^{-1}$ for the excess Gibbs energy of the supercooled liquid mixtures; and the values of 2475 $Jmol^{-1}$ and 3925 $Jmol^{-1}$ for the driving forces in terms of chemical potentials, for limonene and carvone, respectively.

From the observations on the samples having $X = 0.10$ and $X = 0.85$, it follows that, on heating from the fully vitreous state, only one of the components crystallizes – limone in the case of the former, and carvone in the case of the latter composition. These observations indicate that at the temperature of the arrest there is (*metastable) equilibrium* between pure solid component and (supercooled) liquid mixture.

thermodynamic and exothermodynamic correlations

The system NaCl + KCl is a member of the family of binary common-cation alkali halide system, or, in a wider sense, a member of the family of binary common-ion alkali halides. The system C19 + C21 is a member of the family of binary n-alkane systems. In a family of binary systems the components of the member systems belong to a chemically coherent group of substances.

Research on families of systems - rather than isolated systems – quite often offers the opportunity to observe trends and/or properties that are characteristic for the family as a whole. Unnecessary to remark that such trends and characteristics - framed in empirical relations - are useful tools when it comes to the assessment of experimental data and the prediction of properties of member systems for which experimental data are lacking.

excess enthalpy / excess entropy compensation

The treatment above of the region of demixing in NaCl + KCl assumes that, in the experimental temperature range of the data, the excess Gibbs energy is linear in temperature – Equations (2) and (3). If this is really the case, then the excess enthalpy will represent the excess Gibbs energy at zero Kelvin. In Figure 14 experimental equimolar excess data - the excess enthalpy on the axis $T = 0$ K - are plotted for three common-ion alkali halide systems.

From the figure it follows that, within the experimental uncertainties, the assumption of linear behaviour is justified. And, surprisingly, there is a common temperature at which the excess Gibbs energy of the three systems goes through zero.

§(304)

88

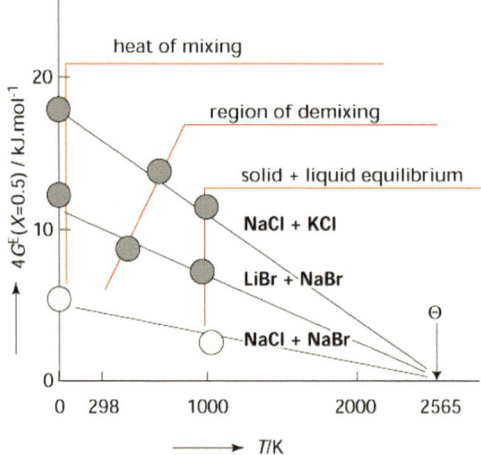

FIG.14.

Binary common-anion alkali halide systems. Representative set of experimental equimolar excess Gibbs energies as a function of temperature

A detailed analysis of the complete group of common-ion alkali halide systems has demonstrated that the excess Gibbs energies of their mixed crystalline state, indeed, are virtually linear in temperature. And moreover, within their experimental uncertainties they comply with the *ABΘ description* (Van der Kemp et al. 1992; Oonk 2001):

$$G^E(T,X) = A \, (1 - T/\Theta) \, X \, (1 - X) \, \{1 + B(1 - 2X)\} \ . \tag{8}$$

The *ABΘ* expression has three *system-dependent parameters*: the *interaction parameter A*, having the dimension of energy; the dimensionless *asymmetry parameter B*; and the *temperature parameter Θ*, which has the dimension of temperature (←212).

The evidence which is shown in Figure 14 is representative of the 21 systems (out of the 52 common-ion alkali halide systems having the NaCl type of structure) with complete *subsolidus miscibility.*

Apart from the linear aspect of the excess functions, the figure reveals that the functions all go through zero at 2565 K. In other terms, the family of systems is characterized by a *common value* for the parameter Θ, which is Θ = 2565 K. In this respect the group of systems has the status of 'class of similar systems in terms of *enthalpy-entropy compensation*' (←212; see e.g. Boots and De Bokx 1989). The temperature of 2565 K is referred to as *compensation temperature*: at this temperature H^E and S^E compensate one another in the sense that $G^E = H^E - TS^E$ becomes zero, or rather goes through zero.

§(304)

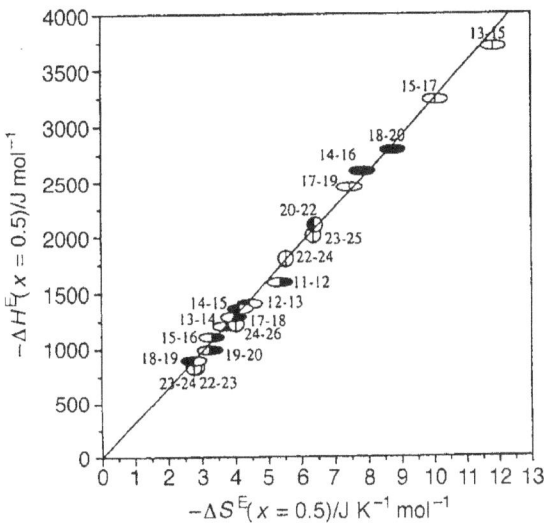

FIG. 15.
For the mixed crystalline rotator I and II forms of binary *n*-alkane systems, equimolar excess enthalpy plotted against equimolar excess entropy (Mondieig et al.1997)

The characteristic that the member systems of a class have a common value for the parameter Θ implies that, for the family as a whole, the graphical representation of excess entropy against excess enthalpy is just a straight line through the origin. For the rotator RI and RII forms of n-alkane systems this is shown by Figure 15, from which it follows that the value of Θ which is characteristic for the class is 320 K (Mondieig et al. 1997; Oonk et al. 1998).

On a general scale and in simple terms, the compensation temperatures are related to the 'hardness' of the materials. Soft materials have low melting temperatures, and their mixed crystals have low compensation temperatures. Hard materials have high melting- and compensation temperatures.
In the log-log diagram, Figure16, compensation temperatures are plotted against melting temperatures – for a number isolated systems and a number of families of systems. The relation between compensation- and (mean, in the case of a family) melting temperature, read the equimolar *EGC temperature*, is

$$\log [\Theta/K] = (1.10 \pm 0.05) \log [T_{EGC}(X{=}0.5)] \ . \tag{9}$$

Each of the two properties under the logarithms in the equation is a property that belongs to the mixed-crystalline state.

§(304)

90

A somewhat more convenient relationship is obtained by replacing the EGC temperature by a property that is related to the melting temperatures, T^o_A and T^o_B, of the components of the system. Van der Kemp et al. (1993) formulated the relationship

$$\Theta = (4.00 \pm 0.16)\, T^o_A T^o_B\, /\, (T^o_A + T^o_B)\,,\qquad\qquad(10)$$

in which the quotient of product of melting points and their sum is taken from the study by Tanaka et al. (1990).

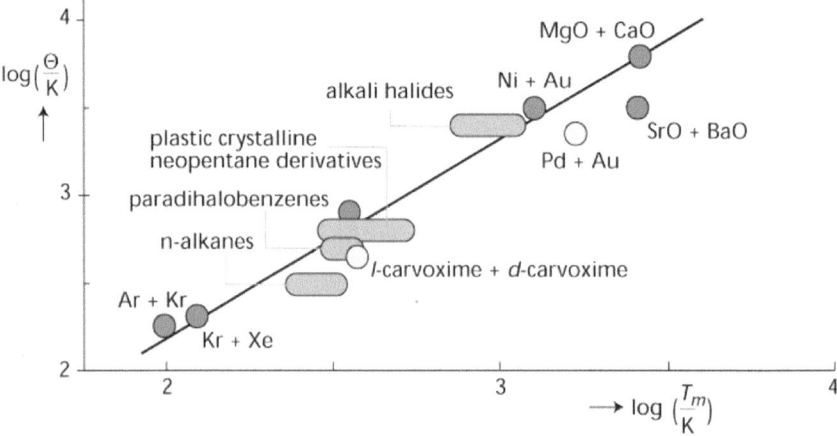

FIG. 16.

Log-log representation of the compensation temperature Θ versus 'melting point' of mixed crystals of equimolar composition. Apart from the alkali halides there are other families of mixed crystals characterized by a common Θ value, such as the *para*-dihalobenzenes (Calvet et al. 1991); the rotator I and 'ordered' forms of *n*-alkanes (Mondieig et al. 1997; Oonk et al. 1998; Rajabalee et al. 2000); a group of plastic-crystalline neopentane derivatives (López et al. 2000). A number of stand-alone systems for which Θ has been or can be calculated: Ar + Kr and Kr + Xe (Walling and Halsey 1958); *laevorotatory*-carvoxime + *dextro*-rotatory carvoxime (Calvet and Oonk 1995); Ni + Au (Sellers and Maak 1966); Pd + Au (Okamoto and Massalski 1985); MgO + CaO and SrO + BaO (Van der Kemp et al. 1994)

For further reading, we refer to the theoretical treatment, in terms of statistical thermodynamics, in Lupis's textbook (Lupis 1983, chapter XV; see also Lupis and Elliot 1967).

mismatch

As follows from Figure 14, the excess enthalpy in the system NaCl + KCl is more than three times the excess enthalpy in NaCl + NaBr. This is plausible, because in NaCl it is 'easier' to replace the Cl⁻ ion (ionic radius 1.81 Å) by the 8% larger Br⁻ ion (1.96 Å), than to replace the Na⁺ ion (1.02 Å) by the 35% larger K⁺ ion (1.38 Å) – ionic radii according to Shannon (1976) for coordination number VI.

In other terms, the parameter A, in Equation (38), is system-dependent: the greater the relative difference (*mismatch*) in size between the units that replace one another, the greater parameter A's value.

For practical reasons it is convenient to replace mismatch in ionic radii by mismatch in molar volume (of the two components of the system)

$$m = \Delta V / V_s,\qquad(11)$$

where V_s is the smaller of the two molar volumes; see Figure 17 in which the A values of 13 common-ion alkali halide systems are plotted as a function of mismatch in molar volume.

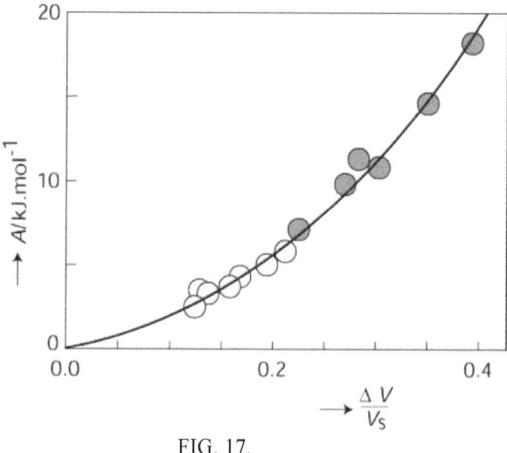

FIG. 17.
Common-ion alkali halide systems. The parameter A of the $AB\Theta$ model Equation (38) as a function of mismatch in molar volume between the two components of the system. Symbols in light grey pertain to systems that show complete miscibility at room temperature; dark grey for systems showing complete subsolidus miscibility along with limited miscibility at room temperture

The curve which is drawn in Figure 17 is given by the equation

$$A = (10.53\ m + 90.68\ m^2)\ \text{kJ mol}^{-1}.\qquad(12)$$

§(304)

The asymmetry parameter B of the $AB\Theta$ model is positive, that is to say if the system $\{(1-X) \text{ mol } A + X \text{ mol } B\}$ is defined such that A is the component with the smaller molar volume. A representative value is $B = 0.2$. Its dependence on m is difficult to assess; an approximation is

$$B = 0.4\,m \tag{13}$$

For the common-ion alkali halide systems the influence of the mismatch parameter is such that, in terms of Equations (12) and (13),

- for $m < 0.19$ the components of the system are completely miscible at $T = 298$ K;
- for $m < 0.44$ there is complete subsolidus miscibility;
- for $m > 0.44$ the solid-liquid phase diagram displays a eutectic three-phase equilibrium;
- for $m > 0.47$ the solubility of the components in one another at $T = 298$ K is less than 0.1 %; for $m > 0.67$ this is less than 1 ppm; and for $m > 0.84$ it is less than 1 ppb.

Like the common-ion alkali systems, mixed crystals in the family of binary n-alkane systems, Cn + C(n+Δn) where Δn positive, represent a class of similar systems in terms of enthalpy-entropy compensation. For these systems the obvious mismatch parameter is the relative difference in *chain length:*

$$m = \Delta n / n_m . \tag{14}$$

For the rotator form RI a system's parameter of the $AB\Theta$ model is related to its mismatch parameter m, with n_m in Equation (14) given by n+0.5Δn, as (Rajabalee et al. 2000)

$$A = (51.09\,m + 376.5\,m^2)\,\text{kJ mol}^{-1} . \tag{15}$$

ionic potential

The phenomenon of partial liquid miscibility, shown by mixtures of silica and metal oxides, plays an important part in petrology (see e.g. Ferguson and Currie 1972) and also in glass technology (see e.g. Rawson 1967). A fact is that the degree of immiscibility is related to the value of the *ionic potential*, z/r, i.e. the ratio of charge and radius of the metal ion, the *network modifier*.
The evidence is such that

- partial immiscibility sets in for z/r values of about 1.6 Å^{-1}; for lower values the region of immiscibilty is metastable, below the liquidus;
- the regions of immiscibility lie at the SiO_2 side of the phase diagram; within each series of given z, they increase with increasing z/r in the direction of both temperature and composition.

§(304)

The observed regularities, as has been shown (Oonk et al. 1976), can be included in one *empirical relation* for the nonlinear part of the Gibbs energy of mixing – by introducing *normalized chemical formulae* for the metal oxides and by choosing a suitable *expression for the excess Gibbs energy.*

Normalized to one oxygen, the formulae for the oxides of the the mono- (I); di- (II); tri- (III); and tetravalent (IV) metals are I_2O; IIO, $III_{2/3}O$; and $IV_{1/2}O$. With these formulae the systems, taking $SiO_2 + IIO$, are defined as {(1-X) times the formula mass of $SiO_2 + X$ times the formula mass of IIO}. It gives all the systems involved the same relation between X and the Si:O ratio, the latter being an important parameter in statistical treatments (see above; and e.g. Charles 1969).

In the usual expression for the Gibbs energy of mixing

$$\Delta_m G(T,X) = RT \{(1-X) \ln(1-X) + X \ln X\} + G^E(T,X) , \tag{16}$$

the excess part is taken as

$$G^E (T,X) \rightarrow G^E (X) = A (n + 1)^{(n+1)} n^{-n} (1-X)^n X . \tag{17}$$

This expression, Equation (17), is such that the maximum of the function is at $X = 1/(n+1)$. In addition, at its maximum the value of the function is equal to the value of the parameter A.

For this model, the expression of the *spinodal* is

$$RT_{spin}(X) = A(n +1)^{(n+1)} n^{-n} X(1-X) \{2n(1-X)^{(n-1)} -n(n-1)X(1-X)^{(n-2)}\} . \tag{18}$$

The X value of the critical point, the maximum of the spinodal is given by

$$X_c = \{(2n +1) - \sqrt{(2n^2-1)}\}/(n +1)^2 . \tag{19}$$

When for given n the value of X_c is introduced in Equation (18), the ratio ρ

$$(\rho =) A(n +1)^{(n+1)} n^{-n}/RT_c$$

is obtained. This ratio is 2.000 for $n = 1$; 2.130 for $n = 2$; 2.159 for $n = 4$; 2.164 for $n = 6$; and 2.166 for $n = 8$.

Using the experimental data for Li, Ba, Sr, Ca, Mg, La, Nd, Sm, Gd, Dy, Y, Er, Yb, U, Sc, and Zr, the following expressions are obtained for the exponent n and the *interaction parameter A*:

$$n -1 = - 0.206 (z/r)^{-1} - 12.57 (z/r)^{-2} ; \tag{20}$$

$$A = \{- 3.21 + 3.82 (z/r) - 0.29 (z/r)^2\} \, kJ\,mol^{-1} . \tag{21}$$

The ionic radii to be used are those given by Shannon (1976).

§(304)

The model contained in Equations (16) and (17) has the mathematical advantage that, thanks to the power n in $(1-X)^n$, the excess Gibbs energy is sharpened at the SiO_2 side – thereby giving rise to a unilaterally positioned, region of demixing. There is some physical realism in the model, in the sense that, at the side of the glass former (SiO_2), the metal ions of the modifier (such as the Sm^{3+} ions from Sm_2O_3) are concentrated in the cavities of the network – the positive (excess) energy effect of their repulsion is increasing with increasing ionic potential z/r.

concluding

time and speed / the system and the investigator

Phase equilibria are studied for many reasons – to verify a physical theory, to determine thermodynamic properties, to serve an industrial goal, to optimize material properties, to reconstruct an event in the past, and so on. People who study phase equilibria are from diverging branches of (materials) science, ranging, say, from chemistry to geology.

A chemist creates his own systems out of a selection of pure substances. As an example, a chemist may take the combination of acetone and chloroform, and decide to study, for a given temperature, the equilibrium between liquid and vapour – in order to calculate, from the measured pressures and the compositions of the phases, the excess Gibbs energy of the liquid mixture as a function of composition. A geologist, on the other hand, frequently investigates samples, e.g. a lunar rock, created by nature. At the start of the investigation, the geologist first has to analyze the rock as to the presence of phases and their chemical composition. After the analysis he/she may have a clue about the conditions of temperature and pressure under which the rock was formed.

The equilibrium between liquid and vapour in the system chloroform + acetone establishes itself at high speed – due to the high mobility of the molecules in each of the two phases. Equilibria that involve mixed solid phases, on the other hand, need much more time, owing to the limited degree of mobility of the structural entities in the solid phases. Equilibrium is reached when there are no more changes – and quite often, the investigator has lost his patience before the equilibrium state has been reached.

To circumvent the problem-of-time-to-reach-equilibrium, the investigator may wish to opt for a method of non-equilibrium. A method, for instance, by which a sample, of given overall composition, is subjected to a programmed rate of heating. If such is the case, the investigator must realize that the sample's course, as a rule, is not a succession of equilibrium states; or in other words that the recordings of the instrument are not a 1 to 1 image of the states of equilibrium. Moreover, the results produced by the instrument not only depend on the rate of heating but also on the thermal history of the sample, in particular how it had been prepared. In addition, if the system to be studied does involve (if it would follow a sequence of equilibrium states) not only a change of state, but also phases of different compositions, one should realize that the change of state is a quasi-instantaneous event, whereas the adjustment of phase composition can be a time-consuming process (in the limiting case a homogeneous mixed crystal can be made to produce liquid of the same composition; ←Exc 301:4).

Spontaneous changes, such as the crystallization of material from a supercooled liquid in a heating experiment, are fascinating and at the same time complicating phenomena. Occasionally, spontaneous changes mislead the investigators in that they run into a false interpretation (→Exc 11).

Each phase, out of a number of phases in equilibrium, is characterized by a certain form. The determination of the forms of the phases is an essential part of phase-equilibrium research. – that is to say, as long as the phases are not simply vapour or (isotropic) liquid. The characterization of forms is a real challenge when polymorphism goes together with isomorphism – especially when monotropism is involved. X-ray diffraction is one of the most powerful tools for the characterization of a form is

from observations to phase diagram / the use and usefulness of thermo

At the start of this section, a distinction was made between the *observation diagram* and the *true phase diagram*. The observation diagram is a representation of the characteristic details of the investigation carried out. The observation diagram leaves room for subjectivity; observation diagrams from different investigations on the same system are never identical. The true phase diagram - the desired goal of the investigation - on the other hand, is unique, and free from errors and misinterpretations.

Ideally the observation diagram is a close image of the true phase diagram: each symbol out of a set of similar symbols in the observation diagram is close to the corresponding curve in the true phase diagram – and so for each set of symbols. Quite frequently there is correspondence (between set of symbols in the observation diagram and the corresponding curve in the true phase diagram) for one set of symbols, and at the same time serious disagreement for another set.

There are a number of simple thermodynamic tools to assess how close an observation diagram reflects the true phase diagram. Tools like the *Konowalow rules;* the *laws for initial slopes;* the *rules for metastable extensions;* and so on. It is unnecessary to remark that these tools preferably should be put into action by the investigator during his investigation, and, at any rate before the publication of 'his' phase diagram.

It is recommended that the investigator, in his publication, should present not only a diagram but also a table with the observed pressure-, temperature-, and phase composition values.

§(304)

96

In thermodynamic phase diagram analysis the parameters of Gibbs energy functions are varied until maximum agreement is obtained between computed phase diagram and set of experimental data. The experimental data are available in the form of a table, or, less favourably, have to be read from a published phase diagram. In most of the cases, the parameters to be adjusted are the constants of the expressions that are used for the excess Gibbs energies – the excess Gibbs energies of the forms that take part in the equilibrium. The output of the computations is the optimized phase diagram, along with the numerical values of the adjusted parameters. The computed phase diagram, at any rate, is free from thermodynamic inconsistencies.

The excess Gibbs energy functions that are used for optimization purposes are manifold, and may or may not be rooted in a physical theory. It is not impossible that two or more diverging functions are capable of reproducing the experimental data with equal precision. And this observation implies that agreement is a sign of mathematical rather than physical power of the function, the model.

In cases where, for the analysis of isobaric equilibria, the Gibbs functions are split up in enthalpy and entropy parts, and both with their own adjustable parameters, it is found that the computed enthalpy and entropy functions, to a greater or lesser extent, disagree with the real enthalpy and entropy functions. The other way round, experimentally available excess enthalpies can be introduced as part of the input data of a phase-diagram analysis, in order to calculate the excess entropies.

In cases where the number of experimental data points is limited, one may opt to create a dummy set of data – a set of data read from either a mathematical or an 'optical' fit of the original data. Occasionally, the agreement between the original data and the computed phase diagram using a dummy set turns out to be better than the agreement obtained with the original data themselves. It is recommended, if not imperative, that one should consult the source of the experimental data, every time a thermodynamic analysis is planned.

And last but not least, a thermodynamic analysis carried out before the publication of experimental data may act as a stimulus to reconsider the experimental methodology and/or the interpretation of observations.

===
Thermodynamics is the most powerful instrument for the investigator to understand his observations on the behaviour of a given system in a given experiment, and to recognize and repair systematic errors. Empirical relationships are useful tools to give a helicopter view of properties displayed by a family of chemically coherent systems.
===

§(304)

EXERCISES

useful equations

In terms of the $AB\Theta$ *model*, and the system being defined as $\{(1\text{-}X) \text{ mol } A + X \text{ mol } B\}$, the equation for the *spinodal* (SPIN) is

Equation (212:19)

$$T_{SPIN}(X) = \frac{2A \cdot X(1-X)[1+3B(1-2X)]}{R + \dfrac{2A}{\Theta}X(1-X)[1+3B(1-2X)]} \ .$$

The equation for the *equal-G curve* (EGC), neglecting heat capacities, is

Equation (212:34)

$$T_{EGC}(X) = \frac{\Delta H(X)}{\Delta S(X)} = \frac{(1-X)\Delta H_A^* + X\Delta H_B^* - H^{E\alpha}(X)}{(1-X)\Delta S_A^* + X\Delta S_B^* - S^{E\alpha}(X)} ,$$

in which the $AB\Theta$ expressions for the excess properties can be substituted, and in which, whenever appropriate, the excess properties of the other state (β) can be introduced.

The equation for the *liquidus*, pertaining to the equilibrium between pure solid A and liquid mixture complying with the $AB\Theta$ model, is given by

Equation (212:42)

$$T_{LIQ}(X) = \frac{\Delta H_A(X)}{\Delta S_A(X)} = \frac{\Delta H_A^* + X^2 \cdot A[1+B(3-4X)]}{\Delta H_A^*/T_A^o - R\ln(1-X) + X^2(A/\Theta)[1+B(3-4X)]} .$$

§(304)

1. *prediction by analogy*

The figures in the matrix which is printed represent the molar volumes of the alkali halides, calculated from the densities given in the *Handbook of Physics and Chemistry*, and expressed in $cm^3 mol^{-1}$. With the exception of CsCl, CsBr, and CsI, which have the *caesium chloride type of structure*, the substances have, at room temperature, the *sodium chloride type of structure*.

	Li	Na	K	Rb	Cs
F	09.84	16.42	23.43	29.37	36.91
Cl	20.50	26.99	37.58	43.19	*42.22*
Br	25.07	32.12	43.27	49.36	*47.93*
I	32.84	40.88	53.04	59.82	*57.61*

The substance CsCl takes a special position in the sense that, at normal pressure, it changes from the caesium chloride type of structure to the sodium chloride type of structure at 743 K with an enthalpy effect of 3.766 $kJ mol^{-1}$.

It is not unreasonable to state that the $AB\Theta$ model, along with the numerical values of its parameters given by Equations (10)-(13), has the capacity of being a good starting point for the prediction of the thermodynamic mixing properties of common-ion alkali halide mixed crystals that have the caesium chloride type of structure.

- First of all and from the trends displayed by the molar volumes, estimate the molar volumes the three substances CsCl, CsBr, and CsI would have if they had the sodium chloride type of structure; and also for CsF if it had the caesium chloride type of structure.

The systems CsF + CsBr and CsF + CsI with their high mismatches certainly will have limited solid-state miscibility. When liquid, the components of the two systems, and also the components of CsF + CsCl, mix with a negative heat effect. In rounded figures, the heats of mixing at equimolar composition, expressed in $J mol^{-1}$, are -300 (F+Cl); -1000 (F+Br) and -1620 (F+I); (see Sangster and Pelton 1987).

- Ignoring any solid-state miscibility, estimate the coordinates of the eutectic point in CsF + CsI. Pure-component data are given below.

In the case of the systems CsF + CsCl, CsCl + CsBr, and CsCl + CsI, the stable phase diagram is the result of a competition between the two mixed-crystalline forms (the following three tasks are rather time-consuming).

§(304)

- For the system CsCl + CsBr with its relatively low mismatch, first infer if the components mix in all proportions at 298 K, and next make a sketch drawing, for $500 \leq (T/K) \leq 1000$, of the phase diagram you expect.

- For CsCl + CsI make a sketch drawing, for $500 \leq (T/K) \leq 1000$, of the phase diagram.

- Out of the three systems with CsCl, the system CsF + CsCl is the most complex. Make a sketch drawing of the phase diagram you expect, again for $500 \leq (T/K) \leq 1000$.

themochemical data

Thermochemical data for the change from solid to liquid. Fusion data from Dworkin and Bredig (1960); transition data from Clark (1959)

	CsF	*CsCl*	*CsBr*	*CsI*
melt. pt (K)	976	918	909	899
ΔH (kJ·mol^{-1})	21.72	20.25	23.60	23.59
trans. pt (K)		742		
ΔH (kJ·mol^{-1})		2.90		

NB. A clever choice should be made of the melting points and melting enthalpies of *metastable forms*.

2. *has minimal Gibbs energy been reached ?*

The seven data triplets - read from the diagram published by Ahtee and Koski (1968) - pertain to the system RbBr + RbI.

T / K	213	233	253	273	293	313	333
X^I	0.13	0.13	0.145	0.16	0.18	0.21	0.26
X^{II}	0.82	0.80	0.78	0.75	0.71	0.64	0.54

It is evident that phase separation in solid solutions of RbBr + RbI at temperatures around 0 °C is slower than the separation around 400 °C in NaCl + KCl. And the more so, because in RbBr + RbI the larger ions, the anions have to change place; in contrast to NaCl + KCl where the cations have to do so. Indeed, Ahtee and Koski had to apply moist to speed up the separation process.

The purpose of this exercise is to assess to what extent the data are representing the states of *minimal Gibbs energy* contained in the $AB\Theta$ description. The following actions are suggested.

§(304)

- From the molar volumes (see Exc 1) calculate the mismatch m between the components, and use m to calculate A and B. From A and B, taking Θ = 2565 K, calculate the coordinates of the critical point of mixing. Adjust A's value to reproduce the experimental critical temperature which is (346 ± 3) K.
- For two or three isothermal sections construct the GX (Gibbs energy versus mole fraction) diagram and use it to determine the mole fractions of the coexistent phases.
- Construct the TX diagram showing the experimental data points along with the binodal following from the isothermal sections and the critical point. Conclusions?

3. *RbBr + RbI as an isolated system*

This time the existence of correlations is ignored and the assumption is made that the data in the table in Exc 2 are representing the true states of minimal Gibbs energy.

- Use the description contained in Equations (2)–(5) to calculate, by means of Equations (5a,b), the values of g_1 and g_2 for each of the seven isothermal sections. By linear least squares determine the values of h_1 and s_1 and h_2 and s_2.
- Transform the result to the $AB\Theta$ description, taking B as the mean of the seven g_2/g_1 quotients.
- With the $AB\Theta$ values, calculate the coordinates of the critical point, and repeat the last two actions of Exc 2.

4. *crossed isotrimorphism*

In Figure 6 each of the three loops is meant to reflect the change of one of the 'normal' solid forms - Op, Mdci, Oi - to the rotator form RI in the case of system C19 + C21; see Figure 4. For the pure components the transition temperatures increase in the order Op→Mdci→Oi.

- Construct the stable phase diagram. And, use vertical lines for the boundaries of each of the two-phase regions between low-temperature forms.

§(304)

5.   *the isothermal transition in C19 + C21*

For the binary n-alkane systems the energetic differences between the forms of the 'normal' solid state are small – and it means that the heat effect of their change to the rotator state RI, as a function of chain length n, can be approximated by one and the same formula (Rajabalee et al. 1999):

$$\Delta H^* = (5.44 - 0.462\,n + 0.0462\,n^2)\ kJ\,mol^{-1}.$$

- Using the temperatures of the isothermal transition, Table 3, and 289.0 K and 302.3 K for the transition temperatures of C19 and C21, respectively, evaluate the transition property $\Delta G^E(X)$.
- Calculate the excess Gibbs energy of the (metastable) form Mdci at equimolar composition.

6. *time for a paradox?*

In the realm of thoughts it is quite easy to create a fantasy system, whose excess properties of the liquid and solid states are such that, in a range of compositions, the mixed-crystalline material does not melt anymore.

As a numerical example, inspired by the properties of the n-alkane systems, a binary system is taken whose liquid state is ideal and whose mixed-crystalline, solid state complies with the $AB\Theta$ model. The properties of that system are as follows. First component: melting point 300 K; heat of melting 10 kJ·mol⁻¹. Second component: 320 K; 12 kJ·mol⁻¹. The constants of the model are $A = 50\ kJ\,mol^{-1}$; $B = 0$; and $\Theta = 300$ K.

NB The large solid state excess enthalpy makes that the heat of melting is negative in the interval $0.305 < X < 0.654$.

- Calculate the position of the *equal-G curve* and also the position of the *spinodal*.
- Draw a sketch of the phase diagram, for $200 \leq (T/K) \leq 400$.

§(304)

7. *the AnΘ model*

As an extension of the model contained in Equations (16) and (17), the system-dependent constant A is made a linear function of temperature:

$$G^E(X) \rightarrow G^E(T,X) = A\,(1 - T/\Theta)\,(n+1)^{(n+1)}\,n^{-n}\,(1-X)^n\,X.$$

The extended model has been used by Hageman & Oonk (1986) to construct a mathematical representation of the binodal curves, based on their experimental region-of-demixing data, for the systems $\{(1-X)\,A + X\,B\}$ with $A = SiO_2$ and $B = SrO$; CaO; MgO; $La_{2/3}O$; and $Y_{2/3}O$. For these systems the relation between Θ and T_c is roughly given by $\log\Theta = 1.04\,\log T_c$.

NB For the construction of the Equations (20) and (21) use was made of experimental data pertaining to the top parts of the regions of demixing.

- For the system $\{(1-X)\,A + X\,B\}$ formulate the expressions for the *spinodal*; component A's *excess chemical potential*; and the *liquidus* for the equilibrium between pure solid A and liquid mixture complying with the *AnΘ* model (ignoring the change with temperature of A's heat of melting).

The expressions to be formulated in Exc. 7 are needed for the following two exercises in 'curve fitting with thermodynamic models'.

8. *Kracek and the heat of fusion of cristobalite*

In his 1930 paper Kracek remarks that "it is desirable to know the exact course of the cristobalite liquidus for alkali silicate mixtures rich in silica" – one of the reasons being that "the curves might be expected to furnish a basis for the calculation of the heat of fusion of cristobalite, a quantity which has not been measured in the calorimeter".

Using the initial slopes of the liquidus curves in the rubidium and caesium systems - the alkali systems having the smallest deviations from ideal behaviour - Kracek arrives at the value of 7.7 kJ·mol^{-1} for cristobalite's heat of fusion ($\leftarrow$302: Equation 20).

Strictly speaking, the value of 7.7 is obtained for $f = 1$, where f is the so-called *Van 't Hoff factor* – the number of species one molecule of the solute is giving rise to (for example, for the system defined by $\{(1-X)\,H_2O + X\,NaCl\}$ the factor is given by $f = 2$, because of the electrolytic dissociation of NaCl into the ions Na^+ and Cl^-). In other terms, for models that correspond to $f = 2$ Kracek's initial slope would imply a heat of fusion of two times 7.7 kJ mol^{-1}, where one mole of SiO_2 has the mass of 60 g. And if the caesium silicate system defined by $\{(1-X)\ SiO_2 + X\ Cs_2O\}$ would, for $X \to 0$, imply the reaction $Cs_2O + n\ SiO_2 \to 2\ Cs^+ + Si_nO_{2n+1}^{2-}$, the factor would be $f = 3$.

The value assessed by Richet et al. (1982) for cristobalite's heat of fusion is 8.92 kJ mol^{-1} with an estimated uncertainty of 1 kJ mol^{-1}.

In this exercise Kracek's definition of the lithium silicate system, i.e. $\{(1-X)\ SiO_2 + X\ Li_2O\}$, is adopted, and the $An\Theta$ model with $\Theta = \infty$ is adopted to assess the heat of fusion of cristobalite, using Kracek's data for the system.

X_{Li2O}	0.0000	0.0583	0.0964	0.1405	0.17 [*])
$t/\ ^\circ C + 273$	1986	1895	1854	1808	1743 [*])

[*]) data point read from figure in original paper

Samples were equilibrated at the selected temperature during 20 minutes, quenched to room temperature, and analyzed under the microscope. The sample having $X = 0.1405$ gave "cristobalite + glass" after equilibration at 1529 °C; and "glass + very rare cristobalite" after equilibration at 1535 °C.

- For $A = 1314$ J mol^{-1} and $n = 8.1$ (which are the values generated by Equations (20) and (21) for $z = 1$ and $r = 0.76$ Å) vary cristobalite's heat of fusion until the best agreement (lowest value for Δ_T), is obtained between the experimental liquidus temperatures and the calculated ones.

9. *an inconvenient result*

For the system $\{(1-X)\ SiO_2 + X\ BaO\}$ the empirical relationships, contained in Equations (20) and (21), produce the following numerical values for the system-dependent parameters: $n = 6.67$, and $A = 1780$ J mol^{-1}. With these values the critical point is calculated at $X_c = 0.084$, and $T_c = 1922$ K; and it means that the region of demixing is metastable, i.e. below the liquidus (cf. Seward et al. 1968).

Experimental liquidus data for the system have been determined by Greig (1927) in the Geophysical Laboratory, Carnegie Institution of Washington. Greig's methodology was the same as the one adopted by Kracek ($\leftarrow$Exc 4). At the time of Greig, the thermoelement was calibrated at the melting points of ice, gold (taken as 1062.6° C ± .8° C), palladium (1549.5° C ± 2° C) and platinum (1755 °C ± 5 °C). Modern values, expressed in kelvin, are 1337.58 (Au), 1825 (Pd), and 2045 (Pt). The melting point of cristobalite was given by Greig as (1723 ± 5) °C; and for the mixed samples he gave the following liquidus temperatures:

X	0.0202	0.0417	0.0647	0.1155	0.1742	0.2071	0.2488
$t\,/\,°C$	1688	1683	1679	1674	1636	1583	1489

- First of all, and after applying thermometric corrections, construct the experimental liquidus curve; and from it derive a *dummy set of experimental data*, such that the new set, apart from the melting point of cristobalite, is composed of ten TX data pairs taken from 1975 K to 1750 K at 25 K intervals.
- Next, taking Richet's value of 8.92 kJ mol^{-1} for the heat of fusion of cristobalite, and ignoring any heat capacities, calculate, for each of the ten dummy data pairs, the value of SiO_2's excess chemical potential in the liquid mixture. From the excess chemical potentials, taking $n = 6.67$, calculate, in terms of the $An\Theta$ model, for each of the ten 'experimental' temperatures the value of $A(T) = A(1-T/\Theta)$. By least squares, make a linear fit of $A(T)$ as a function of T, and from the result derive the values of the constants A and Θ. Use the calculated A and Θ, along with $n = 6.67$ and the heat of fusion of cristobalite, to calculate for the ten 'experimental' mole fractions their corresponding liquidus temperatures; and for the complete result the value of the index Δ_T ($\leftarrow$Exc 7).
- Repeat the calculations, taking two times Richet's value for the heat of fusion of cristobalite ($\leftarrow$Exc 8).
- For the two sets of $An\Theta$ values make a graphical representation - from $X = 0$ to $X = 1$ - of the spinodal curves they give rise to. And, finally, formulate your conclusions.

10. *the system d-limonene + l-carvone by adiabatic calorimetry*

By adiabatic calorimetry a liquid mixture of d-limonene and l-carvone, containing 70 mole% of the latter, is cooled to 130 K, subsequently heated to 185 K and cooled down again to 130 K, and finally heated to 250 K.
- Enumerate the events which successively take place during each of the two heating runs. Whenever appropriate, assess the relative amounts of the vitreous, liquid and crystalline parts of the sample; and also the magnitude of the jump in heat capacity.

§(304)

11. *the system l-limonene + l-carvone by DSC*

In DSC experiments the systems and *l*-limonene + *l*-carvone and *d*-limonene + *l*-carvone show virtually the same behaviour. The set of DSC signals (Figure 9; Table 4) are similar to those found for a series of mixed crystals. The question which arises is this: do limonene and carvone under the circumstances of DSC experiments give rise to a series of *metastable* mixed crystals (first option); or is it so that the observed behaviour is caused by *dual crystallization*, and fully in line with the information contained in Figure 13 (second option)?

- For the first option: in a *TX* diagram plot the onset and end temperatures of melting, that are displayed in Table 4, and also the melting point of pure *l*-limonene (199 K); by hand draw the solidus and liquidus curves; from the diagram and the melting properties of the pure components (the heat of melting of pure limonene is 11.4 kJ mol^{-1}) assess the equimolar value of ΔG^E, the change in excess Gibbs energy on melting. Next, use the heat effects displayed in Table 4 and the heat of melting of pure limonene, to construct a diagram that represents the heat of melting as a function of composition, and from that diagram read the equimolar value of ΔH^E. Finally, and assuming that the liquid state is an ideal mixture, use the equimolar properties to assess the solid state's values of the A and Θ parameters of the $AB\Theta$ model.
- For the second option: use the information contained in Figure 13, to find out if it is possible to give a qualitative interpretation of the behaviour of the mixed samples in DSC experiments, including the absence of any effects for samples having less than 50 mole% of limonene. By a quantitative calculation (lever-rule like) demonstrate that the onset temperature of melting, for a given compostion (take X = 0.8), is consistent with the heat effect registered for that composition.
- Which option do you prefer, and why?

12. *Würflinger's lower loop, an EGC analysis*

The lower loops in the diagrams for C19 + C21, published by Würflinger and Schneider (1973, see Figure 3), Mazee (1957), and Maroncelli et al. (1985),

§(304)

106

all suggest that the two alkanols give rise to a *continuous* (i.e. from X=0 to X=1) *series of mixed crystals* at temperatures below about 9°C. However, in an indirect manner, it can be demonstrated that the course of the lower loop is such that it precludes the existence of a continuous series of mixed crystals.

NB. In an indirect demonstration it is assumed that the statement at hand is true, and subsequently it is shown that it (the statement) is giving rise to contradictory conclusions (for an example ←Exc 004:6).

X	T_{EGC} / K	X	T_{EGC} / K	X	T_{EGC} / K
0.000	294.9	0.376	282.6	0.784	294.7
0.091	285.0	0.475	284.5	0.891	299.3
0.185	282.2	0.576	287.0	1.000	305.0
0.280	281.8	0.679	290.5		

The table gives a set of points on the equal-G curve for the transition from 'normal' solid to rotator, derived from Figure 3.

For the indirect demonstration the following actions are proposed.

- Use the data to construct the *EGC diagram*. Next use the diagram to create a dummy set of 21 data pairs *TX* – from X=0.00 to X=1.00 in steps of 0.05.
- In Equation (6) replace minus $G^{E\alpha}$ by ΔG^E, and use the equation, along with Equation (7), to calculate the value of ΔG^E for each of the (21−2) data pairs.

13. *a delicate technique for the determination of cooling curves*

The table which is printed below is part from Table II in the publication by Campbell and Prodan (1948), who designed "an apparatus for refined thermal analysis", i.e. an experimental set-up for the determination of cooling curves. In their publication one can read the statement:
"The great delicacy of the technique described is proved by the fact that the temperatures of complete solidification and hence the form and the direction of the solidus line are clearly discernible on the cooling curves".

§(304)

The data are for the system $\{(1-X)$ 1,4-dichlorobenzene $+ X$ 1,4-dibromobenzene$\}$ – a key system for the study of mixed crystals. For more information, see also Exc 301:4. For the *pure components*, A=ClCl and B=BrBr, the changes in Gibbs energy on melting are, in $J\,mol^{-1}$,

$$\Delta G^*_A = (18027 - 55.29T/K) \text{ and } \Delta G^*_B = (20387 - 56.56T/K).$$

NB. Observe that the independent information in these expressions is no more than the heat of melting along with the melting point.

X	0.4008	0.4796	0.5706	0.6489	0.7493	0.8501	0.9503	1.0000
liq	63.81	67.05	71.13	74.18	78.26	82.11	85.70	87.30
sol	59.14	62.93	67.09	70.66	75.62	80.09	84.59	87.30

The liquidus temperatures (°C) represent the onset temperature of crystallization, which can be read from the cooling curve with high accuracy. As a matter of fact, Campbell and Prodan's liquidus temperatures have an almost unparalleled accuracy (for a discussion see Van der Linde et al. 2002). The solidus temperatures (°C) represent the temperature at which the tail of the cooling curve starts to develop (in the case of Figure 1 at the start of this section: the point of intersection of the tangent line at the point of inflexion in the tail, at about 550 °C, and the straight line through the first part of the curve after the onset of crystallization).

To assess the value of the statement, printed above in italics, the following 'prior-to-publication-actions' are suggested.

- Make a plot of the liquidus data, and from the initial slope of the liquidus at the BrBr side calculate the intitial slope of the solidus. Use the result, along with the information given, to construct a complete, tentative phase diagram.

- Use the tentative phase diagram to make a guess of the EGC position for the temperature of the second liquidus point in the table. Next, by means of the EGC equation

$$(1-X)\Delta G^*_A(T) + X\Delta G^*_B(T) - \Omega^{sol} X(1-X) = 0,$$

evaluate the numerical value of Ω^{sol}.

§(304)

- From the equality of B's chemical potentials in liquid and solid, and after substitution of their recipes, calculate for the temperature of the second liquidus point, along with the value found for Ω^{sol}, the position of the solidus.

- What is your assessment of the statement printed above in italics?

from § 005

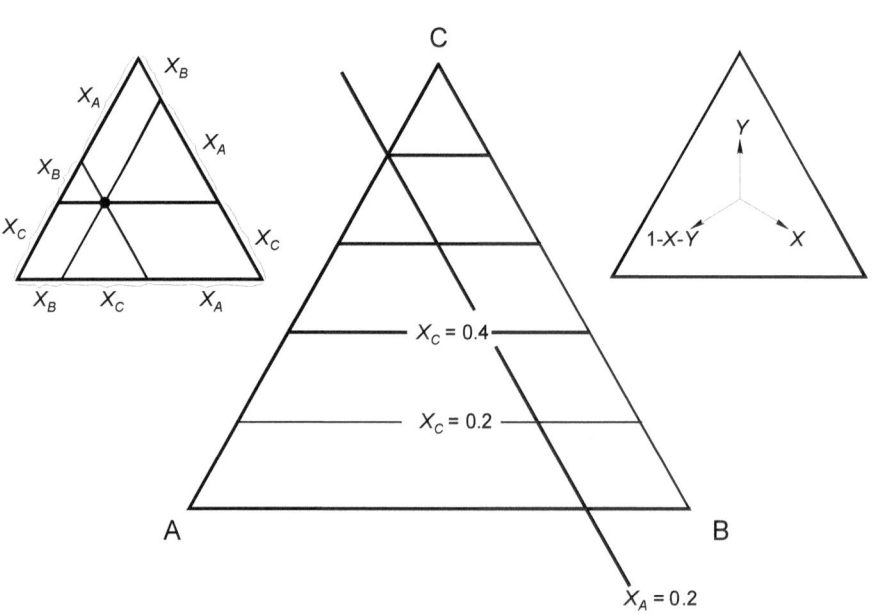

Gibbs' composition triangle

§(304)

==

Binary systems, unlike unary ones, reflect the interaction between different elementary entities. It is plausible to assume that the set of binary subsystems contain almost all of the ingredients needed for the prediction of multicomponent behaviour. In this section the emphasis is on ternary systems.

==

combination of binary excess functions

Let us suppose that (T, P constant) the interaction between A and B, in terms of excess Gibbs energy, is given by

$$G_{AB}^{E} = \Omega_{AB} \cdot X_A \cdot X_B , \tag{1}$$

and such that the interaction is independent of the presence of any other components; and, in addition, such that Ω_{AB} 's value is determined by the ratio X_A/X_B in such a manner that, for the binary (A + B), it can be given as

$$\Omega_{AB} = a + b\left(X_A - X_B\right) . \tag{2}$$

Let us next suppose, that the excess Gibbs energy in the ternary system (A + B + C) can be given as

$$G_{ABC}^{E} = \Omega_{AB} \cdot X_A \cdot X_B + \Omega_{BC} \cdot X_B \cdot X_C + \Omega_{AC} \cdot X_A \cdot X_C ; \tag{3}$$

and ask ourselves if, under the assumptions made, Ω_{AB} will be correctly given by Equation (2).

To find the answer (Y/N), we take a binary system with $X_A = 0.75$ and $X_B = 0.25$, and compare it with a ternary having the same A to B ratio, such that $X_A = 0.375$ and $X_B = 0.125$. According to Equation (2), the value of Ω_{AB} is given by $a + b$ (0.75 – 0.25) = $a + 0.5\ b$.

For the ternary, on the other hand, Equation (2) gives a different value:

$a + b$ (0.375 – 0.125) = $a + 0.25\ b$.

The inconsistency can be repaired by the division by ($X_A + X_B$), i.e. by changing Equation (2) into

110

$$\Omega_{AB} = a + b \ \frac{X_A - X_B}{X_A + X_B}. \tag{4}$$

The example also demonstrates that, if the division by $(X_A + X_B)$ is omitted, the influence of the parameter b is reduced. This comes down to observing that for the ternary mixture an Ω_{AB} would be taken that would correspond to a binary mixture with a lower $\left|(X_A - X_B)\right|$ - i.e. a binary mixture closer to the equimolar composition.

Assuming that the suppositions made are also valid for the interactions between B + C and between A + C, the ternary excess Gibbs energy is given by

$$G_{ABC}^E = \left(a_{AB} + b_{AB} \frac{X_A - X_B}{X_A + X_B}\right) X_A \cdot X_B + \left(a_{BC} + b_{BC} \frac{X_B - X_C}{X_B + X_C}\right) X_B \cdot X_C +$$

$$+ \left(a_{AC} + b_{AC} \frac{X_A - X_C}{X_A + X_C}\right) X_A \cdot X_C \tag{5}$$

The combination of binary excess functions according to Equation (5) can be visualized by Figure 1a – remembering (←005:Exc 9) that, in the composition triangle, the loci of mixtures having the same X_A to X_B ratio are given by straight lines originating from vertex C. The combination according to Equation (5) is referred to as *Kohler's model* (Kohler 1960).

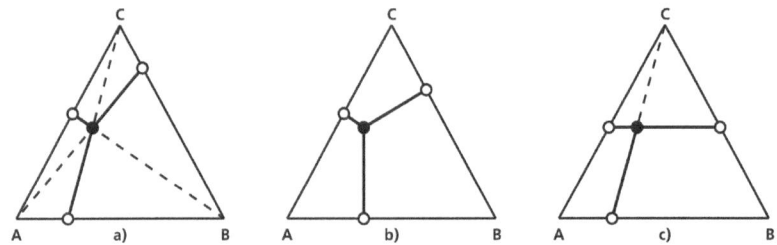

FIG. 1.
Binary excess functions can be combined in many different ways – to yield the excess function for a ternary mixture; see text. The three figures shown are meant to represent the "symmetric" models of Kohler (a) and Muggianu (b); and the "asymmetric" of Toop (c)

§(305)

The combination of "unrepaired" (see above) binary functions to

$$G_{ABC}^{E} = \{a_{AB} + b_{AB}(X_A - X_B)\}X_A \cdot X_B + \{a_{BC} + b_{BC}(X_B - X_C)\}X_B \cdot X_C +$$

$$+ \{a_{AC} + b_{AC}(X_A - X_C)\}X_A \cdot X_C \tag{6}$$

is referred to as *Muggianu's model* (Muggianu et al. 1975).

Kohler's and Muggianu's models are symmetric, in the sense that the three binary subsystems are treated in the same manner. The model by Toop (Toop and Samis 1962), represented by Figure 1 c, is asymmetric. *Toop's model* has been applied to A = MgO; B = CaO; C = SiO_2 by Blander and Pelton (1987): A and B have the same chemical nature; the ionic substances A and B break the network structure of C = SiO_2 ($\leftarrow$ 305). For an exhaustive treatment of the multitude of combination models, the reader is referred to the paper by Wang et al. (1990).

equilibrium between two states in which the components mix in all proportions

Generally the equilibrium states of the isothermal or isobaric equilibrium between two states α and β – in which the three components mix in all proportions – constitute two *surfaces in PXY or TXY space* ($X = X_B$; $Y = X_C$; so that $X_A = 1 - X - Y$). A third surface can be defined which represents the *PXY* or *TXY* conditions for which α and β have equal Gibbs energies. That surface is the *equal-G surface* (EGS) – the ternary equivalent of the binary *equal-G curve* (EGC). The position of the EGS is between the other two surfaces and such that the three surfaces share *stationary points* – the points at which the two phases in equilibrium have equal compositions. A stationary point is a maximum in *PXY/TXY* space, or a minimum or a *saddle point* (which is a maximum in one direction and at the same time a minimum in another direction).

For a simple numerical example (Oonk 1973), we take the case of idealized isothermal equilibrium between liquid and vapour in a hypothetical system. In the *idealized approach* the volume of the liquid is neglected with respect to the volume of the gas, and for the latter the ideal-gas equation is taken. In terms of the variables ln P; X; and Y, the equal-G surface is given by the equation

$$\ln P_{EGS}(X, Y) = (1 - X - Y)\ln P_A^0 + X \ln P_B^0 + Y \ln P_C^0 + \frac{G^{El}(X, Y)}{RT} \tag{7}$$

§(305)

112

which is the logical extension of the binary Equation (212:27). For G^{El}/RT we take the expression

$$G^{El}/RT = (1 - X - Y)X - XY + (1 - X - Y)Y, \qquad (8)$$

and, furthermore, we put $\ln P_A^0 = \ln P_B^0 = \ln P_C^0 = 0$, so that the EGS is given by

$$f(X, Y) \equiv \ln P_{EGS} = (1 - X - Y)X - XY + (1 - X - Y)Y \qquad (9)$$

The function (f) has three binary stationary points and one ternary one, which is a *saddle point* and whose coordinates ($X = Y = f = 0.2$) follow from $\partial f/\partial X = \partial f/\partial Y = 0$. The binary stationary points are maxima at the edges AB and AC, and a minimum at the edge BC.

Except for a few exceptions maybe, binary excess functions have no more than one extremum. It implies, among other things, that binary liquid + vapour equilibria will have no more than one stationary point and that there will be no more than one ternary stationary point (and only so when there is at least one binary stationary point, see also Haase (1950); Serafimov (1967); Serafimov et al. (1971); the other way round, three binary minima do not necessarily involve a ternary one; see below. The maximum possible number of stationary points increases rapidly with increasing number of components: from 1 in binary systems, and (3+1=4) in ternary systems, to (21+35+35+21+7+1=120) in seven-component systems (see also Serafimov 1967).

For the purpose of predicting the characteristics of isobaric ternary behaviour from the binary subsystems, the most convenient expression is an extension of the Equations (212:36, and 37):

$$T_{EGS}(X, Y) = \frac{(1 - X - Y)\Delta H_B^* + X \Delta H_B^* + Y \Delta H_C^* + \Delta G_{EGS}^E(X, Y)}{(1 - X - Y)\Delta S_A^* + X \Delta S_B^* + Y \Delta S_C^*} \qquad (10)$$

To that end, the binary ΔG^E are derived from the experimental data (phase diagram and pure-component transition properties) by means of Equation (212:36, 37); and subsequently combined to ΔG^E (X, Y), to be substituted in Equation (10). In this approach the excess functions are treated as if they were independent of temperature (which, for purpose of predicting ternary equilibria, is the more justified the smaller the temperature range involved).

A speaking example is found in the solid + liquid equilibrium in the system {(1 $-X - Y$) mole of K $+ X$ mole of Rb $+ Y$ mole of Cs}; for details, see Oonk (1973). All of the binary subsystems have a phase diagram with a minimum. The deepest minimum is shown by the binary with the greatest mismatch in atomic size; i.e. K+Cs. The Muggianu-like combination of the binary excess properties results in

$$\Delta G^{E}_{EGS}\left(X,Y\right)=\{\,(\,-469\ +\ 71X\,)\,(1-X-Y\,)\,X-690\,XY+ \\ +\,(\,-2552\ +\ 586\,Y\,)\,(1-X-Y\,)\,Y\,\}\,J\cdot mol^{-1} \tag{11}$$

The minimum of this function and the minimum of the calculated EGS both fall outside the composition triangle: there is no ternary stationary point.

A complete prediction of ternary phase behaviour from binary data has been carried out by Moerkens et al. (1983) on the system 1,4-dichlorobenzene + 1-bromo-4-chlorobenzene + 1,4-dibromobenzene; see Figure 2, which is a copy of Figure 005:11 (note that the parallel lines, drawn in the two-phase region, do not have the status of *tie line*; →Exc 4).

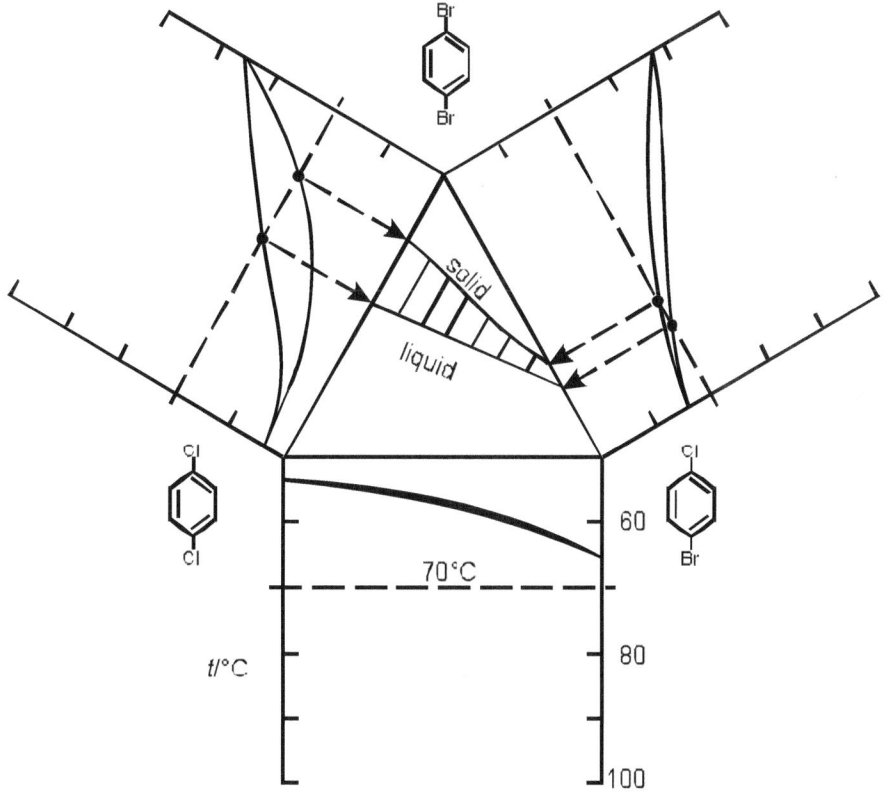

FIG. 2.

Ternary system composed of three 1,4-dihalobenzenes forming mixed crystals. Isothermal section (t = 70 °C) of ternary (solid-liquid) equilibrium, calculated from the information available for the binary subsystems Moerkens et al. 1983); →Exc 4

In Moerkens's analysis the binary solid + liquid equilibria were assessed by means of LIQFIT (← 304; see below); the combination of the binary excess functions was according to Kohler; and the ternary liquidus and solidus surfaces were calculated by means of CALTXY (see below, at the end of this section).

Just to underline the power of the methodology: the mean absolute difference between calculated liquidus temperatures and corresponding experimental liquidus temperatures, measured by Campbell and Prodan (1948), and taken for 39 points inside the composition triangle, is just 0.125 K (see also Van der Linde et al. 2002).

Three calculated isothermal sections, including a bunch of correct tie lines, are shown in Figure 3.

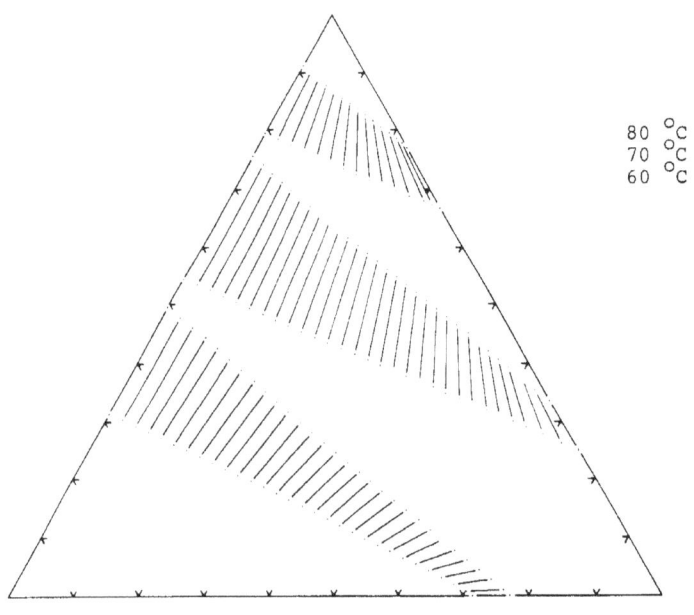

FIG.3.

Ternary system composed of three 1,4-dihalobenzenes forming mixed crystals. Three calculated isothermal sections, including tie lines (Moerkens et al. 1983)

limited solid state miscibility

The components of the system KBr + LiBr + NaBr are isomorphous (*NaCl type of structure*). LiBr and NaBr, and NaBr and KBr show complete *subsolidus miscibility*; their miscibility at room temperature being limited (←304). In the case of LiBr and KBr - owing to the high *mismatch in ionic size* between Li⁺ and K⁺ - solid miscibility is virtually negligible; even at the *eutectic temperature.*

This time, we examine the phase behaviour over a large range of temperature – from room temperature to 1020 K, the melting point of NaBr. The available data are thermodynamic melting properties of the three components; the eutectic phase diagram of KBr + LiBr; and for NaBr + KBr and LiBr + NaBr the solid–liquid loops and the regions of demixing.

The large temperature range makes that the physics of the system cannot be neglected: for this case the temperature dependence of the excess Gibbs energy must be taken into account. Correlations established for the family of alkali halide systems with a common anion or cation (including the three subsystems at hand), reveal that the excess entropy stands in a constant (i.e. within the family system-independent) relation to the excess enthalpy – for the solid mixtures and also for the liquid mixtures (←304; Van der Kemp et al. 1992).

Calculations of the ternary phase behaviour have been carried out by Jacobs et al. (1996), who used the Kohler model for the combination of the binary excess properties (see also Jacobs and Oonk 2006). Besides, the authors report that the combination models of Muggianu et al. (1975) and Colinet (1967) produce the same result within 0.9 mol % or 7 K for the liquidus surface.

The outcome of the computations and their significance, in terms of agreement with ternary experimental data, is shown by Figures 4-7. In Figure 4 a set of experimental liquidus temperatures are shown, along with isothermal section of the liquidus surface calculated for the experimental liquidus temperatures. The six data points, connected in Figure 4 by a straight line, have the same ratio of their KBr and LiBr mole fractions. In Figure 5 it is shown how for compositions along that line the phase behaviour changes as a function of temperature. The section, Figure 5, is referred to as an *isopleth* (originally the term isopleth was introduced to refer to a line of equal composition in a binary phase diagram). Figures 6 and 7 are phase diagrams representing the calculated phase – equilibrium relationships for 750 K and 675 K, respectively.

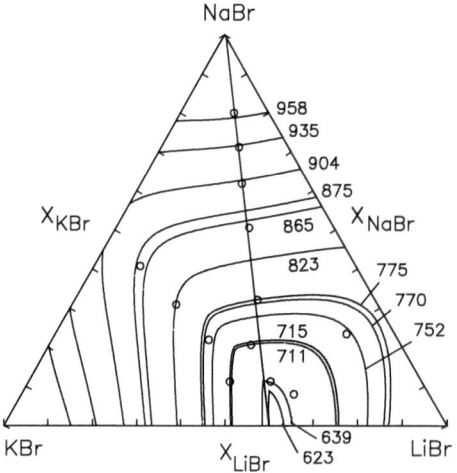

FIG. 4.

Experimental ternary liquidus points, along with isotherms of the liquidus surface calculated –
from binary data – for the experimental liquidus temperatures

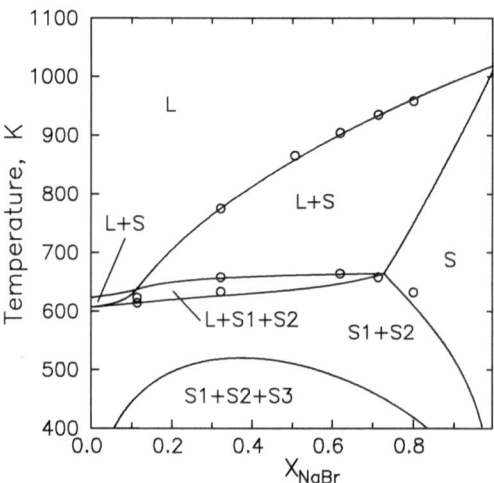

FIG. 5.

Phase behaviour as a function of temperature for the isopleth that corresponds to the straight
line in Figure 4 – characterized by a constant ratio of the LiBr and KBr mole fractions. Open
circles: experimental data read from microcalorimetric recordings

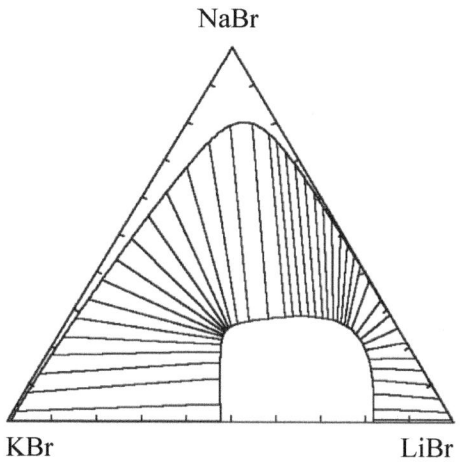

FIG.6.

Ternary phase behaviour calculated for T = 750 K

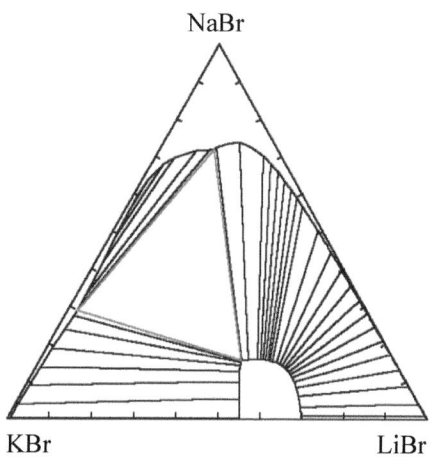

FIG.7.

Ternary phase behaviour calculated for T = 675 K

118

Harker and Tuttle (1955), who studied the decomposition of calcite ($CaCO_3$), magnesite ($MgCO_3$), and dolomite ($Ca_{0.5}Mg_{0.5}CO_3$) and also the limits of solid solution formation between the three carbonates, defined their system of study as a ternary one – $CaO + MgO + CO_2$. In doing so, the system $CaCO_3 + MgCO_3$ becomes a so-called "*binary join*" of the ternary system; and dolomite a ternary compound. This example is to show that the definition of a system leaves some room for a personal interpretation. Indeed, an investigator interested in the system $CaCO_3 + MgCO_3$ on its own will rather refer to it as a binary system – dolomite being a binary compound.

The example also shows that the existence of a ternary compound (e.g. dolomite) cannot be predicted from the binary subsystems ($CaO + MgO$; $CaO + CO_2$; $MgO + CO_2$). Nor is it an easy task to predict the existence of a compound AB (say dolomite) from the thermodynamic properties of the components A and B (calcite and magnesite). The prediction of the existence of compounds is a matter of theoretical chemistry and physics rather than classical thermodynamics.

concluding

The application of classical thermodynamics, for the analysis of phase equilibrium data, coupled with thermochemistry, and the prediction of phase behaviour and thermophysical properties, is an activity which is carried out, in particular, by the members of the *Calphad* community. A special issue of the Claphad journal has been devoted to *software packages:* Tomiska (2002); Yokokawa et al. (2002); Cheynet et al. (2002); Chen et al. (2002); Bale et al. (2002); Davies et al. (2002); and Anderson et al. (2002) – with a foreword written by L. Kaufman, the founding editor of the journal (Kaufman 2002).

A survey of the computer software developed at Utrecht University is given below the *a posteriori* of this section.

==

The examples given demonstrate the power of predicting the phase behaviour of multicomponent systems from the phase behaviour of their binary subsystems and the thermochemical properties of their components.

computer software developed at Utrecht University

The development of methods and software for the thermodynamic analysis of phase diagrams goes back to the sixties of the last century (see References at the end of this work). Typical characteristics of the Utrecht methodology are the frequent use of *equal-G curves* and *equal-G surfaces* and the use of *linear contributions* for the calculation of phase diagrams.

Equal-G curves (EGC) and equal-G surfaces (EGS) direct the treatment of phase equilibria between two mixed forms (such as solid form and liquid) to the intersection of Gibbs functions, rather than to the equality of chemical potentials. The concepts of EGC and EGS are in particular useful when it comes to the *classification of phase diagrams* in terms of stationary points (maxima and minima).

The power of linear contributions lies in the fact that nothing in the world of equilibrium between phases will change if the Gibbs energies of *all of the forms* participating in the equilibrium are extended by terms that are linear in the composition variables, and such that those terms are the same for each of the forms.

The computer routines at Utrecht University were developed in close relation with, and to support the thermodynamic research on families of mixed crystals, such as the common-ion alkali halide systems. Hereafter, a short, incomplete summary is given of the individual routines.

EXTXD is the routine that was developed to derive thermodynamic excess properties from *TX* phase diagrams (Brouwer and Oonk, 1979). There is a version for eutectic systems and one for regions of demixing. Complete listings of the two versions in PASCAL are given in Oonk, Bouwstra, and van Ekeren (1986), and Oonk, Eisinga, and Brouwer (1986), respectively.

LIQFIT (Bouwstra et al. 1980) was designed to derive thermodynamic excess properties of mixed crystals (solid solutions) from the liquidus in *TX* solid-liquid phase diagrams. A complete listing in FORTRAN can be found in Moerkens et al. (1983). A version for Windows, referred to as WINIFIT has been developed in Barcelona by López and López (1998).

Several computer routines have been written for the calculation of binary phase diagrams; either as stand-alone programs or as part of EXTXD and LIQFIT. One of the stand-alone programs is PROPHASE (Van Duijneveldt et al. 1989).

CALTXY (Moerkens et al. 1983, containing a listing in FORTRAN) was designed to calculate ternary phase diagrams pertaining to equilibria between two mixed forms, in particular mixed crystalline solid and liquid.

TXY-CALC (Jacobs 1990; Jacobs et al. 1996) is written in TURBO-PASCAL and designed to calculate thermodynamic properties and phase equilibria in ternary systems (see also Jacobs and Oonk 2006).

All of the developments are reflected in the software package XiPT, which is the successor of TXY-CALC. To date, the program XiPT is used for the construction of a thermodynamic database for materials relevant to geophysics, aiming at predicting thermophysical and thermochemical properties at the extreme conditions prevailing in *planetary mantles.* It includes a number of equations of state, such as the Birch-Murnaghan equation (Birch 1952), our own (Jacobs and Oonk 2000) discussed in section 306 of this work, and the ones by Morse (1929) and Vinet et al. (1986). In addition XiPT includes universal and thermal equations of state based on *lattice vibrations,* in which the *vibrational density of states* is described by a Mie-Grüneisen-Debye model (see Anderson 1995) or Kieffer model (Kieffer 1979); these developments are the subject of section 307.

EXERCISES

1. *ternary azeotropes*

Knowledge about the presence of azeotropes and their coordinates is of vital importance for the separation of substances by means of *distillation.* And because the experimental determination of the position of an azeotrope is a time-consuming task, much attention has been given to the prediction of azeotropic data (see CRC *Handbook of Chemistry and Physics*; and for the purpose of this exercise Malesiński 1965, and Haase 1950, 1956).

The most elementary EGS formula - related to Equation (10) - to predict the coordinates of a *ternary azeotrope* from the coordinates of the three binary subsystems, has the form

$$T_{EGS}(X,Y) = (1-X-Y)T^o_A + XT^o_B + YT^o_C + (1-X-Y)X \cdot \Theta_{AB} + (1-X-Y)Y\Theta_{AC}$$

$$+ XY\Theta_{BC.} . \quad (I)$$

The binary interaction parameters follow from the coordinates of the azeotropes in the binaries; in the case of A + B:

$$\Theta_{AB} = [T_{az} - (1-X_{az})T^o_A - X_{az}T^o_B] / [X_{az}(1-X_{az})] . \quad (II)$$

- What are the assumptions on which the derivation of (I) and (II) from Equation (10) is based?

§(305)

The data in the table below have been taken from Ohe's (1989) compilation of *vapour-liquid equilibrium data*. They all are valid for $P = 760$ Torr.

The tasks indicated by the bullets below the table are not interdependent.

system (1-X)A+XB	reference	T^o_A / °C	T^o_B / °C	T_{az} / °C	X_{az}
acetone+chloroform	Reinders & de Minjer (1940)	56.1	61.7	64.3	0.649
acetone+methanol	Uchida et al. (1950)	56.2	64.8	55.4	0.20
chloroform+methanol	Nagata (1962)	61.7	64.8	53.5	0.35
benzene+methanol	Nagata (1969)	80.0	64.7	58.2	0.60

- Taking 56.1°C for acetone's *normal boiling point,* calculate the coordinates of the saddle point in (1-X-Y) acetone + X chloroform + Y methanol.

In the system benzene+chloroform+methanol, the binaries benzene+methanol and chloroform+methanol have minimum azeotropes The binary benzene+chloroform, on the other hand, is zeotropic (does not have a stationary point); its equimolar EGC temperature is 72.2°C (as follows from the diagram by Nagata 1962).

- Is it likely that the system benzene+chloroform+methanol has a *ternary stationary point*?

To appreciate that pinpointing the position of a ternary stationary point requires considerable time and effort, a hypothetical system (1-X-Y) A + X methanol(M) + Y benzene(B) is defined as follows. Substance A's boiling point is equal to the azeotropic temperature in the system benzene+methanol. The interaction parameters of the binaries A+methanol and A+benzene are such that the two systems just fail to have a minimum azeotrope: $\Theta_{AM} = -6$ K; and $\Theta_{AB} = -21$ K. The interaction parameter in M+B, as follows from the data in the table, has the value of -52.6 K.

- As a first step, calculate the XY coordinates of the ternary stationary point. Next, in order to establish its nature and to observe the flatness of the region around it, you are invited to calculate a number of temperatures of the equal-G surface, and to construct a series of isotherms.In a zone of the XY plane, having a width of $\Delta Y = 0.25$ and extending from A's boiling point to the minimum azeotrope in benzene+methanol, calculate a number of EGS temperatures - to three (!) decimal places - and use them to draw by hand the isotherms for the Celsius temperatures from 58.200 to 58.400 at 0.050 intervals.

§(305)

2. *solubility of mixed crystals*

The issue addressed in this exercise is the isothermal, isobaric equilibrium between mixed crystals formed by the substances A and B and their solutions in a liquid solvent C.

In a Gibbs triangle with top vertex C, the solubility curve - the locus of the equilibrium compositions of the liquid phase - is a curve inside the triangle, the base AB being the locus of the equilibrium compositions of the solid phase.

- To start with, write down the $f = M - N$ *system formulation*. To avoid excessive decoration of symbols, use the symbol X for B's mole fraction in the solid and the symbols m_A and m_B for the mole fractions of A and B in the liquid, respectively.

- Assuming ideal liquid mixing, and with $G^E = X(1-X)\ [g_1 + g_2\ (1- 2X)]$ for the solid state, derive the formulae for m_A and m_B as a function of X. Introduce the solubilities of the pure substances, m_A^o and m_B^o, in order to eliminate their solid and liquid molar Gibbs energies (the four G^* quantities).

- For $m_A^o = 0.6$; $m_B^o = 0.4$; $g_1 = 3000$ J mol^{-1}; $g_2 = 600$ J mol^{-1}; and $T = 300$ K, calculate m_A and m_B, for X values from $X = 0.1$ to $X = 0.9$, at intervals of 0.1. Construct the triangular phase diagram, including the tie lines for the calculated equilibrium states.

- Transform the triangular phase diagram into a rectangular one, such that the horizontal axis is for $X = X_B / (X_A + X_B)$ - for A and B mole fractions in solid and liquid - and the vertical axis for $Y = m_A + m_B$.

NB. The rectangular XY diagram, which is an analogue of the so-called *Lippmann diagram* (Lippmann 1980; Königsberger and Gamsjäger 1990; Gamsjäger et al. 2000), has the appearance of the PX diagram for the binary equilibrium between liquid (or mixed-crystalline solid) and vapour ($\leftarrow$005; Exc 006:13).

3. *isothermal section with binary and ternary compound*

This exercise is about the construction of an isobaric, isothermal section ($P = 760$ Torr; $T = 900$ K) for a hypothetical, idealized ternary system $\{(1-X-Y)$ A $+ X$ B $+ Y$ C$\}$ which has a binary compound AC and a ternary compound ABC.

The designation 'idealized' is used here to state that 1) liquid mixing is ideal; 2) solid miscibility is absent; and 3) heat capacities are negligible.

The fusion of the compounds can be seen as a chemical reaction:

AC (sol) → A (liq) + C (liq);

ABC (sol) → A (liq) + B (liq) + C (liq).

Accordingly, one can use the concept of *equilibrium constant K* to describe the equilibrium between compound and liquid mixture (←007):

$$\ln (K / K_o) = - (\Delta H^o / R) (1/T - 1/T_o). \tag{I}$$

The melting point T_o of the ternary compound is 1000 K, and its heat of fusion, to change into 1 mol A + 1 mol B + 1 mol C, is $\Delta H^o = 2500 \, R$ K, R being the gas constant.

- First formulate K / K_o for the fusion of the ternary compound. Next, calculate and subsequently plot the ternary section of the compound's liquidus surface.

-

Clue. Take X as the independent variable and solve, for given X, the quadratic equation in Y.

- In the constructed ternary diagram, introduce the sections of the liquidus surfaces for compound AC, and component B, making use of the following information: 1) in the binary subsystem A + C at 900 K, and at the side of C, A's mole fraction in the-liquid-in-equilibrium-with-solid-AC is 0.2; and 2) in the binary subsystem B + C at 900 K, B's mole fraction in the liquid-in-equilibrium-with-solid-B is 0.5.

- Finally, out of the diagram with the sections of the three liquidus surfaces, construct the stable isobaric, isothermal phase diagram – with its single-phase fields, two-phase regions with tie lines, and three-phase triangles. At 900 K, the components A and C are liquid.

4. *an amusing faux pas*

The lines that are drawn in the two-phase region in Figure 2 are not all, and not at all, *tie lines*; apparently they have been drawn as a manner of marking the field (hatching). The true tie lines are shown in Figure 3. For a comment the reader is referred to the review written by J.C. Wheeler, Physics Today 62(2009)62-64.

- Explain in simple wording that it is impossible that, in a ternary two-phase region like the one in Figure 2, all tie lines are parallel to each other.

5. *the idealized isobaric ternary liquid+vapour equilibrium*

As a follow-up of the binary description in § 210, the case is considered where in the system $\{(1-X-Y)$ mol A $+ X$ mol B $+ Y$ mol C$\}$, and under isobaric circumstances, an ideal liquid mixture is in equilibrium with an ideal gas mixture. Part of the binary description is reproduced here:

The change in Gibbs energy of component A, is given by

$$\Delta G_A^*(T) = -\Delta S_A^*(T - T_A^o) \ ,$$

in which ΔS_A^* is the *entropy of vaporization*, the quotient of the heat of vaporization and boiling point temperature. Similarly for component B,

$$\Delta G_B^*(T) = -\Delta S_B^*(T - T_B^o) \ .$$

After substitution of the *recipes for* the *chemical potentials* into the equilibrium equations, the following relations are obtained.

$$(1 - X^{liq}) = (1 - X^{vap})\exp(\Delta G_A^* / RT) \ ;$$

$$X^{liq} \ = \ X^{vap}\exp(\Delta G_B^* / RT) \ .$$

By addition of the last two equations, X^{liq} is eliminated and X^{vap} is obtained as a function of temperature:

$$X^{vap}(T) = \frac{1 - \exp(\Delta G_A^* / RT)}{\exp(\Delta G_B^* / RT) - \exp(\Delta G_A^* / RT)} \ ,$$

which is, in other words, the formula for the *vaporus*. The formula for the *liquidus* is obtained by combining the fourth and the fifth eaquation:

$$X^{liq}(T) = \frac{1 - \exp(\Delta G_A^* / RT)}{\exp(\Delta G_B^* / RT) - \exp(\Delta G_A^* / RT)} \ \exp(\Delta G_B^* / RT) \ .$$

- For the ternary case write down the *system formulation*, and see that there are two independent variables.

- For the two independent variables T and Y^{liq}, formulate the equations by means of which X^{liq}, X^{vap}, and Y^{vap} can be solved.

- Demonstrate that, in isothermal sections for the idealized case at hand, the liquidi and vapori are straight lines.

§(305)

6. *no more than Antoine constants*

The three *homologs* n-hexane (C_6H_{14}), n-heptane (C_7H_{16}), and n-octane (C_8H_{18}) intuitively form quasi-ideal liquid mixtures. It means that only pure-component data are needed to perform an approximate calculation of the ternary liquid+vapour equilibrium states. This time, the pure-component data are just the constants of the *Antoine equation* for vapour pressure as a function of temperature:

$$\ln(P/\text{Torr}) = A - B\{(T/K) - C\}^{-1}.$$

Substance	A	B	C
n-hexane	15.85	2706	49
n-heptane	15.89	2917	56
n-octane	15.94	3120	63

- For each of the three substances: calculate its *normal boiling point* (i.e. at $P = 760$ Torr) and its heat of vaporization at the nbp (←Exc 110:5).

Using the results obtained in Exc 5, and for $P = 760$ Torr carry out the following tasks.

- For $T = 380$ K, and for $Y^{liq} = 0.45$; 0.55; 0.65; and 0.75, calculate the compositions of the coexisting phases and use them to construct the triangular isothermal section; i.e. the complete liquidus and vaporus, along with the four tie lines.

- In a *cylinder-with-piston* experiment, an amount of equimolar liquid mixture is heated from room temperature to $T = 400$ K. Calculate the temperature at which the liquid starts to boil, and, next, the temperature at which the last amount of liquid changes into vapour (dew point).

7. *linear contributions at work*

This exercise is about a hypothetical liquid system $\{(1-X-Y)\text{ mol } A + X \text{ mol } B + Y \text{ mol } C\}$, whose phase behaviour resembles the behaviour displayed by the triangular diagram which is shown (see next page), and which represents the equilibrium between two liquid phases, at 1 atm and 30 °C, in the system methylcyclohexane (A) + n-hexane (B) + methanol (C); (←005).

§(305)

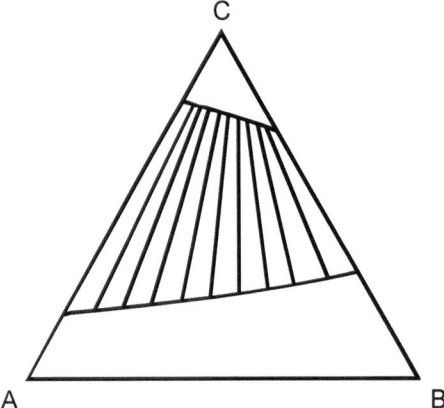

The ternary excess Gibbs energy function of the hypothetical system is a Muggianu type of combination of the excess functions of the binary subsystems AC and BC. The full Gibbs energy of the system, which is studied under isobaric conditions at $T = 300$ K, is represented by the expression

$$G(T,X,Y) = RT \, LN(X,Y) + (1{-}X{-}Y)Y \, a_{AC} + XY\{a_{BC} + (X - Y)b_{BC}\} + AX + BY,$$

with $LN(X,Y) = (1{-}X{-}Y) \ln (1{-}X{-}Y) + X \ln X + Y \ln Y$.

The values of the binary excess parameters are set at $a_{AC} = 5760$ J mol^{-1}; $a_{BC} = 5360$ J mol^{-1}; and $b_{BC} = -220$ J mol^{-1}. These values are such that the liquid+liquid two-phase region in the binary AC extends from $Y = 0.2$ to $Y = 0.8$; and in the binary BC from $Y = 0.3$ to $Y = 0.75$.

   The main purpose of this exercise is to give an idea of how the/a ternary phase diagram can be calculated by the use of the linear contributions $AX + BY$. To that end the following actions are suggested:

- To start with, and to have an idea of the extension of the expected two-phase region in the central part of the triangle, calculate the Gibbs energies along the line $Y = 1 - 2X$ (the perpendicular bisector on AB). Carry out two manners of calculation: the first with $A = B = 0$; and the second with $A = 0$, and B's value taken such that $G(Y = 1) = G(Y = 0)$. Construct the plots of G versus Y, and from them read the Y- values of the points of contact of the double tangent. From the result make an educated guess of the course in the composition triangle of the upper and the lower boundary of the two-phase region.

From the *system formulation*

$$f = M \, [X^I, Y^I, X^{II}, Y^{II}] - N \, [\mu_A^I = \mu_A^{II}; \mu_B^I = \mu_B^{II}; \mu_C^I = \mu_C^{II}] = 4 - 3 = 1,$$

where the superscripts *I* and *II* denote the phases rich in A+B and the ones rich in C, respectively, it follows that the equilibrium values of Y^I, X^{II}, and Y^{II} follow from the choice of X^I.

§(305

In the calculations to be carried out, the Gibbs energies of phase I are calculated with fixed X^I for Y^I values around the equilibrium value expected for the latter; and the Gibbs energies for phase II for X^{II} and Y^{II} values around the equilibrium values expected. For example, for $X^I = 0.2$, the equilibrium values expected are in the vicinity of 0.22 for Y^I; 0.05 for X^{II}; and 0.79 for Y^{II}.

In two or three rounds of calculations the values of the parameters A and B have to be adjusted such that the two local minima of the Gibbs function have the same function value. In that situation the two minima have the X and Y values of the two phases in equilibrium.

- First round. For $X = 0.2$ and $Y = 0.22$, calculate starting values for A and B from the conditions $(\partial G/\partial X) = 0$ and $(\partial G/\partial Y) = 0$. Using the values obtained for A and B, calculate Gibbs energies - in six significant figures and in mole-fraction steps of 0.01 - along the line $X = 0.2$ around $Y = 0.22$; next do the same for X and Y values of points on a lattice around $X = 0.05$; $Y = 0.79$. From the two sets of calculated Gibbs energies read the coordinates of the two local minima, and so to three decimal places.

- Second round. Use the difference between the Gibbs energies of the two local minima found in the first round, to adjust the value of the parameter B, such that, for the coordinates of the minima, the Gibbs energies will be the same. Using the adapted value for B and the original one for A, calculate Gibbs energies - in seven significant figures and steps of 0.001 - and from them determine the coordinates of the two local minima to four decimal places.

8. *the incidence of symmetry*

The triangular diagram which is shown (see next page) is the phase diagram of a hypothetical system in which only one *form* is present and in which three miscibility gaps make their appearance.

The case was used as an example in the paper by Jacobs and Oonk (2006) on the calculation of ternary phase diagrams by using linear contributions. The dashed lines in the figure are related to the strategy of dealing with the task.

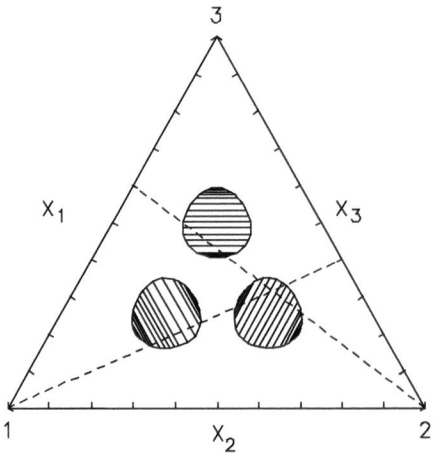

The Gibbs energy of the form is given by the expression

$$G(T,X,Y) = RT \ \text{LN} \ (X,Y) + (10^5 \ \text{Jmol}^{-1}) \ (1-X-Y) \ X \ Y,$$

with $\text{LN} \ (X,Y) = (1-X-Y) \ln (1-X-Y) + X \ln X + Y \ln Y.$

In the expression, the molar Gibbs energies of the pure components are omitted: they can be considered as the sum of a constant and a linear contribution term $AX + BY$, which have no effect on the equilibrium compositions to be calculated.

The phase diagram printed above, with $X_2 = X$ and $X_3 = Y$, represents the computational result for the temperature of 1450 K.

After the removal of the properties of the pure components, and due to the high symmetry of the system, the *system formulation* for each of the three separated two-phase equilibria, at isothermal, isobaric conditions, is just a case of one equilibrium condition acting upon two variables. For the upper of the three equilibria shown, and for which a phase of composition (X^I, Y) is in equilibrium with a phase of composition (X^{II}, Y):

$$f = \text{M} \ [X^I, Y] - \text{N} \ [(\partial G/\partial X)_Y = 0] = 2 - 1 = 1.$$

- By constructing the cross-section of the Gibbs function for $Y = 0.5$, appreciate the system formulation, and determine the X coordinates of the coexisting phases.

- Through the second derivative of the Gibbs function with respect to X, determine the coordinates of the upper and lower critical points of the upper miscibility gap.

§ 306 HIGH PRESSURE, HIGH TEMPERATURE

===

The greatest challenge in phase theoretical classical thermodynamics is the computational prediction of phase behaviour and thermophysical material properties for conditions that prevail in the interior of the earth. Conditions that are far away from the pressure and temperature values at the surface of the earth, and for that reason rather inaccessible to experimental investigation.

===

introduction

This section (§ 306) and the following one (§ 307) have the character of a discovery journey: the sections give an account of the efforts undertaken by the authors of this book and their colleagues to construct a reliable thermodynamic structure for the analysis and the prediction of thermodynamic properties and phase equilibria under conditions of high pressure and high temperature. For that matter, a frequent use is made of the "we" form: "we argued that" rather than for instance "Jacobs and Oonk (2001) argued that".

In this section a start is made with the construction of an *equation of state*. Numerous equations of state for solids are available in the literature and detailed overviews can be found in the books by for example Anderson (1995) and Poirier (2000). Rather than detailing all specific characteristics of different equations of state, this section is limited to the construction of a simple mathematical framework to derive the Gibbs energy of a substance. The substance MgO is taken as a vehicle for demonstrating the characteristics of an equation of state we used for describing thermodynamic properties of mantle materials. The mathematical framework is easily extended to mixtures and can be used as a tool for describing thermodynamic properties in pressure-temperature-composition space. Because of its mathematical nature problems may arise when thermodynamic properties are extrapolated to extreme conditions prevailing in the Earth mantle.

In § 307 the models of Einstein, Debye, and Kieffer based on *lattice vibrations* make their appearance. This time the substance aluminium is taken to introduce the models. The substance Mg_2SiO_4, which is one of the major constituents of the mantle of the earth, is the subject, again, of a thermodynamic analysis.

The equation of state constructed in this section forms a part of a thermodynamic framework enabling the derivation of the Gibbs energy function and in section (§307) the Helmholtz energy function. From these two functions all thermodynamic properties, such as volume, thermal expansivity and heat capacity can be derived, and from the Gibbs energy also the equilibria between different phase assemblages in a mineralogical system. The thermodynamic formalism in this framework must fulfil at least four criteria to be robust enough for application at high pressure-temperature conditions.

The first, trivial, criterion is that thermodynamic properties derived from it should be accurate enough to represent measured data, preferably to within experimental uncertainty. This condition particularly applies to data measured in the low pressure regime, which are not affected by problems such as pressure calibration and pressure effects on the emf of thermocouples.

The equation of state in the formalism is used to extrapolate thermodynamic properties to high temperature and high pressure conditions, for which experimental data are scant. Such extrapolation should be reliable; a second criterion, which should be fulfilled. However, the judgement of reliability by the investigator using the thermodynamic formalism is a cumbersome and often a subjective matter. On the other hand there is a weaker criterion, which can be used as an indication for reliability. One could state that the minimum criterion for a thermodynamic formalism to be reliable is that thermodynamic properties derived from the formalism are well-behaved, without the occurrence of physically unrealistic behaviour, such as negative thermal expansivity at high pressure or an increase of isothermal bulk modulus with temperature. It should be noted that this criterion of 'well-behaving behaviour' of thermodynamic properties cannot replace the criterion of reliability. To proof if an extrapolation by a thermodynamic description is reliable without experimental data, requires additional information. At present missing information can be supplied in the form of results obtained by first-principles calculations, which have shown to be powerful tools to obtain values for thermodynamic properties.

A third criterion is that the thermodynamic formalism should be reasonably fast enough to construct the phase behaviour and thermodynamic properties associated with these phases in complicated systems encountered in planetary interiors. In that way the formalism can also be made available for application in mantle convection research to investigate the dynamics and thermal history of planets in particular the Earth.

The main direct observables giving knowledge about the interior of the Earth are the velocities of sound waves, determined in the field of seismology. A thermodynamic formalism should therefore be able to compare calculated results with experimental results obtained by seismology. This is a fourth criterion which should be fulfilled by a thermodynamic formalism.

In this section we shall construct a thermodynamic framework applicable to pure substances and to stoichiometric compounds. The resulting formalism for the Gibbs energy can be applied to derive thermodynamic properties and the phase behaviour of mixtures, such as solid solution phases in the same way as has been discussed in the preceding sections.

the Gibbs energy

In practice, the *(molar) Gibbs energy* of a substance, in a given form, at arbitrary conditions of temperature and pressure, is obtained by integration over temperature followed by integration over pressure – starting from a reference temperature T_0 and a reference pressure P^o ($\leftarrow$301):

$$G(T,P) = \Delta_f H^\circ + \int_{T_0}^{T} C_p^\circ dT - T \left\{ S^\circ + \int_{T_0}^{T} \frac{C_p^\circ}{T} dT \right\} + \int_{P_0}^{P} V dP \,. \tag{1}$$

In this expression the property $\Delta_f H^\circ$ is the *heat of formation* and S° the *absolute entropy*, both at T_0 and P°.

The heat capacity, as a rule is described by a *polynomial function*, such as

$$C_p^\circ(T) = C_1 + C_2 \cdot T + C_3 \cdot T^{-1} + C_4 \cdot T^{-2} + C_5 \cdot T^2 + C_6 \cdot \ln T \,. \tag{2}$$

Consequently and necessarily, the volume of the substance, in the form it has taken, must be known as a function of temperature *and* pressure.

equations of state

In the context of this work, *equations of state* (EOS) are mathematical expressions that give the relationship between a system's molar volume and its temperature and pressure. A well-known example is the *Van de Waals equation* for a *fluid substance* (←206):

$$(V - b)\left(P + \frac{a}{V^2} \right) = RT \tag{3}$$

Another example of EOS is the family of equations named after Birch and Murnaghan – equations that are widely used in geophysics (see Anderson 1995). A member of the family is the so-called *third-order Birch-Murnaghan equation*, which is formulated as

$$P = P^\circ + \frac{3}{2} K^\circ \left[\left(\frac{V}{V^\circ} \right)^{-7/3} - \left(\frac{V}{V^\circ} \right)^{-5/3} \right] \left\{ 1 - \frac{3}{4}\left(4 - \left(K^\circ \right)' \right) \left[\left(\frac{V}{V^\circ} \right)^{-2/3} - 1 \right] \right\}, \tag{4}$$

where the open circle in the superscript refers to a reference pressure, P°, and where K' is the symbol for the partial derivative with respect to pressure of the isothermal *bulk modulus* K. Generally, the properties at the reference pressure, having the superscript $^\circ$ are functions of the temperature only.

The two equations, Equations (3) and (4), are rooted in a physical model – rather than in a law of nature. As a result, their degree of successfulness will depend on the nature of the application, the material at hand, and the circumstances of temperature and pressure. To illustrate this, Equation (4) has been used to calculate the molar entropy of MgO, at 298.15 K, as a function of pressure. To that end, the necessary data for MgO were taken from the database constructed by Saxena et al. (1993). The result of the computational exercise is shown in Figure 1 (from Jacobs & Oonk 2000).

132

The fact that the computed entropy, Figure 1, is going through a minimum would mean, by virtue of the *Maxwell equation* ($\leftarrow$107) for the change of entropy with pressure [$(\partial S/\partial P)_T = -(\partial V/\partial T)_P$], that the volume of the material is going through a maximum, its *thermal expansivity* becoming negative, the substance starts to shrink on heating. Although there is no law, which forbids that the thermal expansivity changes from positive to negative, such a thing is not obvious for substances, like MgO, that have a *close-packed structure.*

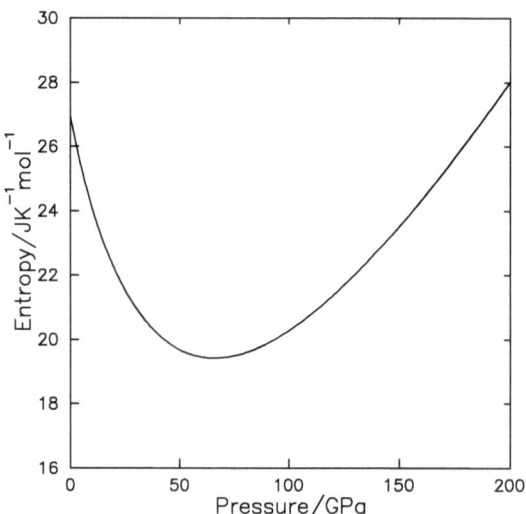

FIG. 1.
Calculated entropy curve for MgO at 298.15 K generated by the third-order Birch-Murnaghan equation of state, Equation (4), using the database by Saxena et al. (1993)

an empirical conditioner

From Figure 1, and other indications of anomalous behaviour, we observed that some kind of *magic relationship* (cf. $\leftarrow$206) would be needed to stabilize the thermodynamic rigour of the computational systems. With these things in mind, we decided to study the stabilizing possibilities of the empirical relationship which was discovered by Grover, Getting and Kennedy (Grover et al. 1973).

From shock-wave and static compression measurements on a variety of metals, Grover and co-workers found "a nearly precise linear relation" between the logarithm of bulk modulus and volume – up to volume changes of 40%. Their relationship can be written as

$$V(P,T) = V^\circ(T_0) + b \cdot \ln\left(\frac{K(P,T)}{K^\circ(T_0)}\right), \tag{5}$$

in which the open-circle superscript of *V* and *K* refers to *standard pressure* (1 bar), and T_0 is a *reference temperature*. The model parameter '*b*' has the same dimension as volume, m³/mol. A graphical illustration of the relationship is shown in Figure 2 – for MgO and the forsterite form of Mg_2SiO_4, taken as $Mg_{2/3}Si_{1/3}O_{4/3}$.

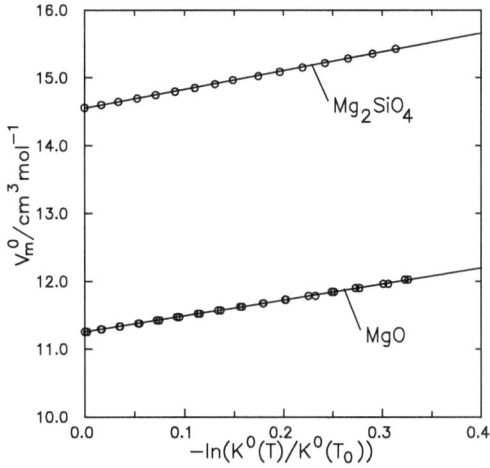

FIG. 2.
Showing the 'nearly precise linear relation' (Grover et al.1973) between volume and logarithm of bulk modulus. NB. The data for each of the two materials cover the range from 300 to 1800 K

Equation (5), which can be rewritten to give

$$K(P,T) = -V\left(\frac{\partial P}{\partial V}\right)_T = K^\circ(T_0) * \exp\left(\frac{V - V^\circ(T_0)}{b}\right), \qquad (6)$$

corresponds to a *differential equation* between pressure and volume. The solution of this equation is found by writing the exponential in Equation (6) in a *power series* of volume resulting in an expression for the pressure:

$$P = P^\circ - K^\circ(T_0)\left\{\exp\left(-\frac{V^\circ(T_0)}{b}\right)\right\}\left[\ln\left(\frac{V}{V^\circ}\right) + \sum_{j=1}^{\infty}\left(\frac{b^{-j}\left(V^j - (V^\circ)^j\right)}{j * j!}\right)\right] \qquad (7)$$

in which $V^\circ(T_0) = V(P^\circ, T_0)$ and $V^\circ = V(P^\circ, T)$.

Equation (7), as a result, is the equation of state stabilized by the empirical relationship, Equation (5), between volume and bulk modulus.

Gibbs energy in terms of the stabilized equation of state

In terms of the equation of state, Equation (7), the last term in Equation (1) takes the form

$$\int_{P^\circ}^{P} V\,\mathrm{d}P = \int_{P^\circ,V^\circ}^{P,V} \mathrm{d}(PV) - \int_{V^\circ}^{V} P\,\mathrm{d}V = V(P-P^\circ) + K^\circ(T_0)\left\{\exp\left(-\frac{V^\circ(T_0)}{b}\right)\right\}*$$

$$*\left\{V*\ln\left(\frac{V}{V^\circ}\right) - (V-V^\circ) + \sum_{j=1}^{\infty}\left[\frac{b^{-j}}{j*j!}\left(\frac{V^{j+1}-(V^\circ)^{j+1}}{j+1} - (V^\circ)^j(V-V^\circ)\right)\right]\right\}. \qquad (8)$$

Owing to the fact that the (material-dependent) constant b is negative, the terms of the power series are alternately negative and positive. Figure 3 shows how the calculated Gibbs energy behaves as a function of n, the number of terms of the power series. The example, which is representative of MgO, for $T = 2000$ K and $P = 50$ GPa, gives the difference between the Gibbs energy calculated for $n = n$ and the value calculated for $n = 50$. At $n = 34$ the difference is reduced to 1 picojoule per mole – a precision that allows the numerical calculation of the first and the higher-order derivatives of the Gibbs energy. This is elaborated in the following paragraphs for MgO; to begin with the two first partial derivatives, i.e. volume and entropy (for details and references, see Jacobs and Oonk 2000).

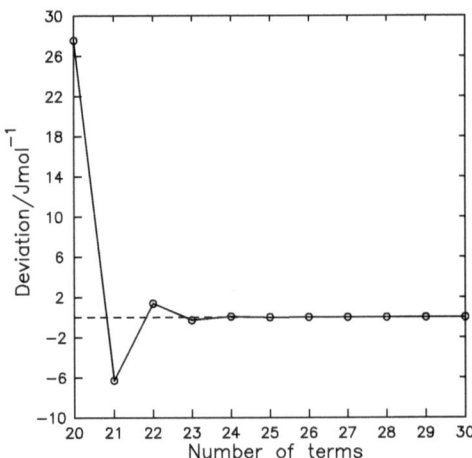

FIG. 3.
Deviation of the calculated Gibbs energy from the final value, as a function of the number of terms (n) in the summation of Equation (8). Substance MgO; $T = 2000$ K; $P = 50$ GPa

volume and entropy

For the calculation of molar volume at $T = 298.15$ K as a function of pressure, the input is needed of just three numbers. These are the values of molar volume and bulk modulus at ambient conditions, i.e. $V^o(T_0)$ and $K^o(T_0)$, and the value of the parameter *b*. In respective order: 11.25×10^{-6} $m^3 mol^{-1}$; (161.5 ± 0.6) GPa; and $(-2.359 \pm 0.003) \times 10^{-6}$ $m^3 mol^{-1}$. The result of the computation is shown in Figure 4. The calculated result is in agreement with the available experimental data – from static compression and shock compression experiments.

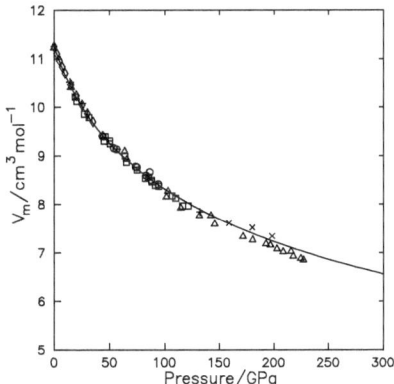

FIG. 4.
Calculated molar volume of MgO at $T = 298.15$ K as a function of pressure; along with experimental data

Calculated entropies as a function of pressure are shown in Figure 5 for a set of selected temperatures – using the value of 26.94 $J K^{-1} mol^{-1}$ for the entropy at ambient conditions. This time the anomaly that is seen in Figure 1 is absent. Experimental data are not available.

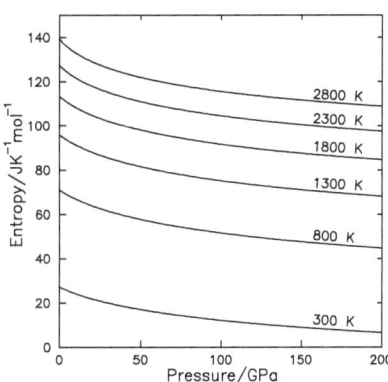

FIG. 5.
Calculated isotherms of MgO's molar entropy as a function of pressure. Thanks to the stabilized equation of state, the anomaly shown in Figure 1 has been removed

§(306)

second and higher-order properties

The thermodynamic properties related to the second-order partial derivatives are: the heat capacity at constant pressure (C_P); the isothermal compressibility (κ) or its inverse, the isothermal bulk modulus (K); and the cubic expansion coefficient, or expansivity (α).

As follows from the description so far: the heat capacity at ambient pressure and the expansivity at ambient pressure, both as a function of temperature, serve as the input for the Gibbs energy function. That is to say, before the equation of state comes into action. The same holds true for the isothermal bulk modulus at ambient conditions; its value is 161.5 GPa; see above.

For MgO the optimized expression, in terms of Equation (301:20), for the molar heat capacity at $P = 0.1$ MPa as a function of temperature is

$$\frac{C_P^o}{\mathrm{J\cdot K^{-1}\cdot mol^{-1}}} = 180.7689 + 8.7874\times10^{-3}\left(\frac{T}{K}\right) - 1.5081\times10^4\left(\frac{K}{T}\right) +$$

$$+ 6.0367\times10^5\left(\frac{K}{T}\right)^2 + 6.1868\times10^{-7}\left(\frac{T}{K}\right)^2 - 18.0218\ \mathrm{ln}\left(\frac{T}{K}\right). \quad (9)$$

For the thermal expansivity, in terms of Equation (301:21),

$$\frac{\alpha^o}{K^{-1}} = 4.5248\times10^{-5} + 8.4711\times10^{-10}\left(\frac{T}{K}\right) - 4.1959\times10^{-3}\left(\frac{K}{T}\right) +$$

$$+ 2.4984\times10^{-12}\left(\frac{T}{K}\right)^2 \quad (10)$$

Experimental data for non-ambient conditions are available for the thermal expansivity, but not for the heat capacity. In Figures 6 and 7 it is shown how the thermal expansivity behaves as a function of pressure and temperature, respectively; and also how the agreement is between the experimental data and the results obtained through the equation of state.

As regards the third- and fourth-order partial derivatives of the Gibbs energy, experimental data are available for the isothermal derivatives with respect to pressure. The third-order derivative is related to the pressure derivative of the isothermal bulk modulus – through Equation (6):

$$K' = \left(\frac{\partial K}{\partial P}\right)_T = -\frac{V}{b} \quad . \quad (11)$$

V in Equation (11) denotes molar volume. For MgO at ambient conditions the property is calculated as $K' = 4.769 \pm 0.005$. Literature values range from 4.13 ± 0.09 to 5.40 ± 0.20.

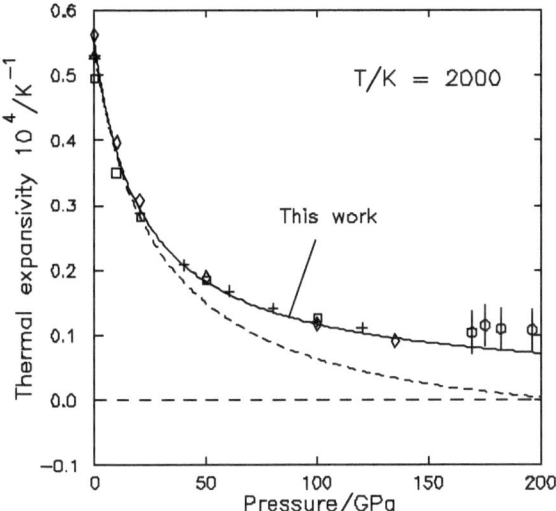

FIG. 6.
The thermal expansivity of MgO as a function of pressure at *T* = 2000 K.
Calculated curve, along with experimental data

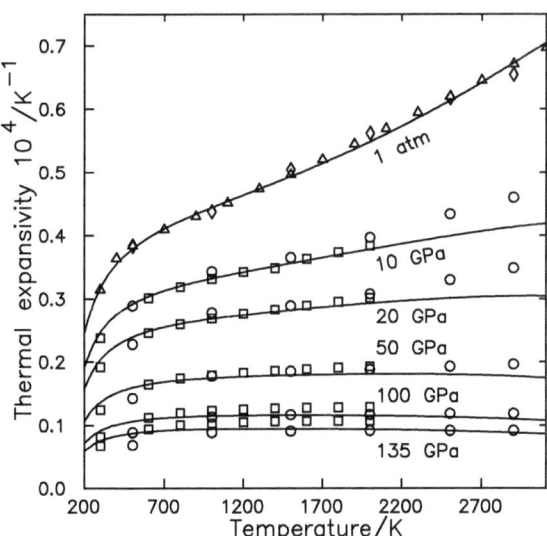

FIG. 7.
Thermal expansivity of MgO as a function of temperature: calculated isobars
along with experimental data

The fourth-order derivative with respect to pressure is related to the second derivative of the bulk modulus:

$$K'' = \left(\frac{\partial^2 K}{\partial P^2}\right)_T = \frac{V_m}{b * K}. \tag{12}$$

For ambient conditions K'' is calculated as $K'' = (-0.0295 \pm 0.0001)$ GPa^{-1}. Due to b's sign K'' invariably is negative, which is a favourable characteristic: Stacey et al. (1981) pointed out that an EOS, when applied to terrestrial data, becomes useful if the bulk modulus has a concave nature. The experimental determination of K'' is rather cumbersome. In spite of that, from the scarce experimental data, the values of (-0.0289 ± 0.0143) GPa^{-1} and (-0.022 ± 0.004) GPa^{-1} can be derived.

Because the parameter 'b' is now related to the pressure derivative of bulk modulus it is possible to write the equation of state expressed in Equation (7) in its final form. The result is:

$$P = P^\circ - K^\circ(T_0)\exp\left(K'_{ref}\right)\left\{\ln\left(\frac{V}{V^\circ(T)}\right) + \sum_{j=1}^{\infty}\left[\frac{(-1)^j}{j \cdot j!}K'_{ref}\frac{V^j - (V^\circ(T))^j}{(V^\circ(T_0))^j}\right]\right\}; \tag{13}$$

$$K'_{ref} = \left(\frac{\partial K}{\partial P}\right)_{P^\circ, T_0}. \tag{14}$$

The 'ref' in Equations (13 and (14)) denotes the reference condition. The final expression for isothermal bulk modulus is:

$$K = K^\circ(T_0)\exp\left[-K'_{ref}\frac{V - V^\circ(T_0)}{V^\circ(T_0)}\right]. \tag{15}$$

Finally, thermal expansivity can be derived by taking the temperature derivative of the left-hand and right-hand side of Equation (13) at constant pressure. The result is:

$$\alpha(P,T) = \alpha^\circ(T)\exp\left[K'_{ref}\frac{V - V^\circ(T)}{V^\circ(T_0)}\right]. \tag{16}$$

Equation (16) indicates that the sign of thermal expansivity only depends on the sign of the one-bar thermal expansivity. It is therefore a positive property in pressure-temperature space, a characteristic we were searching for in the beginning of this chapter.

Another interesting characteristic is the behaviour of the pressure derivative of isothermal bulk modulus in pressure-temperature space. From Equation (11) it is observed that:

$$K' = \frac{V}{V^\circ(T_0)}K'_{ref}. \tag{17}$$

Equation (17) indicates that K' decreases with pressure and increases with temperature. This is also a characteristic we were searching for. Its temperature derivative is written as:

$$\left(\frac{\partial K'}{\partial T}\right)_P = \frac{\alpha V K'_{ref}}{V^\circ(T_0)}.$$ (18)

For MgO this derivative ranges between 1.5×10^{-4} to 3.5×10^{-4} for temperatures between 300 K and 3000 K. These values compare quite well with values obtained with the vibrational model established in §307.

summarizing for MgO

In terms of the approach outlined above, the thermodynamic properties of MgO as a function of temperature and pressure, in the range from ambient conditions up to 3100 K and 200 GPa, can be calculated from a set of 15 numerical values. These 15 values are composed of

- the three values at 298.15 K and 1 bar for the enthalpy of formation, the absolute entropy and volume; these values are, in respective order: -601241 $J\,mol^{-1}$; 26.924 $J\,K^{-1}\,mol^{-1}$; and 11.25×10^{-6} $m^3\,mol^{-1}$;
- the two equation of state parameters, Equation (5): $K^\circ(T_0) = 161.5$ GPa; and $b = -2.359 \times 10^{-6}$ $m^3\,mol^{-1}$;
- the six constants in Equation (9) for the heat capacity at constant pressure, at 1 bar, as a function of temperature;
- the four constants in Equation (10) for the cubic expansion coefficient, at 1 bar pressure, as a function of temperature.

thermodynamic analysis in terms of the equation of state

A thermodynamic analysis, in terms of the equation of state, of a given substance in a given form, comes down to using all of the available experimental data to determine the numerical values of the material-dependent properties in Equations (1) and (8). These are

- the heat of formation ($\Delta_f H^\circ$), the absolute entropy (S°), and the volume (V°) of the material at 298.15 K and 1 bar (0.1 GPa);
- the parameters of the polynomial expression for the heat capacity at constant pressure (C_p°), at 1 bar;
- the parameters of the polynomial expression for the cubic expansion coefficient (α°), at 1 bar;
- the parameters that define the equation of state: K° and b.

140

For many cases the values of the heat of formation, the absolute entropy, and the volume at 298.15 K and 1 bar are known, and can be found in thermodynamic tables. For those cases in which (part of) these properties are not known, the establishment of their values is part of the analysis.

The parameters of the polynomial expressions for the heat capacity at constant pressure and the cubic expansion coefficient, or thermal expansivity, are derived from calorimetric and volumetric data, respectively, as a function of temperature, and at ambient pressure.

The parameters that define the equation of state are derived from available experimental data on volume and bulk modulus. Apart from volume as a function of temperature at 1 bar pressure, volume data also may include information on volume as a function of pressure.

application to Mg_2SiO_4, solid solutions

It is generally accepted that $(Mg,Fe)_2SiO_4$, containing 10 mole % of the iron component, is the major constituent material in the *earth's transition zone*. The zone, which is located at depths between 400 and 520 km in the earth's upper mantle, is characterized by a transformation of the low-pressure form, olivine, to the forms wadsleyite and ringwoodite, during displacement of material to deeper regions, see Figure 8.

The mineralogical name of the form of Fe_2SiO_4 stable at ambient conditions is *fayalite*. *Olivine* is the general term used by geologists to refer to the complete series of mixed crystals of forsterite, Mg_2SiO_4, and fayalite.

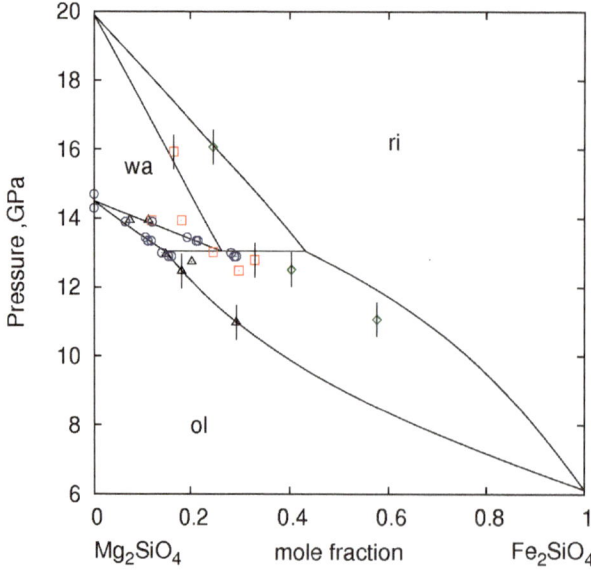

FIG. 8.
The system Mg_2SiO_4 + Fe_2SiO_4. Pressure versus composition phase diagram, valid for $T = 1673$ K. Experimental data are from Bertka and Fei (1999), circles, and Frost (2003), triangles olivine, squares wadsleyite and diamonds ringwoodite

§(306)

To explain the experimentally observed seismic jumps, associated with sound waves propagating through the zone, a thermodynamic framework is required to predict the thermophysical properties of the three mixed-crystalline forms for conditions of temperature and pressure prevailing in the zone. To achieve that, thermodynamic descriptions for the Gibbs energies of the *end members* in the three forms should be available. Once these are available the Gibbs energies of the three mixed-crystalline forms olivine, wadsleyite and ringwoodite can be constructed using phase diagram data.

For the end members forsterite and ringwoodite, we demonstrated in Jacobs and Oonk (2001) that the formulation of the equation of state, Equation (7) is incomplete. A general characteristic of the linear relationship formulated by Equation (5) is derived by using the identity:

$$\left(\frac{\partial K}{\partial T}\right)_V = \left(\frac{\partial K}{\partial T}\right)_P + \alpha K \left(\frac{\partial K}{\partial P}\right)_T. \tag{19}$$

From the linear relationship written in the form of Equation (6), it follows that $(\partial K/\partial T)_P = \alpha K V/b$. Combination of the latter identity with Equation (11) for $(\partial K/\partial P)_T$ and Equation (19) leads to $(\partial K/\partial T)_V = 0$, which is the general characteristic of our equation of state, Equation (7). Experimental data for forsterite and ringwoodite indicate non-zero values for $(\partial K/\partial T)_V$. To cover the experimental data in pressure-temperature space for these end members we used a mathematical extension by writing the molar volume in Equation (5) as:

$$V_m(T,P) = V_m^o(T_0) + a \cdot (T - T_0) + b \cdot \ln\left(\frac{K(T,P)}{K^o(T_0)}\right). \tag{20}$$

Although it is also possible to extend Equation (5) by writing the parameter b linearly dependent in temperature, mathematical expressions for the derivatives $(\partial K/\partial T)_P$ and $(\partial K/\partial T)_V$ resulting from Equation (20) are simpler; for example the latter is

$$\left(\frac{\partial K}{\partial T}\right)_V = -\frac{a \cdot K}{b}. \tag{21}$$

The introduction of the constant a has as consequence that in Equations (8), (9) and (10), the term $K^o(T_0) \cdot \exp(-V^o(T)/b)$ is replaced by the term $K^o(T_0) \cdot \exp(-V^o(T)/b) \cdot \exp(a \cdot (T_0 - T)/b)$.

For forsterite we find values for a typically between -2.691×10^{-10} m$^3 \cdot$K$^{-1} \cdot$mol^{-1} and -2.217×10^{-10} m$^3 \cdot$K$^{-1} \cdot$mol^{-1} depending on which thermal expansion dataset is used. Applying Equation (21), the average value of $(\partial K/\partial T)_V$ in the temperature range between 300K - 2000 K and 0.1 MPa pressure is -0.003 GPa/K which is about 14 % of the average value of $(\partial K/\partial T)_P$.

In Jacobs and Oonk (2005) we have combined mathematical expressions for the Gibbs energies of the end members in terms of the equation of state constants a and b given in Equation (20). These expressions were used in a general expression for the Gibbs energy for describing the *mixing behaviour* of *solid solutions*. For a solid solution of a particular form the Gibbs energy referred to one-site mixing, $(Mg,Fe)(SiO_4)_{1/2}$, is:

$$G(T,P,x_j) = \sum_{j=1}^{2} x_j G_j^*(T,P) + RT \sum_{j=1}^{2} x_j \ln(x_j) + G^E(T,P,x_j). \tag{22}$$

We showed that for the ringwoodite form of Mg_2SiO_4 a significant range of uncertainty is present in data for thermal expansivity and volume at ambient conditions, probably due to a hydration effect. For this form we found that thermodynamic analyses, based on polynomial expressions for 1 bar pressure, do not allow favouring one thermal expansivity data set in particular and that the effect on the description of the phase diagram of the mixed crystalline phases is insignificant. However, the effect on calculated bulk sound velocities at transition zone conditions is significant, calling for more tightly constrained models, such as those discussed in the next section.

some reflections using isochoric heat capacity

The isochoric heat capacity, C_V, is a property, not directly accessible to experimentation. In this section its role has been limited to the derivation of isothermal bulk modulus from experimentally determined adiabatic bulk modulus data. However, the significance of C_V becomes clearer when molecular calculations, performed using properties in the microscopic world, are coupled with thermodynamic macroscopic observables. Because energetic effects depend on atomic distances, it is more natural for these methods to start from the *Helmholtz energy*, from which C_V can be derived by differentiating twice with respect to temperature. Such calculations, rooted in *first principles*, show characteristic behaviour of this property. To enhance a comparison between thermodynamic properties calculated by our formalism and those derived by first principle calculations, C_V is of crucial importance.

To elucidate this point in more detail we return to MgO and investigate its behavior using the isochoric heat capacity in pressure-temperature space. Because MgO is a substance abundant in the Earth's lower mantle, we calculate isochoric heat capacity as function of temperature up to pressures prevailing at the core-mantle boundary, about 135 GPa.

Figure 9 illustrates that at 1 bar pressure a limiting value is approached for four different thermodynamic calculations including our own one, marked as Jacobs & Oonk (2000). This limiting value is somewhat lower than $6R$, the *Dulong-Petit value* for MgO (Dulong & Petit, 1819).

At higher pressures our own calculated C_V progressively increases with pressure and temperature. As we shall see in §307, this increase could be explained in terms of *intrinsic anharmonic behaviour*. However, some difficulties arise when these calculations are compared with first-principles calculations. A *molecular dynamics* study, including anharmonic effects, by Inbar and Cohen (1995) indicates that intrinsic anharmonicity is important at low pressures, but that it decreases at high pressures. Their results indicate that MgO behaves quasi-harmonically at lower mantle pressures and therefore C_V will approach the 6R limit at high pressures and temperatures. This behaviour is not captured by our formalism. Figure 9 shows C_V behaviour calculated using three thermodynamic databases popular in geophysics, which are based on polynomial parameterizations of 1 bar properties. The result obtained with the database of Fei et al. (1991) and that of Saxena (1996), both employing a Birch-Murnaghan equation of state, show similar behaviour as our own result. That obtained by using the database of Fabrichnaya et al. (2004), employing a Murnaghan equation of state, shows different behaviour. From Figure 9 it can be concluded that the behaviour of C_V is quite different in different thermodynamic databases and that no database results in a proper convergence to the 6R limit, hampering a fair comparison with results obtained by first-principles calculations. Figure 9 evidences that parameterization studies, using polynomial functions for the description of 1 bar thermodynamic properties, simultaneously covering the majority of the available experimental data in pressure-temperature space, do not necessarily imply a reliable description at the extreme conditions in the lower mantle. This calls for a thermodynamic formalism that incorporates, somehow, physical effects such as anharmonicity. This is the reason why we have included in section §307 methods enabling a more convenient comparison with first principles calculations.

perovskite, MgSiO₃

> To complete the description of the Mg_2SiO_4 phase diagram, a thermodynamic analysis of perovskite is needed. General consensus exists in the geophysical community that perovskite is the most abundant material in the Earth. For that reason it is important to constrain its thermophysical properties as much as possible. However for this phase only a limited amount of one-bar data are available because it may either transform to enstatite (Knittle et al., 1986) or become amorphous (Paris et al., 1990, and Wang et al., 1994).
The isobaric heat capacity has been determined by Akaogi and Ito (1993) using Differential Scanning Calorimetry in the temperature range between 140 K and 295 K. Adiabatic bulk modulus is established at ambient conditions by Yeganeh-Haeri (1994) and Sinogeikin et al. (2004). One-bar volume has been determined up to 400 K by Ross and Hazen (1989) and Wang et al. (1994). A thermodynamic analysis based on polynomial functions for one bar properties, therefore, heavily relies on volume data at high pressure and temperature, such as measured by Funamori et al. (1996), Utsumi et al. (1995) and Wang et al. (1994).
Although the available data constrain the model parameters to a specific extent, one-bar parameterizations for heat capacity, thermal expansivity and bulk modulus might be not constrained well enough to allow reliable extrapolations to high *P-T* conditions. For this reason we postpone a thermodynamic description until we develop models that incorporate more physics, and which are constrained by experimental data on microscopic physical properties. These models are discussed in the next section where we apply a vibrational model to $MgSiO_3$ perovskite. And therewith we establish a parameterization in line with the methodology of the present section.

144

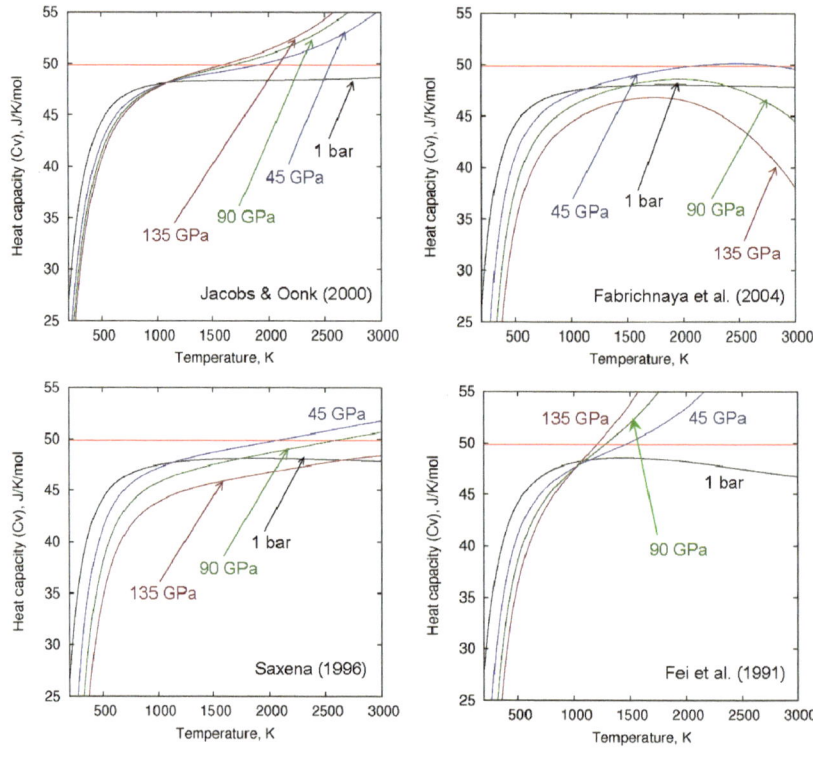

FIG. 9.
Isochoric heat capacity calculated from four different thermodynamic
descriptions of MgO as function of temperature and pressure. Our own
calculated C_V is given by that of Jacobs and Oonk (2000). The horizontal line
in the plots denotes the Petit-Dulong (1819) value

*The "nearly precise linear relation" between logarithm of bulk modulus and
volume has been transformed to an equation of state, with the help of which
thermodynamic properties of materials under high pressure can be derived from
thermodynamic data obtained at ambient pressure as a function of temperature.
The substance MgO has been taken to investigate the thermodynamic rigour of the
equation of state. Subsequently, the methodology has been applied to the three
polymorphs - forsterite(fo), wadsleyite(wa), and ringwoodite(ri) - of Mg₂SiO₄. Some
reflections were made regarding the coupling of the results of the methodology with
results from microscopic theories. The thermodynamic analysis of perovskite
requires incorporating more physics into the thermodynamic modelling.*

§ 307 TOWARDS VIBRATIONAL MODELS

==

A thermodynamic formalism is discussed, which, in a self-consistent way, enables the derivation of thermodynamic properties, the calculation of the shear modulus (→308), and the calculation of phase diagrams, for conditions of pressure and temperature that prevail in the interior of the earth. That formalism is based on the assumption that the Helmholtz energy of an insulator material can be taken as the sum of two contributions, one of them being related to the static lattice and the other to lattice vibrations. The vibrational models by Einstein and Debye are applied to aluminium. The model introduced by Kieffer is applied to the forsterite form of Mg_2SiO_4.

==

introduction: the Helmholtz energy; the formalism

In a sense, the preceding sections of this work have been written from a chemical background, where the investigator has control over *temperature* and *pressure – the natural variables* of the *Gibbs energy* (←107). In this section, the central role of the Gibbs energy is taken over by the *Helmholtz energy* (*A*), whose natural variables are *temperature* and *volume*. The Helmholtz energy is composed of *energy* (*U*) and *entropy* (*S*):

$$A = U - TS, \tag{1}$$

and the *fundamental equation* (←107) for its change is

$$dA = -SdT - PdV. \tag{2}$$

From an experimental, geophysical standpoint the Helmholtz energy has the advantage that its natural variable *V* is more accessible to experimental determination than pressure *P*. Geophysical *equations of state* give pressure as a function of volume, rather than volume as a function of pressure.

From a theoretical standpoint the Helmholtz energy has to be preferred because of the fact that *lattice vibrations*, which make their appearance in this section, mainly depend on volume.

146

The Helmholtz energy is linked to the *partition function* and, therefore, amenable to *statistical mechanical* treatment. And because pressure is directly derived from the Helmholtz energy:

$$P = -(\partial A/\partial V)_T,\tag{3}$$

the additional advantage is that pressure can be partitioned in its physical contributions.

The formalism which is detailed in this section is based on the assumption that, for an *insulator material*, such as Mg_2SiO_4, the lattice vibrations give an independent contribution to the Helmholtz energy. The remaining contribution comes from a hypothetical lattice in which vibrations are absent: the *static lattice.* In other terms, the Helmholtz energy of the material is the sum of two contributions, one from the static lattice and the other from *lattice vibrations*:

$$A(T,V) = U^{st}(V) + A^{vib}(T,V).\tag{4}$$

Equation (4) reflects the fact that the static lattice has zero entropy: its energy depends on volume only. In the following, the contribution of the static lattice is split up into two parts, such that our basic equation for the Helmholtz energy is

$$A(T,V) = U^{ref}(V_0^{st}) + U^{st}(V) + A^{vib}(T,V).\tag{5}$$

In the equation $U^{ref}(V_0^{st})$ represents the energy contribution at zero Kelvin and zero pressure for a substance in which there is no vibrational motion of the atoms. The term is adjusted such that the *enthalpy of formation* of the material at 298.15 K and 1 bar pressure calculated by the formalism based on Equation (5) is equal to the value reported in e.g. the *JANAF thermochemical tables* compiled by Chase et al. (1985).

The volume of the static crystal lattice at zero pressure and zero Kelvin, V_0^{st}, is not experimentally accessible because in a real crystal lattice vibrational motions exist at these conditions.

The term $U^{st}(V)$ represents the change of the Helmholtz energy resulting from a change in volume of the static lattice from V_0^{st} to V. This change can be caused by compression due to an increase in pressure or by an expansion due to an increase in temperature. The third term in Equation (5) represents the change in Helmholtz energy due to lattice vibrations.

NB. Equation (5) may be expanded by including additional physical contributions, such as contributions due to electronic and magnetic effects or due to lambda transitions.

§(307)

properties of the static lattice

The second term on the right-hand side of Equation (5) is calculated with an *equation of state*, such as the one derived in § 306, or one of the equations of state that are frequently used in geophysics.

A comparison of the equations of state used in geophysics has been made by Cohen et al. (2000). They showed, for low and high compressible materials, covering the compression range of substances in the Earth, that *Vinet's equation of state* (Vinet et al. 1987) produces more accurate results than several other ones, such as a *Birch-Murnaghan equation of state* (Birch, 1952). Vinet's equation of state is mathematically simpler than the one we have used in § 306; for that reason we have used it to assess thermodynamic properties of mantle materials. Jacobs and Schmid-Fetzer (2010) showed that Keane's (1954) equation of state is more successful than Vinet's for representing thermodynamic properties of aluminium.

Vinet's equation is expressed as:

$$P^{st}(V) = 3K_0^{st} \left[\left(\frac{V}{V_0^{st}} \right)^{-2/3} - \left(\frac{V}{V_0^{st}} \right)^{-1/3} \right] \exp \left\{ \frac{3}{2} \left(K_0^{'st} - 1 \right) \left[1 - \left(\frac{V}{V_0^{st}} \right)^{1/3} \right] \right\}, \quad (6)$$

where K_0^{st} represents the *isothermal bulk modulus* of the static lattice and $K_0^{'st}$ its pressure derivative. Keane's equation of state is expressed as:

$$\frac{P^{st}(V)}{K_0^{st}} = \frac{K_0^{'st}}{\left(K_\infty^{'st} \right)^2} \left[\left(\frac{V}{V_0^{st}} \right)^{-K_\infty^{'st}} - 1 \right] + \left(\frac{K_0^{'st}}{K_\infty^{'st}} - 1 \right) \ln \left(\frac{V}{V_0^{st}} \right) \quad (7)$$

Relative to Vinet's equation of state Equation (7) is characterized by an additional property, $K_\infty^{'st}$, the pressure derivative of the bulk modulus at infinite pressure. In the following we take Vinet's equation of state to illustrate properties of the static lattice. We use Keane's equation of state in an example for deriving thermodynamic properties of aluminium in pressure-temperature space.

For $K_0^{'st} \neq 1$ the energy $U^{st}(V)$ is found by integration of Equation (6):

$$U^{st}(V) = \frac{4K_0^{st}V_0^{st}}{\left(K_0^{'st} - 1 \right)^2} \left\{ 1 + \left[\frac{3}{2} \left(K_0^{'st} - 1 \right) \left(1 - \left(\frac{V}{V_0^{st}} \right)^{1/3} \right) - 1 \right] \exp \left[\frac{3}{2} \left(K_0^{'st} - 1 \right) \left(1 - \left(\frac{V}{V_0^{st}} \right)^{1/3} \right) \right] \right\}$$

$$(8)$$

§(307)

And for $K_0^{'st} = 1$ it is given by:

$$U^{st}(V) = -9K_0^{st}V_0^{st}\left[\left(\frac{V}{V_0^{st}}\right)^{1/3} - \frac{1}{2}\left(\frac{V}{V_0^{st}}\right)^{2/3} - \frac{1}{2}\right].$$ (9)

Next, the *total pressure* derived from Equation (5) is:

$$P(T,V) = P^{st}(V) + P^{vib}(T,V).$$ (10)

The properties V_0^{st}, K_0^{st} and $K_0^{'st}$ and $U^{ref}(V_0^{st})$ can be obtained from a least-squares optimization of available experimental data. Alternatively they can be predicted by *ab initio static calculations*. At the condition $P = 0$ and $T = 0$, lattice vibrations are present and the volume of the crystal is denoted by V_0. Equations (6) and (10) lead to the relation:

$$0 = 3K_0^{st}\left[\left(\frac{V_0}{V_0^{st}}\right)^{-2/3} - \left(\frac{V_0}{V_0^{st}}\right)^{-1/3}\right]\exp\left\{\frac{3}{2}\left(K_0^{'st} - 1\right)\left[1 - \left(\frac{V_0}{V_0^{st}}\right)^{-2/3}\right]\right\} + P^{vib}(0,V_0)$$ (11)

Generally the *zero-point pressure*, $P^{vib}(0,V_0)$, is a positive property. For instance within the framework of the *Debye model* (Debye 1912) this vibrational contribution to the pressure is given by:

$$P^{vib}(0,V_0) = \frac{9nR\theta\gamma}{8V_0},$$ (12)

where n represents the *number of atoms in a formula unit* ($n = 7$ for Mg_2SiO_4); R the *gas constant*; θ the *characteristic Debye temperature*; and γ the *Grüneisen parameter* (see below, Equation 15) (Grüneisen 1926). Because all properties in Equation (12) are positive, it follows that the vibrational contribution to pressure is positive; and consequently, as follows from Equation (11), the *equation-of-state-pressure* for the static lattice, $P^{st}(V_0)$, is negative.

The work of Kiefer et al. (2001) on the wadsleyite form of Mg_2SiO_4 is instructive to get an impression of the magnitude of the properties V_0^{st}, V_0, and $P^{vib}(0,V_0)$. They applied a method based on *density functional theory* and symmetry-preserving, variable cell-shape structure relaxation to obtain the values of: the isothermal bulk modulus (180.8 GPa); its pressure derivative (4.34); and the volume of the static lattice (39.700 $cm^3 mol^{-1}$). From Equations (11) and (12) and the values 975 K for the Debye temperature and 1.32 for the Grüneisen parameter it follows that the real physical volume at zero Kelvin, V_0, is 40.175 $cm^3 mol^{-1}$, which is about 1.2% larger than V_0^{st}. An independent way to confirm this value of V_0 is offered by Born and Huang (1966), who derived the approximate expression:

$$\frac{V - V_0^{st}}{V_0^{st}} = \frac{\gamma U^{vib}}{V K_0^{st}}, \tag{13}$$

where U^{vib} represents the vibrational contribution to the energy. In the Debye model at $T=0$ and $P=0$, this contribution equals:

$$U^{vib}(0, V_0) = \frac{9}{8} n R \theta(0, V_0). \tag{14}$$

From Equations (13) and (14), V_0 can be deduced analytically to have the value of 40.16 $cm^3 mol^{-1}$. This value deviates only 0.04% from the value calculated by Equations (11) and (12). The reason why V_0 differs from V_0^{st} by 1.2% is that the vibrational contribution, $P^{vib}(0, V_0)$, to the total pressure has the significant value of 2.1 GPa; in Equation (11) that value must be counter-balanced by $P^{st}(V_0)$.

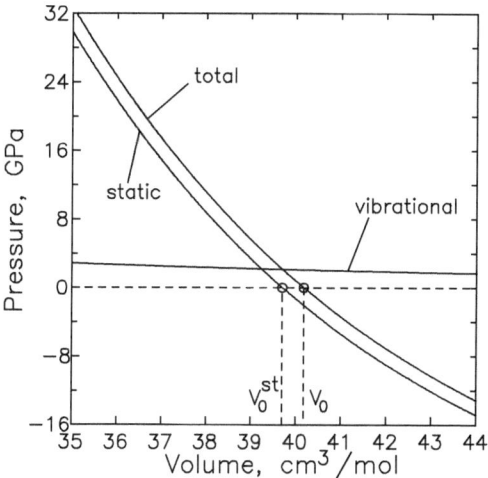

FIG. 1.
Plot of the pressure contributions as a function of volume for wadsleyite, Mg_2SiO_4. The intersection of the static contribution to pressure, labelled 'static', with the line $P = 0$ results in the volume of the static lattice V_0^{st}. The intersection of the total pressure curve, labelled 'total', and the line $P = 0$, the external pressure, results in the volume V_0 and is the solution of Equation (11). At this volume the vibrational contribution, labelled 'vibrational' counter-balances the static contribution

Figure 1 illustrates the contributions of the static lattice and the lattice vibrations to the total pressure. The fact that V_0 is larger than V_0^{st} can also be understood using the energy expression, Equation (8).

The energy of the static lattice as a function of volume reveals the asymmetric curve plotted in Figure 2. At $P=0$ and $T=0$ the crystal of wadsleyite, Mg_2SiO_4, is in equilibrium with the external environment when its energy is minimal, by virtue of Equations (1) and (3). At this condition, the volume at which attractive and repulsive forces in the static crystal are in equilibrium is V_0^{st}. Lattice vibrations are present at zero Kelvin and their presence shifts the total energy to higher values; and the minimum is displaced to a higher volume. This displacement of the minimum from V_0^{st} is caused by the positive value of the Grüneisen parameter. This can be illustrated as follows. As we shall see further on in this section, the definition of the Grüneisen parameter is:

$$\gamma = -\left(\frac{\partial \ln v}{\partial \ln V}\right)_T = -\left(\frac{\partial \ln \theta}{\partial \ln V}\right)_T. \tag{15}$$

The reason for the second term in Equation (15) is that in Debye's theory the vibrational *cut-off frequency* is related to the Debye temperature as: $\theta = hv/k$. Integration of Equation (15), assuming that γ is independent of volume, results in:

$$\theta(0,V) = \theta(0,V_0) \cdot \left(\frac{V}{V_0}\right)^{-\gamma}. \tag{16}$$

The reason that V_0 appears in Equation (16) instead of V_0^{st} is that lattice vibrations take place at the real physical volume of the crystal.

The total energy at $P = 0$ and $T = 0$ follows from Equation (5) and it is written as:

$$U(0,V) = U^{ref} + U^{st}(V) + \frac{9}{8} nR\, \theta(0,V_0) \cdot \left(\frac{V}{V_0}\right)^{-\gamma}. \tag{17}$$

We have derived the value of $U^{ref} = -2.224269$ MJ mol^{-1} from the experimental value for the heat of formation at 975 K measured by Akaogi et al. (1989) and by making use of the complete Debye expressions, as is discussed further on in this section. For the present discussion this value is not crucial because a different value merely displaces the two curves in Figure 2 to different energies in the same manner.

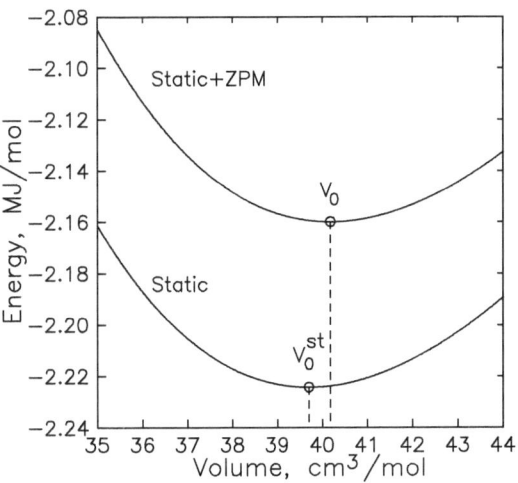

FIG. 2.

Plot of energy versus volume at $P = 0$, $T = 0$ for wadsleyite, Mg_2SiO_4. The curve labelled 'static' represents the energy ignoring vibrational motions. The curve is steeper to the left of V_0^{st} because the average repulsion energy of the atoms is shorter in range compared to the attraction energy. The curve labelled 'Static+ZPM' means that *Zero-Point-Motion* of the atoms is included in the energy. Quantum mechanical theory requires that lattice vibrations are present at these conditions. Lattice vibrations not only increase the total energy of the substance, but also the volume relative to the static lattice

The second term on the right-hand side of Equation (17) is represented by Equation (8). When the Grüneisen parameter, γ, is zero, the Debye temperature does not depend on volume, and the third term on the right-hand side of Equation (17) is a constant. In this case the minimum of the total energy is located at the same volume as the minimum of the static lattice energy. Using Equation (17) it can be easily demonstrated that the value 1.32 for the Grüneisen parameter established by Kiefer et al. (2001) leads to a larger value of V_0 relative to V_0^{st}. It also demonstrates that the magnitude of the displacement of the volume, from V_0^{st} to V_0, is determined by the volume dependence of the vibrational frequencies.

Kiefer et al. (2001) applied their static *ab-initio method* in combination with a Debye model to obtain the value of 40.28 $cm^3 mol^{-1}$ for the volume at ambient conditions, a value, which we confirmed by using the complete Debye expression for the vibrational contribution to pressure. The ambient volume thus deviates from V_0 by only 0.3%. In our formalism, V_0^{st} is necessary in Equations (6)-(9) and V_0 is needed to determine θ and to determine vibrational frequencies. The static volume V_0^{st} is calculated from Equations (11) and (12), once the θ, γ, K_0^{st}, $K_0^{'st}$, and the real volume of the crystal, V_0, are known. These properties result from a process in which experimental data are fitted.

properties of lattice vibrations: Einstein model

The temperature dependence of the heat capacity was a hot debate at the turn of the 19th century, before the advent of quantum mechanics. At that time classical mechanics predicted that *heat capacity at constant volume*, C_V, is a constant equal to 3nR, which is also known as the *Dulong-Petit value*. In his ground-breaking paper Einstein (1906) showed, by using a *monochromatic oscillator model*, in which the atoms vibrate independently with the same frequency, that C_V for solid substances is not a constant, but that it changes with temperature, being zero at zero Kelvin and approaching the Dulong-Petit limit at infinite temperature. The heat capacity predicted from Einstein's model is in reasonable accord with experimental observation, although a refinement is necessary at temperatures approaching zero Kelvin. This refinement is possible by using the *vibrational model* developed by Debye (1912) (which we shall discuss further on). There are three reasons for discussing the Einstein model. The first is that expressions for vibrational contributions to thermodynamic properties are simple enough to be calculated analytically. The second is that calculated thermodynamic properties are fairly accurate for metallic elements. The third reason is that we shall use these expressions in models in which the *vibrational density of states* (VDoS) differs from that of a monochromatic oscillator.

To derive the formulas for thermodynamic properties using Einstein's theory, we depart from the *partition function* Z^{vib}, using the *harmonic approximation* in which *Hooke's law* is obeyed:

$$Z^{vib} = \sum_{n=0}^{\infty} \exp\left(\frac{-\varepsilon_n}{kT}\right) = \sum_{n=0}^{\infty} \exp\left(-\left(n+\frac{1}{2}\right)\frac{h\nu}{kT}\right) = \frac{\exp\left(-\dfrac{h\nu}{2kT}\right)}{1-\exp\left(-\dfrac{h\nu}{kT}\right)}, \tag{18}$$

where ε_n denotes the *energy of* the *oscillator*, n a *quantum number* indicating the energy level, h *Planck's constant*, ν the *frequency*, and k *Boltzmann's constant*. In the following we write:

$$x = \frac{h\nu}{kT}. \tag{19}$$

The partition function becomes:

$$Z^{vib} = \frac{\exp(-x/2)}{1-\exp(-x)}, \tag{20}$$

and

$$\ln\left(Z^{vib}\right) = -\frac{x}{2} - \ln\left(1 - e^{-x}\right). \tag{21}$$

Thermodynamic properties are derived from the partition function using *statistical mechanics*, and the expressions are summarized as follows.

Helmholtz energy: $A^{vib} = -kT \ln Z^{vib}$;   (22)

Energy : $U^{vib} = kT^2 \left(\dfrac{\partial \ln Z^{vib}}{\partial T}\right)_V$;   (23)

Heat capacity: $C_V^{vib} = \left(\dfrac{\partial U^{vib}}{\partial T}\right)_V$;   (24)

Entropy: $S^{vib} = k \ln Z^{vib} + kT \left(\dfrac{\partial \ln Z^{vib}}{\partial T}\right)_V$;   (25)

Thermal pressure: $P^{vib} = kT \left(\dfrac{\partial \ln Z^{vib}}{\partial V}\right)_T$.   (26)

In the harmonic approximation it is assumed that the frequency of each oscillating atom does not depend on volume and temperature. In real systems thermodynamic properties are more accurately described using the *quasi-harmonic approximation* in which it is assumed that the frequency depends on volume. The volume dependence of the frequency can be elucidated by making isothermal measurements of the frequency as a function of pressure. For instance for MgO, it has been demonstrated by Chopelas (1996), by *Raman spectroscopy* at different pressures, that the frequency depends on volume. The volume dependence of the frequency is described by the *Grüneisen parameter;* and to extract it from the pressure-frequency measurements use is made of its definition:

$$\gamma = -\left(\frac{\partial \ln v}{\partial \ln V}\right)_T = -\frac{V}{v}\left(\frac{\partial v}{\partial V}\right)_T = -\frac{V}{v}\left(\frac{\partial v}{\partial P}\right)_T\left(\frac{\partial P}{\partial V}\right)_T = +\frac{K}{v}\left(\frac{\partial v}{\partial P}\right)_T, \tag{27}$$

where V denotes volume, P pressure, K isothermal bulk modulus, and v frequency.

§(307)

Equation (27) indicates that additional measurements of isothermal bulk modulus and volume are necessary to obtain a value for the Grüneisen parameter. Several expressions are available in the literature to describe the vibrational frequency as a function of volume. One description we used for materials in the Earth is the expansion of the logarithm of the frequency in a *Taylor series* (see e.g. Gillet et al. (2000)) and truncate it after the second term.

$$\ln\left(\frac{v(V)}{v_0}\right) = -\gamma_0 \ln\left(\frac{V}{V_0}\right) - \frac{1}{2}q_0 \cdot \gamma_0 \ln^2\left(\frac{V}{V_0}\right).$$
(28)

In Equation (28) v_0 represents the vibrational frequency at zero pressure and zero temperature. At these conditions, the Grüneisen parameter and the *mode parameter* q are given by:

$$\gamma_0 = -\left(\frac{\partial \ln v}{\partial \ln V}\right)_{T=0,P=0} ; \quad q_0 = \left(\frac{\partial \ln \gamma}{\partial \ln V}\right)_{T=0,P=0}.$$
(29)

As a result, the Grüneisen parameter is a function of volume and it can be written as:

$$\gamma(V) = \gamma_0 + q_0 \cdot \gamma_0 \ln\left(\frac{V}{V_0}\right).$$
(30)

Equation (30) indicates that the Grüneisen parameter may become zero at $V/V_0 = \exp(-1/q_0)$ resulting in a maximum of the frequency versus volume (or pressure) curve. To avoid this situation more terms may be included in Equation (30). An expression eliminating this problem is used by Jacobs and Schmid-Fetzer (2010) in their application of vibrational models to aluminium and iron to high compressions. They write the Grüneisen parameter as:

$$\gamma(V) = \gamma_\infty + (\gamma_0 - \gamma_\infty)\left(\frac{V}{V_0}\right)^m$$
(31)

In Equation (31) γ_∞ represents the Grüneisen parameter at infinite pressure. Equation (31) results in the volume dependence of the vibrational frequency as follows:

$$\ln\left(\frac{v(V)}{v_0}\right) = -\gamma_\infty \ln\left(\frac{V}{V_0}\right) - \frac{\gamma_0 - \gamma_\infty}{m}\left[\left(\frac{V}{V_0}\right)^m - 1\right] \quad \text{and } m \neq 0$$
(32a)

$$\ln\left(\frac{v(V)}{v_0}\right) = -\gamma_0 \ln\left(\frac{V}{V_0}\right) \quad \text{and } m = 0$$
(32b)

§(307)

The Grüneisen and mode q parameters at 0 K and 0 Pa, are defined by:

$$\gamma_0 = -\left(\frac{\partial \ln v}{\partial \ln V}\right)_{T=0,P=0} \quad, \quad q_0 = \left(\frac{\partial \ln \gamma}{\partial \ln V}\right)_{T=0,P=0} = \frac{m(\gamma_0 - \gamma_\infty)}{\gamma_0} \tag{33}$$

In the following we assume a quasi-harmonic monochromatic oscillator model in which $\gamma \neq 0$. The Einstein expressions for thermodynamic properties including static and vibrational contributions are obtained by using Equations (20) to (26) and $x = hv/kT$. They are summarized below.

Helmholtz energy:

$$A = U^{ref} + U^{st}(V) + A^{vib} = U^{ref} + U^{st}(V) + 3nRT\left[\frac{x}{2} + \ln(1 - \exp(-x))\right]; \tag{34}$$

Energy:

$$U = U^{ref} + U^{st}(V) + U^{vib} = U^{ref} + U^{st}(V) + 3nRT\left(\frac{x}{2} + \frac{x}{\exp(x) - 1}\right); \tag{35}$$

Entropy:
$$S = S^{vib} = 3nR\left[-\ln(1 - \exp(-x)) + \frac{x}{\exp(x) - 1}\right]; \tag{36}$$

Heat capacity:
$$C_V = C_V^{vib} = \frac{3nR\,x^2\,\exp(x)}{(\exp(x) - 1)^2}; \tag{37}$$

Pressure:
$$P = P^{st}(V) + P^{vib} = P^{st}(V) + \frac{3nRT\gamma}{V}\left(\frac{x}{2} + \frac{x}{\exp(x) - 1}\right); \tag{38}$$

Isothermal bulk modulus:
$$K = K^{st}(V) + K^{vib} = K^{st}(V) - V\left(\frac{\partial P^{vib}}{\partial V}\right)_T \tag{39}$$

$$K = K^{st}(V) + \frac{3nRT}{V}\left\{\left(\frac{1}{2} + \frac{1}{\exp(x) - 1}\right)\cdot x\cdot\left(\gamma^2 + \gamma - V\left(\frac{\partial \gamma}{\partial V}\right)_T\right) - \frac{\gamma^2 x^2 \exp(x)}{(\exp(x) - 1)^2}\right\}; \tag{40}$$

Thermal expansivity:
$$\alpha = \frac{1}{K}\left(\frac{\partial P^{vib}}{\partial T}\right)_V = \frac{3nR\gamma}{KV}\left\{\frac{x^2 \exp(x)}{(\exp(x) - 1)^2}\right\}; \tag{41}$$

Isobaric heat capacity:
$$C_P = C_V + \alpha^2 KVT; \tag{42}$$

Enthalpy:
$$H(P,T) = U^{st}(V) + U^{vib}(T,V) + \left(P^{st}(V) + P^{vib}(T,V)\right)\cdot V; \tag{43}$$

Gibbs energy:
$$G(P,T) = H(P,T) - T\cdot S^{vib}(T,V). \tag{44}$$

Note that in the expression for thermal expansivity, the total isothermal bulk modulus is used, which is the sum of the contributions of the static lattice and lattice vibrations. The expressions above are for one mole of molecular formula unit, the unit consisting of n atoms. Because the atoms in one mole of a monatomic substance have $3N_A$ degrees of vibrational freedom all properties have been multiplied with $3nN_A$ in the derivation from Equations (20)-(26). All expressions (34)-(41) can be written in the more familiar *Einstein characteristic temperature*, θ^E, by using $x = \theta^E/T$.

calculation of thermodynamic properties

There are eleven parameters for an Einstein model needed to determine all thermodynamic properties of a solid substance. That also applies to the Debye model. These model parameters are:

1. V_0^{st} Volume of the static lattice
2. K_0^{st} Isothermal bulk modulus of the static lattice
3. $K_0^{'st}$ Pressure derivative of isothermal bulk modulus of the static lattice
4. V_0 Real (measurable) volume of the lattice at zero Kelvin and zero pressure
5. v_0 or θ_0^E Einstein frequency or Einstein temperature at zero Kelvin and zero pressure
6. γ_0 Grüneisen parameter at zero Kelvin and zero pressure
7. γ_∞ Grüneisen parameter at zero Kelvin and infinite pressure
8. q_0 (or m)   Mode q-parameter (or m) at zero Kelvin and zero pressure
9. U^{ref} Reference energy
10. β_{el} Electronic coefficient in the free electron gas theory
11. γ_{el} Electronic Grüneisen parameter in free electron gas theory

The eleven model parameters are obtained by fitting experimental data to these parameters in a least-squares optimization. Equation (11) shows that V_0^{st} can be obtained once all other parameters are fixed to specific values. As we shall see in the example for aluminium, the *electronic parameters* β_{el} and γ_{el} can be obtained from low temperature heat capacity data and in an optimization process they can be kept constant. That results in a total of 8 fitting parameters in the Einstein and Debye model. Each calculation of a thermodynamic property at a specific pressure and temperature starts with the determination of the volume by using Equation (38). This is achieved in an iterative process by changing the volume until the total pressure matches the desired pressure. In each step of volume the frequency is calculated using Equation (28) or (32) and the Grüneisen parameter by Equation (30) or (31). Once volume has been established, all other thermodynamic properties are calculated from their analytical expressions, Equations (34)-(44).

example: aluminium

As an example to demonstrate the accuracy of the model we use the thermodynamic analysis of aluminium performed by Jacobs and Schmid-Fetzer (2010). We modified their description such that thermal pressure decreases with volume at constant temperature by setting $q_0=1$ in equation (33). We fitted all available experimental data to the model parameters and the results are given in Table 1.

Table 1:
Properties of aluminium resulting from an Einstein and a Debye model. These properties were calculated by a least-squares optimization of experimental data. The zero Kelvin, zero pressure Einstein temperature calculated from the frequency v_0 is 283.45 K and the Debye temperature is 396.70 K. The electronic coefficient is 13.5×10^{-4} $J K^{-2} mol^{-1}$ and the electronic Grüneisen parameter is 2/3. Equation (32) is used for the frequency with $\gamma_\infty=0$, and $q_0=1$. Keane's equation of state (7) has been used for the static lattice

U^{ref}	V_0^{st}	K_0^{st}	K_0^{st}	K_∞^{st}	V_0	v_0	γ_0
kJ/mol	cm³/mol	GPa			cm³/mol	cm⁻¹	
Einstein model							
-8.253(7)	9.770(3)	81.74(9)	5.30(2)	2.34(6)	9.865(3)	196.61(20)	2.157(8)
Debye model							
-8.346(7)	9.766(3)	81.77(9)	5.30(2)	2.34(6)	9.867(3)	275.53(20)	2.166(7)

Above we have started to develop the model for an insulator material. It means that Equation (5) does not contain an electronic contribution to the Helmholtz energy. Because aluminium is not an insulating material, the electronic contribution should be added. To keep our discussion short and because the effect of the electronic contribution to thermodynamic properties is fairly small we have assumed in our calculations a simple *free electron gas model*. For its contribution to the Helmholtz energy we follow Brown and McQueen (1986) by writing:

$$A^{el}(T) = -\frac{1}{2} \beta_{el} \left(\frac{V}{V_0} \right)^{\gamma_{el}} T^2 . \tag{45}$$

In the free electron gas model, the electronic Grüneisen parameter has the value of 2/3. We add this contribution to Equation (5) under the assumption that the electronic effect is independent of the static lattice and the vibrational contributions to the Helmholtz energy.

Phillips (1959) established a value 13.5×10^{-4} $JK^{-2}mol^{-1}$ for the *electronic coefficient*, β_{el}, by using Debye's model and their measured low-temperature heat capacity data. A detailed discussion on the free electron gas model and its volume dependence is given by Gopal (1966). From Equation (45) it follows that the electronic contribution to heat capacity is nearly linear in temperature.

Figure 3 shows that heat capacity data at temperatures above room temperature are quite well represented by the Einstein model, but that deviations occur below this temperature.

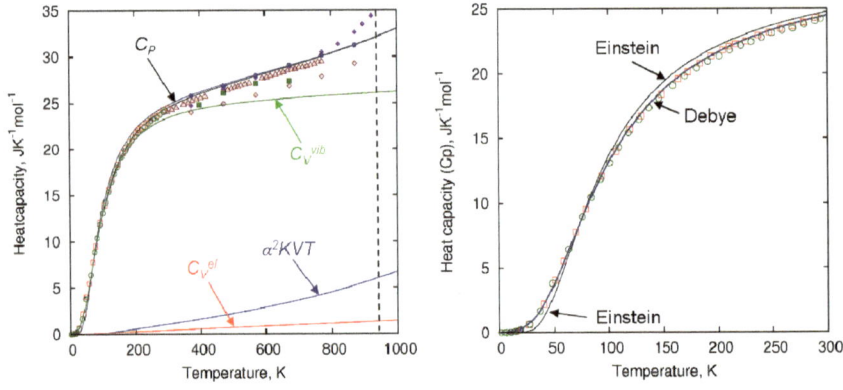

FIG. 3.

Heat capacity of aluminium at 1 bar pressure using Einstein's and Debye's models. Experimental data are from: Circle: Downie and Martin (1980); Triangle: Leadbetter (1968); Solid diamond: Pochapsky (1953); Square: Giauque and Meads (1941); Open diamond: Avramescu (1939); Solid square: Quinney and Taylor (1937); Solid circle: Eastman et al. (1924). The vertical dashed line represents the melting temperature of aluminium

Our Einstein model underestimates heat capacity below 75 K and overestimates it above this temperature. The effects counterbalance each other to some extent resulting in an entropy value of 28.26 $JK^{-1}mol^{-1}$ at room temperature in accordance with the experimental value derived by Downie and Martin (1980), 28.24 ± 0.1 $JK^{-1}mol^{-1}$. At temperatures below about 50 K the calculated heat capacity shows an exponential behaviour, a typical characteristic of the Einstein model. Experimental data however show a T^3 behaviour, a feature as we shall see further on, is better represented with the Debye model. The exponential behaviour of heat capacity at low temperatures is, as Einstein already noticed, due to the simplification that atoms vibrate independently with the same frequency.

Physically it is more plausible that inside the crystal the motion of atoms affect the motions of other atoms. The atoms will in that case vibrate with a range of possible frequencies. Figure 3 shows the relative magnitudes of the different contributions to the isobaric heat capacity; the electronic contribution is small, but cannot be neglected. Figure 4 shows that thermal expansivity is described well in a large range of temperature. At low-temperature the same kind of behaviour is present in thermal expansivity as in the heat capacity.

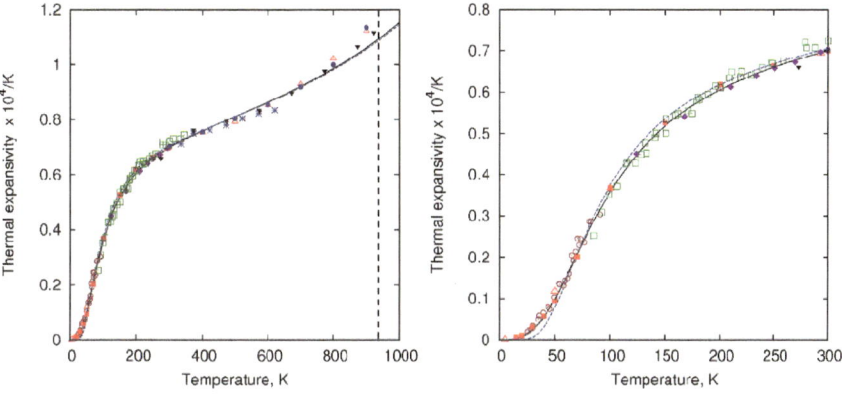

FIG. 4.

Calculated thermal expansivity of aluminium at 1 bar pressure using Einstein's (dashed curve) and Debye's (solid curve) models. Experimental data are from: Solid circle: Pathak and Vasavada (1970); Circle: Fraser and Hollis-Hallett (1965); Solid circle: Nicklow and Young (1963); Solid square: Fraser and Hollis-Hallett (1961); Open square: Strelkov and Novikova (1957); Inverse solid triangle: Wilson (1941); Asterisk: Honda and Okubo (1924)

Figure 5 shows that the *adiabatic bulk modulus* (K_S) and the isothermal bulk modulus are quite well represented with such a simple model for the vibrational density of states.

Because the electronic contribution affects heat capacity at constant volume, C_V, it also affects the adiabatic bulk modulus through the equation ($\leftarrow$Exc 301:28)

$$K_S = K \left(C_P/C_V \right). \tag{46}$$

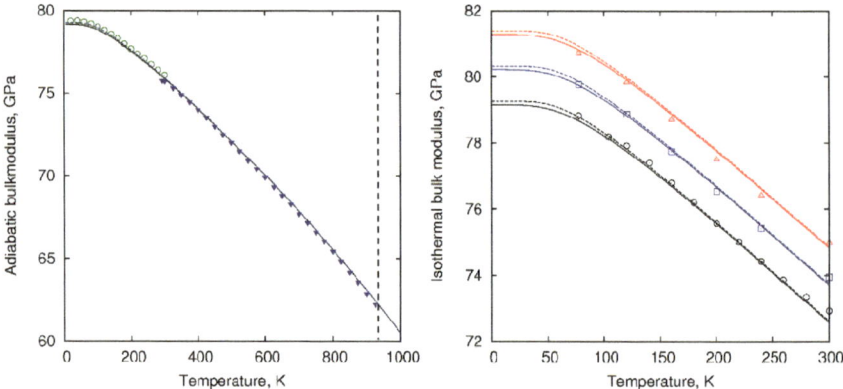

FIG. 5.
Left: calculated adiabatic bulk modulus of aluminium at 1 bar pressure using Einstein's (dashed curve) and Debye's (solid curve) models. Experimental data are from: Triangle: Gehrlich and Fisher (1969); Circle: Kamm and Alers (1964). The vertical dashed line represents the melting temperature.
Right: calculated isothermal bulk modulus and experimental data of Ho and Ruoff (1969) at 1 bar, 0.2 GPa, and 0.4 GPa. Einstein's model (dashed curves), Debye's model (solid curves)

Despite its simplification, the break-through of Einstein's theory is that it was shown for the first time that the behaviour of the heat capacity is the result of the *quantization of energy.* And notwithstanding its physical limitation, it is clear that the Einstein model represents experimental data fairly accurate despite its small number of parameters. An additional advantage is that the parameters in the model are physical properties, which in principle can be measured, or predicted by ab initio methods.

For the application of not only the Einstein model but also for other vibrational models, such as the Debye model, it is important to constrain the thermodynamic analysis with measurements of thermodynamic properties as a function of pressure, such as volume; see Figure 6. This puts tighter constraints on the model parameters.

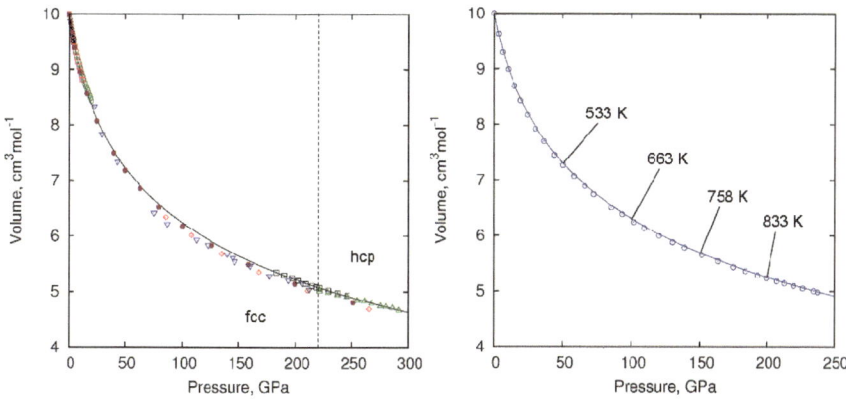

FIG. 6.

Left: calculated volume-pressure isotherm of aluminium at 300 K. Experimental data are from: Square and Triangle: Akahama et al. (2004); Circle: Ming et al. (1986); Solid square: Syassen and Holzapfel (1978); Asterisk: Vaidya and Kennedy (1970); deduced from shock-wave data: Solid circle: Wang et al. (2000); Inverse triangle: Greene et al. (1994); Diamond: Nellis et al. (1988).

Right: calculated isentrope with experimental shock-wave data from Davis (2006).

The results using Einstein's model and Debye's model are indistinguishable on the scale of the two plots

thermodynamic properties in the Debye model

So far, we found that the Einstein model for aluminium represents the experimental thermodynamic data well, except for heat capacity data at low temperature. Debye developed a model in which the atoms do not vibrate independently, but with a range of frequencies between zero and a *cut-off frequency*, v_D. The cut-off frequency is associated with the vibration having the largest *restoring force*. The *density of states* in his model behaves parabolic in frequency. The vibrational contribution to each thermodynamic property is derived from the vibrational density of states and the Einstein expression for that property. Generally, in the Debye model each thermodynamic property M can be written as:

$$M(v) = \int_0^{v_D} M_E(v) \cdot g(v)\, dv = \frac{9nN_A}{v_D^3} \int_0^{v_D} M_E(v) \cdot v^2\, dv \,, \qquad (47)$$

where M_E denotes the expression for the same property in the Einstein model. Equation (47) is valid for a substance consisting of n atoms per *formula unit*. To obtain shorter formulas we make use of the definition Equation (19) to obtain:

§(307)

$$M(x) = \frac{9nN_A}{X_D^3} \int_0^{X_D} M_E(x) \cdot x^2 dx . \tag{48}$$

Using Equation (48) and the Einstein expressions, Equations (34) to (41) for the vibrational contributions, the vibrational contribution to thermodynamic properties in the Debye formulation are as follows:

Helmholtz energy:
$$A^{vib}(T,V) = \frac{9nRT}{X_D^3} \left[\frac{1}{8} X_D^4 + \int_0^{X_D} x^2 \ln(1 - e^{-x}) dx \right]; \tag{49}$$

Energy:
$$U^{vib}(T,V) = 9nRT \left[\frac{X_D}{8} + \frac{1}{X_D^3} \int_0^{X_D} \frac{x^3}{e^x - 1} dx \right]; \tag{50}$$

Entropy:
$$S^{vib}(T,V) = \frac{9nR}{X_D^3} \left[\int_0^{X_D} \frac{x^3}{e^x - 1} dx - \int_0^{X_D} x^2 \ln(1 - e^{-x}) dx \right]; \tag{51}$$

Heat capacity:
$$C_V^{vib}(T,V) = \frac{9nR}{X_D^3} \int_0^{X_D} \frac{x^4 e^x}{\left(e^x - 1\right)^2} dx ; \tag{52}$$

Thermal pressure:
$$P^{vib}(T,V) = \frac{9nRT\gamma}{X_D^3 V} \left[\frac{X_D^4}{8} + \int_0^{X_D} \frac{x^3}{e^x - 1} dx \right]; \tag{53}$$

Isothermal bulk modulus:

$$K^{vib}(T,V) = \frac{9nRT}{X_D^3 V} \left\{ \left(\gamma^2 + \gamma - V \left(\frac{\partial \gamma}{\partial V} \right)_T \right) \left[\frac{X_D^4}{8} + \int_0^{X_D} \frac{x^3}{e^x - 1} dx \right] - \int_0^{X_D} \frac{\gamma^2 x^4 e^x}{\left(e^x - 1\right)^2} dx \right\}; \tag{54}$$

Thermal expansivity:
$$\alpha = \frac{9nR\gamma}{KX_D^3 V} \int_0^{X_D} \frac{x^4 e^x}{\left(e^x - 1\right)^2} dx , \tag{55}$$

with:
$$K(T,V) = K^{st}(V) + K^{vib}(T,V) . \tag{56}$$

The calculation of thermodynamic properties using the Debye model differs from that for the Einstein model, inasmuch as the integrals given in Equations (49) - (55) have to be evaluated numerically. The computationally most efficient way to do this is to tabulate the integral for C_V as function of x, from which the other integrals follow from:

$$\int_0^{X_D} \frac{x^3}{e^x - 1} dx = \frac{1}{4} \frac{X_D^{\ 4}}{e^{X_D} - 1} + \frac{1}{4} \int_0^{X_D} \frac{x^4 e^x}{\left(e^x - 1\right)^2} dx \text{, and} \tag{57}$$

$$\int_0^{X_D} x^2 \ln(1 - e^{-x}) dx = \frac{1}{3} X_D^3 \ln(1 - e^{-X_D}) - \frac{1}{3} \int_0^{X_D} \frac{x^3}{e^x - 1} dx . \tag{58}$$

The calculation of thermodynamic properties proceeds in the same way as has been explained above for the Einstein model. The result of the thermodynamic analysis of the experimental data is included in Table 1. Thermal expansivity, bulk modulus and volume-pressure data are represented by the Einstein and the Debye model with comparable accuracy. Figure 3 and Figure 4 demonstrate that the Debye model leads to a superior description of the heat capacity and thermal expansivity at low temperature. That is also demonstrated in Figure 7: the exponential behaviour of the low-temperature heat capacity in the Einstein model does not allow an accurate description of experimental data. From the low temperature heat capacity it is possible to derive the electronic coefficient given by Equation (45). Taking the low-temperature limit of Debye's model results in:

$$\underset{T \to 0}{Lim} C_P = \underset{T \to 0}{Lim} C_V = \beta_{el} \cdot T + \frac{12\pi^4 R}{5\theta_D^3} \cdot T^3 . \tag{59}$$

Figure 7 clearly shows that the linear term due to the electronic contribution to heat capacity is necessary to describe the low-temperature data. Phillips (1959) derived a value of 13.5×10^{-4} J K^{-2} mol^{-1} for the electronic coefficient and a value $\theta_D = 427.7$ K for the *Debye temperature*, matching the data plotted in the right frame of Figure 7. Table 1 shows that our value for the Debye temperature is lower by about 30 K. Therefore our calculated heat capacity is steeper at low temperatures. The left frame of Figure 7 shows that because of this artefact in our Debye model the heat capacity is not described to all fine details between about 8 K and 30 K. The reason for the discrepancy is that the vibrational density of states (VDoS) resulting from a Debye model is too crude to describe a realistic vibrational density of states depicted in Figure 8. We have used the expression for the parabolic VDoS in Debye's model in Equation (47) and (48):

$$g(\nu) = \frac{9nN_A}{\nu_D^3} \nu^2 \delta . \tag{60}$$

The *Kronecker* δ equals one for all frequencies below ν_D and zero above ν_D. The surface underneath the VDoS gives the *total degree of vibrational freedom*, $3N_A$ for one mole of metallic element ($n = 1$). The VDoS itself expresses the density of vibrational freedom in a specific frequency range.

164

Despite the small artefact in heat capacity and the crude representation of the VDoS, the Debye model results in an amazingly accurate representation of thermodynamic properties in large ranges of pressure and temperature.

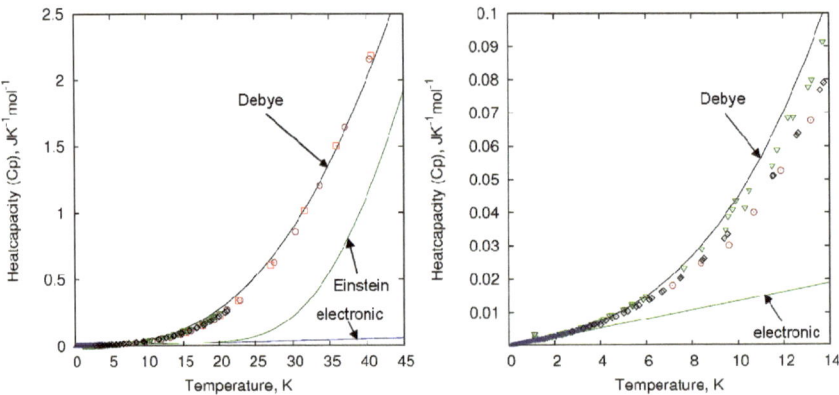

FIG. 7.
Calculated heat capacity of Al at low temperature using Einstein's and Debye's model. Experimental data are from: Circle: Downie and Martins (1980); Diamond: Berg (1968); Triangle: Phillips (1959); Square: Giauque and Meads (1941); Inverse triangle: Kok and Keesom (1937)

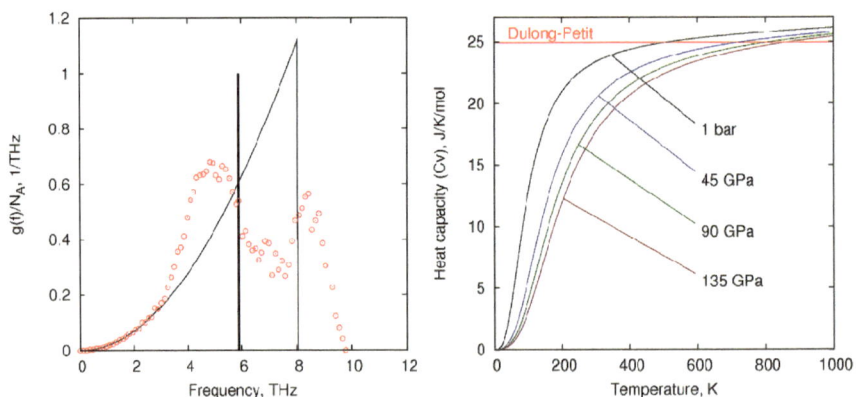

FIG. 8.
Left: The vibrational density of states of a Debye model is parabolic and that of an Einstein model is monochromatic. Experimental data are from Kresh et al. (2008). The conversion from frequency in cm^{-1} in Table 1 to frequency in Hz is a factor $100c$, with c the speed of light. Calculations were performed at 300 K and 1 bar pressure.
Right: Isochoric heat capacity of aluminium calculated with the Debye model does not show anomalous behaviour in P-T space. C_V breaks through the Dulong-Petit limit due to the electronic contribution

Jacobs and Schmid-Fetzer (2010) investigated in a detailed analysis the effects of the expression for the equation of state of the static lattice and that for the Grüneisen parameter on thermodynamic properties. They arrived at the conclusion that thermodynamic properties are represented to within experimental accuracy in the complete pressure-temperature stability range of aluminium. Figure 8 shows that the Debye model does not suffer from anomalous behaviour in isochoric heat capacity, which we discussed in §306 for MgO. The behaviour of the heat capacity is described in terms of physical properties.

density of states in Kieffer's model

As a next step, we turn to the model introduced by Kieffer (1979). Kieffer's model is developed for substances, such as Mg_2SiO_4, that have *densities of states* resulting from *acoustic and optic modes*. For such substances the Debye model often fails: more details of the *phonon spectrum* are needed to obtain an accurate description of their thermodynamic properties.

With Z the number of molecules in the *primitive cell*, a total of $3nZ$ normal modes of vibration are present, three of which are acoustic. The *total degree of vibrational freedom* is $3nN_A$, where n is the number of atoms in a *formula unit*, and N_A *Avogadro's number*. Therefore the degree of freedom for acoustic vibrations is $3N_A/Z$. The number of remaining optic oscillators is $3nN_A - 3N_A/Z = (3n - 3/Z)N_A$.

The optic vibrations are represented as continua of Einstein oscillators with a lower and an upper *cut-off frequency* v_{l_j} and v_{u_j} respectively.

The density of states $g(v)$ is written as:

$$g(v) = \frac{3N_A}{Z} \left(\frac{2}{\pi}\right)^3 \sum_{i=1}^{3} \left(\frac{\arcsin^2(v/v_i)}{\left(v_i^2 - v^2\right)^{1/2}} \cdot \delta_i \right) + \left(3n - \frac{3}{Z}\right) \cdot N_A \sum_{j=1}^{N_{OC}} \frac{f_j}{v_{u_j} - v_{l_j}} \cdot \delta_j. \qquad (61)$$

In Equation (60) N_{OC} represents the number of optic modes. The *Kronecker delta function* δ is zero when the frequency v exceeds the cut-off frequency of a particular mode otherwise it equals one.

The first term on the right-hand side of Equation (61) represents the contributions due to acoustic lattice vibrations. They are characterized by constant group velocities, and the dispersive character of these vibrations is described with a simple sinusoidal relation resulting in the arcsine term. Three acoustic modes are defined, one longitudinal and two transverse ones.

The values for v_i are determined by the cut-off frequencies of the modes. Each of the modes is characterized, at long wavelengths, by a directionally averaged sound velocity.

The second term describes optic modes of vibration, which may interact with electromagnetic radiation, and f_j denotes the fraction of the total number of optic oscillators in mode j. Finally, for Mg_2SiO_4 the number of atoms per molecule is $n = 7$.

By making use of $x_j = hv_j/kT$, the vibrational contribution to the Helmholtz energy is given by:

$$A^{vib}(T,V) = \frac{3N_A}{Z}\left(\frac{2}{\pi}\right)^3 \sum_{i=1}^{3}\left[\int_0^{x_i}\frac{x_i \arcsin^2\left(x/x_i\right)}{\left(x_i^2 - x^2\right)^{1/2}}A_{E,i}^{vib}dx\right] + \left(3n - \frac{3}{Z}\right)N_A \sum_{j=1}^{N_\infty}\left[\int_{x_{l_j}}^{x_{u_j}}\frac{f_j}{x_{u_j} - x_{l_j}}A_{E,j}^{vib}dx\right]$$

(62)

where $A_{E,i}^{vib}$ represents the vibrational contribution to the Helmholtz energy of an Einstein oscillator for mode i given by the term A^{vib} (per oscillator) in Equation (34).

application of Kieffer's model to Mg_2SiO_4 forsterite

Forsterite is not the simplest substance to illustrate Kieffer's formalism, because, as is well-known, it shows a deviation from quasi-harmonic behaviour. This deviation, also known as *intrinsic anharmonic behaviour* will be treated further on in this section. Here we illustrate that the Einstein and Debye models are not sufficiently accurate to describe the thermodynamic properties, in particular the heat capacity, and that, to improve the description, more information about the *vibrational density of states* (VDoS) is required. To achieve this, we limit our description to temperatures well below 1460 K, the temperature at which the heat capacity at constant volume becomes larger than the *Dulong-Petit limit.*

At these temperatures, say below 700 K, it is reasonably safe to apply the quasi-harmonic approximation to calculate forsterite's thermodynamic properties.

Because Kieffer's model takes into account more vibrational modes than the models of Einstein and Debye, it obviously has more parameters. On the other hand, it is constrained by more experimental data: by *Raman* and *infrared spectroscopy* data and longitudinal and transverse *sound wave velocity* data.

For the Einstein and Debye models we followed the same procedure as for aluminium, detailed above. A survey of the experimental data available for forsterite is given by Jacobs and de Jong (2005). The result of a least-squares optimization process of the available data is given in Table 2.

For the application of Kieffer's model we started from the representation of the VDoS given by Chopelas (1990). Forsterite has 84 normal modes of vibration, such as is determined by the relation $3nZ$, with $Z = 4$ the number of molecules in the primitive cell and $n = 7$, the number of atoms in one formula unit. Hofmeister (1987) determined that 35 modes are infrared active, that 36 modes are Raman active and that 10 modes are spectroscopically inactive. The remaining 3 modes are acoustic.

Table 2:
Optimized properties for Mg_2SiO_4 (forsterite) using the Einstein and Debye models in the quasi harmonic approximation. Einstein temperature is 535.73 K, and the Debye temperature is 762.47 K. Eqn. (28) has been used from the frequency expression. Vinet's equation of state (6) has been used for the static lattice

U^{ref}	V_0^{st}	K_0^{st}	$K_0^{'st}$	V_0	v_0	γ_0	q_0
kJ/mol	cm^3/mol	GPa		cm^3/mol	cm^{-1}		
Einstein model							
-2.2426(50)	43.122(20)	135.7(9)	4.66(6)	43.502(5)	371.7(15)	1.09(7)	1.57(25)
Debye model							
-2.2452(50)	43.104(20)	135.6(9)	4.63(6)	43.511(5)	529 0(13)	1 09(6)	1 48(20)

By means of *factor group analysis* Chopelas (1990) determined the total number of modes for each vibrational motion type by a method described by Fately et al. (1971). The frequency intervals of each motion type have been established by Hofmeister (1987) and Chopelas (1990). Because it is impractical to incorporate all 84 vibrational normal modes into a thermodynamic description, the modes are grouped in vibrational motion types, such as symmetric bending motions in SiO_4 units, $v_2(SiO_4)$ in Table 3. Chopelas (1990, 2000) showed that thermodynamic properties are not too sensitive to details of the VDoS, a property, which we already observed for the Debye model of aluminium. The grouping into motion types offers an elegant reduction of the complexity of the 84 modes of vibrations.

The VDoS on the left in Figure 9 is a simplified model density of states at ambient conditions based on spectroscopic measurements relative to a *lattice dynamics calculation* of Price et al. (1987) shown in the right frame of the Figure. The representation of experimental thermodynamic data for the three models is not significantly different except for the heat capacity.

Table 3:
Properties for Mg_2SiO_4 (forsterite) resulting from an optimization process, using Kieffer's model including intrinsic anharmonicity. Debye temperature is 784.5 K at zero Kelvin. NB. For temperatures below 700 K, computations with and without including intrinsic anharmonicity do not give significantly different results. Vinet's equation of state (6) has been used for the static lattice

Mode type	Fraction of oscillators	motion type	Frequency range, cm^{-1}	$\gamma_{j,0}$	$q_{j,0}$	$a_j \times 10^5$ K^{-1}
Mg_2SiO_4 (forsterite)						
AC1	1/84	TA	0.00-101.4(2)	1.05(5)	1.70(90)	-4.94(5)
AC2	1/84	TA	0.00-102.4(2)	1.05(5)	1.70(90)	-4.94(5)
AC3	1/84	LA	0.00-172.7(1)	1.66(2)	0.78(30)	-1.55(5)
OC1	2/84	$T(SiO_4)$	106.4(60)-147.6(60)	1.02(4)	2.33(17)	0.00
OC2	7/84	$T(SiO_4)$	167.9(63)-229.1(63)	2.21(9)	0.80(16)	0.00
OC3	12/84	$T[M(2)O_6]$	229.2(30)-355.1(30)	1.76(6)	2.49(5)	-2.31(6)
OC4	12/84	$T[M(1)O_6]$	279.7(30)-414.2(30)	1.22(6)	2.49(5)	-1.59(6)
OC5	12/84	$R(SiO_4)$	305.8(30)-474.1(30)	1.42(6)	2.49(5)	-0.84(6)
OC6	8/84	$v_2(SiO_4)$	409.7(34)-510.7(34)	0.56(9)	2.49(5)	-2.95(6)
OC7	12/84	$v_4(SiO_4)$	507.0(34)-647.2(34)	0.56(8)	-2.78(6)	-0.84(8)
OC8	4/84	$v_1(SiO_4)$	841.5(34)-842.5(34)	0.41(9)	-3.83(4)	-0.84(8)
OC9	3/84	$v_3(SiO_4)$	871.5(34)-872.5(34)	0.39(9)	-3.83(4)	-0.84(8)
OC10	4/84	$v_3(SiO_4)$	919.5(34)-920.5(34)	0.32(9)	-3.83(4)	-0.84(8)
OC11	5/84	$v_3(SiO_4)$	978.5(34)-979.5(34)	0.56(9)	-3.83(4)	-0.84(8)

K_0^{st} /GPa	$K_0^{'st}$	V_0 /cm³/mol	V_0^{st} /cm³/mol	U^{ref} /MJ/mol	Z
134.8(1)	4.74(3)	43.476(48)	4.093(48)	-2.2533(40)	4

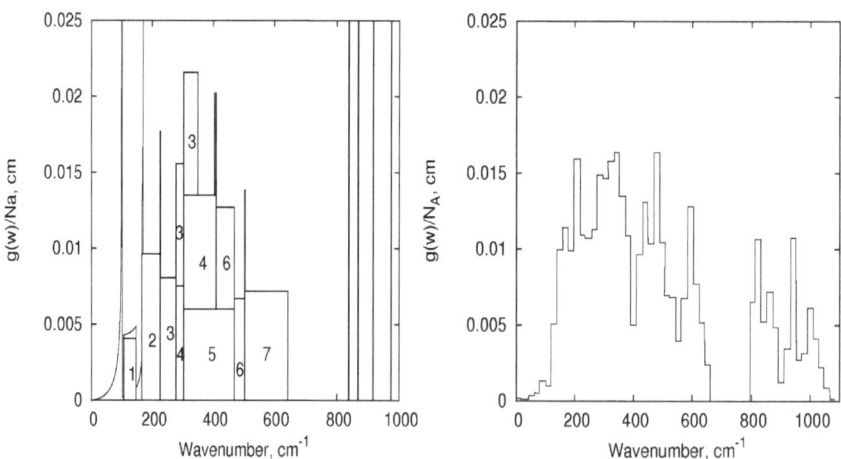

FIG. 9.
Vibrational density of states (VDoS) of forsterite at ambient conditions. The VDoS in the left frame is a simplification and the numbers correspond to the optic motions in Table 3. The VDoS in the right frame has been derived by Price et al. (1987) using a lattice dynamics model

This is more clearly demonstrated in Figure 10, which indicates that the Einstein and Debye models perform rather poor for this property. The absolute entropy of forsterite is represented well by Kieffer's model only. Because the description of the Gibbs energy is strongly affected by the description of the entropy, Kieffer's model for the VDoS is better suitable to describe phase boundaries prevailing at conditions of the Earth mantle. This is especially the case when phase boundaries and thermodynamic data together are included in a thermodynamic analysis. In such a case Kieffer's model is better suitable to point to inconsistencies in either thermodynamic data or phase boundary data. From Figure 10 we also make the observation that Kieffer's model is better suitable relative to a Debye or Einstein model to indicate deviations from quasi-harmonic behaviour. In other words, it opens the possibility to pinpoint more accurately physical effects leading to such a deviation. Further on we discuss a deviation from quasi-harmonic behaviour due to *intrinsic anharmonicity.*

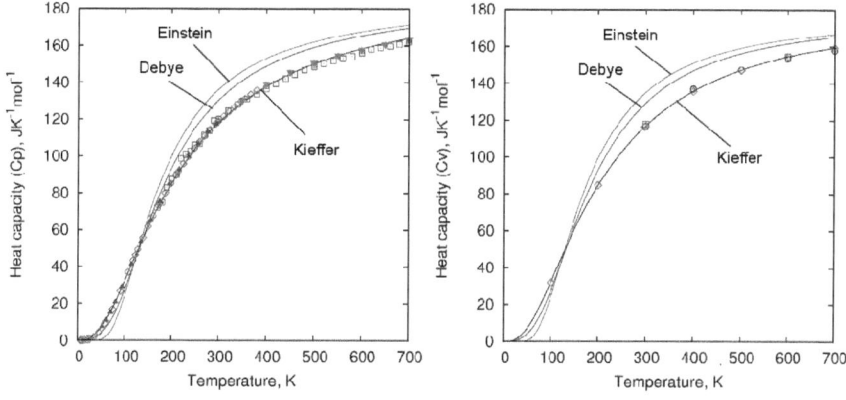

FIG. 10.

Left: calculated low-temperature heat capacity, C_P, of forsterite Mg_2SiO_4 at 1 bar pressure using the models of Einstein, Debye, and Kieffer. Experimental data are from: Square: Ashida et al. (1987); Inverse solid triangle: Watanabe (1982); Diamond: Robie et al. (1982); Solid triangle: Kelley (1943).
Right: calculated low temperature heat capacity, C_V, of forsterite Mg_2SiO_4 at 1 bar pressure according to the three models. Data points are derived from: Circle: Cynn et al. (1996); Square: Gillet et al. (1991); Diamond: Chopelas (1990)

insufficiency of the quasi-harmonic model

For forsterite, Mg_2SiO_4, there is strong evidence that the *quasi-harmonic model* is not valid. Experimental work on isobaric heat capacity, thermal expansion and bulk modulus reveals that forsterite shows *intrinsic anharmonic behaviour* of the isochoric heat capacity, C_V, (see e.g. Jacobs and Oonk, 2001, and references therein).

§(307)

Figures 11a-c show available data for heat capacity, adiabatic bulk modulus and thermal expansivity. Figure 11c demonstrates that thermal expansivity is relatively uncertain compared to heat capacity and bulk modulus. From the data presented in Figure 11a-c it is possible to derive the heat capacity at constant volume, C_V, by fitting them to polynomial expressions. To calculate C_V we combine Equations (46), for the relation between adiabatic and isothermal bulk modulus, and Equation (63) for the difference between C_p and C_V ($\leftarrow$Exc 107:10)

$$C_P - C_V = \alpha^2 K \cdot V \cdot T ,$$ (63)

and derive

$$C_V = \frac{C_P^2}{\alpha^2 K_S V T + C_P} .$$ (64)

The result of the calculation is plotted in Figure 11d.

Two issues are associated with this plot. First, the heat capacity C_V may become higher than the $3nR$ Dulong-Petit limit - independent of which experimental dataset is chosen for the thermal expansivity. This implies that a vibrational model employing the quasi-harmonic approximation is insufficient to describe C_V up to the melting point. Figure 11d also demonstrates that heat capacity C_V may differ by about 7% at the melting point, depending on which thermal expansivity investigation is taken in the thermodynamic analysis.

Secondly, thermal expansivity has a considerable impact on the *'Clapeyron slope'* of the *phase boundary* between forsterite and wadsleyite. This phase boundary has been the subject of several experimental investigations, and resulted in different locations and slopes.

To address these issues, we required a model which more precisely constrains thermal expansivity than the 'simple' parameterizations given above. For this reason we employed Kieffer's vibrational model in which thermodynamic properties are constrained with additional data such as infrared and Raman spectroscopic data.

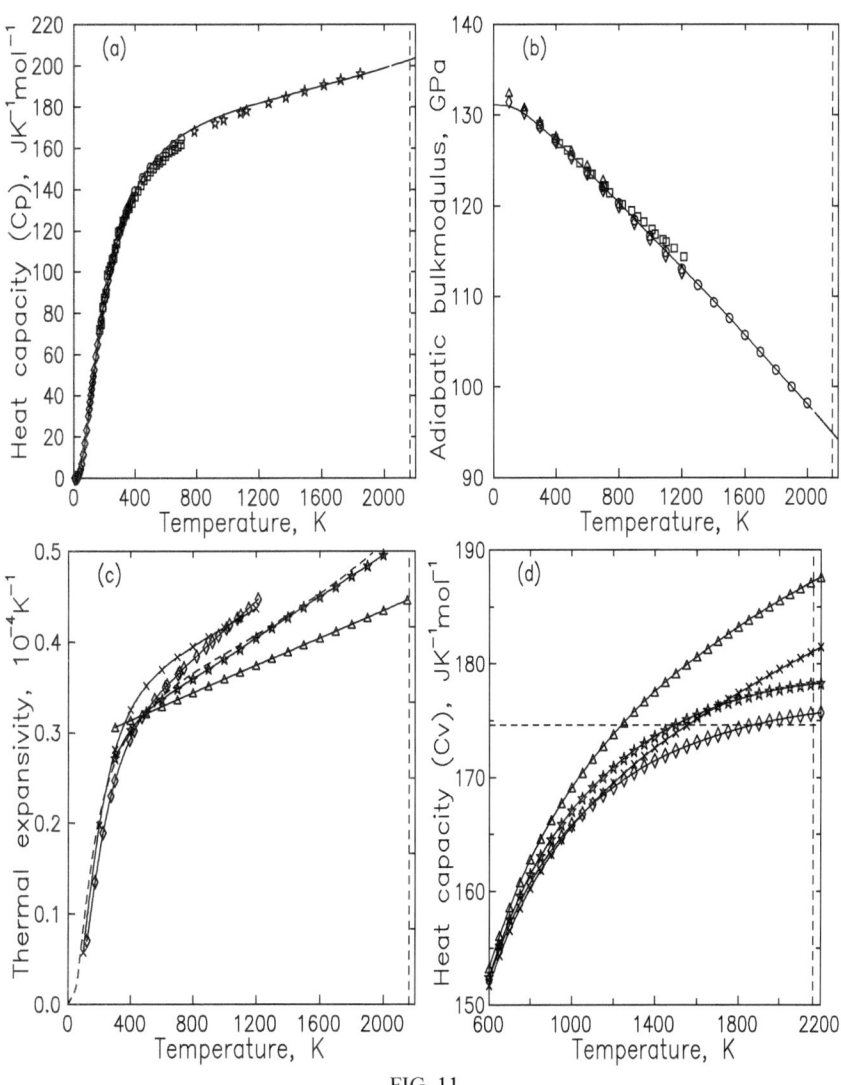

FIG. 11.

Experimental thermodynamic properties for forsterite, Mg_2SiO_4. (a) Isobaric heat capacity measured by Robie et al. (1982), $\Diamond$; Watanabe (1982), $\times$; Ashida et al. (1987), $\Box$. The solid curve has been calculated with the values in Table 3. (b) Adiabatic bulk modulus measured by Manghnani et al. (1981), $\Diamond$; Suzuki et al. (1983), $\Box$; Isaak et al. (1989), O. (c) Thermal expansivity measured by Suzuki et al. (1984), $\Diamond$; Matsui et al. (1985), $\times$; Kajiyoshi et al. (1986), $\star$; Bouhifd et al. (1996), Δ. The data of Suzuki (1975) and Gillet (1991) have been omitted for transparency. The dashed curve has been calculated with the values in Table 3. (d) Isochoric heat capacity has been derived from the experimental data and Equation (63). The symbols have the same meaning as in frame (c) the horizontal dashed line denotes the $3nR$ Dulong-Petit limit

172

One possible explanation for the behaviour in C_V is found in the phenomenon of *intrinsic anharmonicity*. According to Gillet et al. (1991; 1997) intrinsic anharmonicity is related to a strong phonon coupling in the microscopic vibrations of the atomic arrays that build up the crystal lattice. They showed experimentally that the frequencies of the vibrational modes do not only depend on volume but also on temperature, and such that the temperature derivative of the frequency at constant volume is non-zero. This is illustrated as follows. The *anharmonicity parameter* for a specific mode of vibration, i, is defined as:

$$a_i = \left(\frac{\partial \ln \nu_i}{\partial T} \right)_V. \tag{65}$$

Because measurements of vibrational frequencies at constant volume are impracticable to perform, we write Equation (65) in a different form using a classical thermodynamic relationship:

$$a_i = \left(\frac{\partial \ln \nu_i}{\partial T} \right)_V = \left(\frac{\partial \ln \nu_i}{\partial T} \right)_P + \alpha K \left(\frac{\partial \ln \nu_i}{\partial P} \right)_T \tag{66}$$

$$= \left(\frac{\partial \ln \nu_i}{\partial V} \right)_P \left(\frac{\partial V}{\partial T} \right)_P + \alpha K \left(\frac{\partial \ln \nu_i}{\partial V} \right)_T \left(\frac{\partial V}{\partial P} \right)_T$$

$$= \alpha \left(\frac{\partial \ln \nu_i}{\partial \ln V} \right)_P - \alpha \left(\frac{\partial \ln \nu_i}{\partial \ln V} \right)_T. \tag{67}$$

We use the following definition of the Grüneisen parameter along the different paths in PT space:

$$\gamma_{i,P} = -\left(\frac{\partial \ln \nu_i}{\partial \ln V} \right)_P \quad \text{and} \quad \gamma_{i,T} = -\left(\frac{\partial \ln \nu_i}{\partial \ln V} \right)_T, \tag{68}$$

The anharmonicity parameter is then written as:

$$a_i = \alpha \left(\gamma_{i,T} - \gamma_{i,P} \right). \tag{69}$$

In the quasi-harmonic approximation, all a_i are zero and $\gamma_{iP} = \gamma_{iT}$ indicating that the vibrational frequency is depending just on the volume. Equations (68) and (69) indicate that for the determination of the anharmonicity parameters isothermal and isobaric measurements are necessary.

It is customary to construct $\ln(\nu)$-$\ln(V)$ plots to determine the Grüneisen- and the mode anharmonicity parameters. Figure 12 shows an example of two translational modes $T[Mg(1)O_6]$ at different vibrational frequencies.

Each mode shows a discontinuity in the slope $d\ln\nu/d\ln V$ indicating that γ_{iP} differs from γ_{iT}. The experimental points were connected with straight lines enabling the calculation of the Grüneisen parameters using Equation (68). For the mode with wave number 306 cm^{-1} at ambient conditions the values of γ_{iT} are 1.73 or 1.53 if the data of Chopelas (1990) or those of Wang et al. (1993) are used. The value for γ_{iP} derived from the measurements of Gillet et al. (1997) is 2.53. Using a value for thermal expansivity of 2.67×10^{-5} K^{-1} these values result in anharmonicity parameter of -2.14×10^{-5} K^{-1} and -2.67×10^{-5} K^{-1}, respectively, depending on which dataset is used for the pressure direction. For the mode having a frequency of 331 cm^{-1} at ambient conditions the value for the anharmonicity parameter can be derived in the same way: -1.46×10^{-5} K^{-1}.

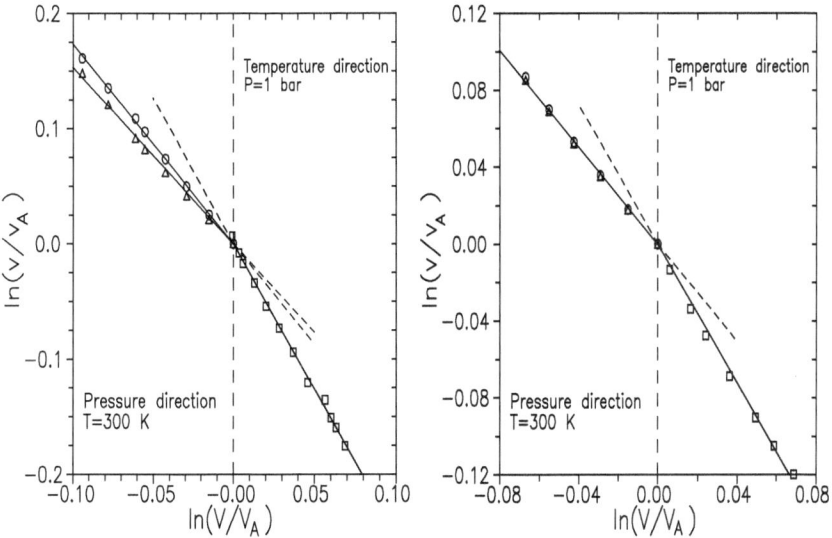

FIG. 12.

Calculated logarithmic frequency-volume plots for two translational modes T[Mg(2)O$_6$] with peak wave numbers of 306 (left frame) and 331 (right frame) cm^{-1} at room conditions. The pressure range is between 1 bar and 15 GPa and the temperature range is between 300 and 2000 K. The experimental data in the pressure direction are from Chopelas (1990), O and Wang et al. (1993), Δ. Experimental data in the temperature direction are from Gillet et al. (1997), □. The dashed lines are extrapolations of the experimental data. The vertical lines indicate ambient conditions. V_A and ν_A indicate volume and frequency at ambient conditions

The construction of Figure 12 is not trivial. For the conversion of the frequency-pressure data of Chopelas and Wang et al. to frequency-volume data an equation of state must be employed. For the conversion of the frequency-temperature data of Gillet et al. accurate thermal expansivity data at 1 bar pressure are required. As can be expected from Figure 12, the uncertainty in thermal expansivity contributes significantly to the uncertainty in the value for γ_{iP}, about 10%. Consequently, the uncertainties in γ_{iT} and γ_{iP} propagate in the uncertainty of the anharmonicity parameters. Figure 13 shows that uncertainties in anharmonicity parameters for vibrational modes in forsterite can be significant, especially in the lattice modes at lower frequencies. It also illustrates that the anharmonicity parameters differ significantly from zero, having negative values.

To constrain more accurately mode Grüneisen and anharmonicity parameters it is useful to employ a vibrational model, which enables pinpointing consistency between the disparate experimental datasets. In the next paragraph we discuss the construction of a vibrational model including intrinsic anharmonicity.

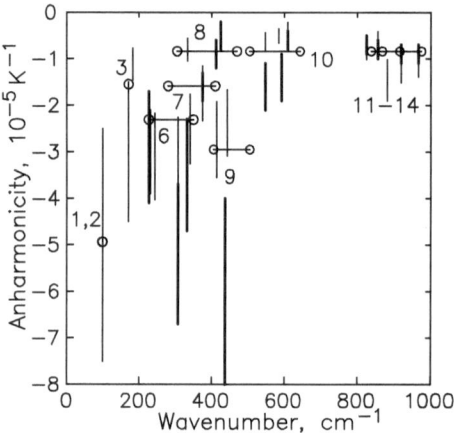

FIG. 13.

Calculated mode anharmonicity parameters for the fourteen mode types given in Table 3. The thin vertical lines are Raman spectroscopic experimental data from Gillet et al. (1991) and the thick vertical lines are from Gillet et al. (1997). The vertical lines indicate the uncertainty in the experiments. Circles indicate cut-off frequencies of acoustic and optic mode types given in Table 3. We found mode anharmonicity in the Raman inactive $T[Mg(1)O_6]$ mode but no mode anharmonicity in $T[SiO_4]$ modes (mode 4 and 5)

constructing a model including intrinsic anharmonicity

Intrinsic anharmonic effects are ignored in the quasi-harmonic approximation. In the *anharmonic phonon theory* presented by Wallace (1972), these effects are associated with the asymmetry of the vibrational potential. Winkler and Dove (1992) showed by molecular dynamics simulations applied to $MgSiO_3$ perovskite that the departure from the quasi-harmonic approximation results in vibrational frequencies that depend not only on volume but also on temperature. On rigorous lattice dynamic theoretical grounds, Wallace (1972) demonstrated that it is incorrect to substitute the volume and temperature dependent frequency of an arbitrary vibrational mode i, $v_i(V,T)$, in the expression for the Helmholtz energy based on the quasi-harmonic model. Instead, he showed that, to first order, the correct *entropy* as a function of volume and temperature is obtained when the actual measured $v_i(V,T)$ is substituted in the quasi-harmonic expression for the vibrational entropy. This is what we shall refer to as the *"Wallace theorem"*; its application is the key issue we discuss here.

As demonstrated by Gillet et al. (1997), the theorem has the drawback that the Helmholtz energy must be calculated by numerical integration of the entropy. Because all thermodynamic properties of a substance are built up of partial derivatives of the Helmholtz energy, its non-analyticity results in a large computational effort relative to the quasi-harmonic model. The effort becomes cumbersome when thermodynamic properties must be calculated in many points of pressure-temperature-composition space, as for instance is the case in geodynamic mantle convection modelling. As a result, the theorem has not been widely applied. Here we show how this difficulty can be circumvented; read, reduced to a method that requires a computational effort comparable to that for the quasi-harmonic model.

We made the method operational as an *inversion technique*, enabling us to detect - within the framework of a chosen model for the vibrational density of states - inconsistencies in specific experimental data,

According to Wallace it is to first order correct to substitute the volume and temperature dependent frequency into the quasi-harmonic expression for the entropy. The entropy of an Einstein oscillator is given by the relation, Equation (36):

$$S(T,V) = k\left(\frac{x}{\exp(x)-1} - \ln(1-\exp(-x))\right), \tag{70}$$

where k denotes Boltzmann's constant, $x = hv/kT$, h Planck's constant and v the frequency of the oscillator.

The Helmholtz energy follows from the integration of the entropy:

$$A^{vib}(T,V) = -k\int_0^T \left(\frac{x}{\exp(x)-1} - \ln(1-\exp(-x))\right)dT + A^{vib}(0,V).$$ (71)

According to Gillet (1991), the temperature dependent frequency can be written as:

$$v(T,V) = v(V)\exp(aT),$$ (72)

where a represents the anharmonicity parameter and $v(V)$ is determined by Equation (28). The combination of Equations (71) and (72) leads to an inconvenient numerical integration to calculate the Helmholtz energy in each point of temperature-volume space. In fact only the heat capacity at constant volume, $C_V = T(\partial S/\partial T)_V$, and the temperature derivative of thermal pressure, $(\partial P^{vib}/\partial T)_V = (\partial S/\partial V)_T$, can be evaluated analytically. For Mg_2SiO_4, Gillet et al. (1991, 1997) did not carry out this numerical integration based on the theorem of Wallace. However for $MgSiO_3$ perovskite, Gillet et al. (2000) presented a numerical calculation scheme based on it.

In the following we express the frequency as:

$$v(T,V) = v(V)(1+aT).$$ (73)

For Mg_2SiO_4, the anharmonicity parameter for each mode of vibration is a negative number and its value is of the order -3×10^{-5} K^{-1}. This indicates that Equation (73) offers a physically realistic description for temperatures up to the melting point, 2271 K. Equation (73) is also a first order approximation of Equation (72). Below we demonstrate that the advantage of Equation (73) is that its combination with Equation (71) results in an expression for the Helmholtz energy which can be evaluated with less computational effort compared to the numerical integration resulting from the combination of Equations (71) and (72).

The integral in Equation (71) can be changed into an integral over x by using Equation (73), leading to:

$$A^{vib}(T,V) = +k\int_\infty^x T(1+aT)\left[\frac{1}{\exp(x)-1} - \frac{\ln(1-\exp(-x))}{x}\right]dx + A^{vib}(0,V).$$ (74)

The $T(1+aT)$ term in Equation (74) can be expressed in x by using the identity $x = hv/kT$ and Equation (73), resulting in:

$$T(1+aT) = \frac{y}{(1-ay)^2}, \text{ where } y = \frac{kx}{hv(V)}.$$ (75)

Equation (75) is approximated with the power series:

§(307)

$$T(1+aT) = y \sum_{j=0}^{m} \left((j+1)(a\,y)^j \right). \tag{76}$$

Because a is a small negative number, for forsterite typically of the order -3×10^{-5} K^{-1}, the power series in Equation (76) converges rapidly to the result calculated by Equation (75) with a small number of terms. For instance, at temperatures close to the melting point of forsterite, y attains values of about 2000. Taking $m = 2$ in Equation (76), the difference between results calculated by Equations (75) and (76) is less than 0.1%.

Equation (76) has two advantages. The first is that the integral of Equation (71) is now expressed in x only. The second is that the volume dependent frequency $v(V)$ can be placed outside the integral operator because the integration in Equation (71) is over constant volume by the definition of the entropy $S = -(\partial A/\partial T)_V$. Inserting Equation (76) into Equation (74) leads to:

$$A^{vib}(T,V) = k \sum_{j=0}^{m} \left[a^j \left(\frac{h v(V)}{k} \right)^{j+1} \frac{\ln(1-\exp(-x))}{x^{j+1}} \right] + k \sum_{j=1}^{m} \left[j a^j \left(\frac{h v(V)}{k} \right)^{j+1} \int_{\infty}^{x} \frac{dx}{x^{j+1}(\exp(x)-1)} \right]$$

$$+ \frac{1}{2} kT \frac{x}{1+aT} \quad , \tag{77}$$

where use has been made of:

$$\int_{\infty}^{x} \frac{\ln(1-\exp(-x))}{x^{j+1}} dx = -\frac{1}{j} \frac{\ln(1-\exp(-x))}{x^j} + \frac{1}{j} \int_{\infty}^{x} \frac{dx}{x^j(\exp(x)-1)} \quad .$$

The *zero-point energy* contribution to the Helmholtz energy, $A^{vib}(0,V)$ in Equation (74), cannot be determined from the integration of the entropy. However, Oganov and Dorogukupets (2004) showed for MgO that the anharmonic contribution to the zero-point energy is about 0.17% of the harmonic zero-point energy. For Mg_2SiO_4 we calculate that the contribution of the harmonic zero-point energy is less than 3% of the total Helmholtz energy. Because the contribution of the anharmonic zero-point energy to the Helmholtz energy is possibly of the order 0.005% we assumed in Equation (77) that the zero-point energy contribution to the Helmholtz energy can be described by the quasi-harmonic model.

By using the identity $x = h v/kT$ and Equation (73), the final expression for the Helmholtz energy contribution for an Einstein oscillator becomes:

178

$$A^{vib}(T,V) = kT \sum_{j=0}^{m} \left[\frac{(aT)^j}{(1+aT)^{j+1}} \ln(1-\exp(-x)) \right] + kT \sum_{j=1}^{m} \left[j \frac{(aT)^j x^{j+1}}{(1+aT)^{j+1}} \int_{\infty}^{x} \frac{dx}{x^{j+1}(\exp(x)-1)} \right] +$$
$$+ \frac{1}{2} kT \frac{x}{1+aT} . \tag{78}$$

The integral expressions in the second term on the right hand side can be easily tabulated as a function of only x. Once these tables exist, the Helmholtz energy can be calculated with comparable computational effort as for the quasi-harmonic model. The advantage of Equation (78) is that only a small number of terms are necessary in the summations to reach the accuracy in thermodynamic properties comparable with that of the numerical integration, Equation (71). This can be demonstrated by deriving the entropy from Equation (78):

$$S(T,V) = k \left(\frac{x}{\exp(x)-1} - \ln(1-\exp(-x)) \right) \sum_{j=0}^{m} \left[\frac{(j+1)(aT)^j}{(1+aT)^{j+2}} \right]. \tag{79}$$

When $m \to \infty$, this expression reduces to Equation (70).

The summation term in Equation (79) is a measure for the error made when a specific number of terms, m, are taken into account. For example when $m = 2$ and $a = -3 \times 10^{-5}$ K^{-1}, the entropy at 2000 K calculated by Equation (79) deviates from that calculated by Equation (70) by only 0.1%. This is about the error in entropy at ambient conditions for Mg_2SiO_4 derived from experimental data (0.11%).

Table 4 summarizes the deviation between thermodynamic properties derived from Equation (78) and calculated from the numerical integration, Equation (71). In the construction of Table 4, we have selected the shear vibrational mode of Mg_2SiO_4 because the deviations are largest at small frequencies combined with large negative values for the anharmonicity parameter. Table 4 indicates that the deviation for the entropy at 2000 K increases from 0.1% to 0.7 % when a decreases from -3×10^{-5} K^{-1} to -4.9×10^{-5} K^{-1} when $m = 2$. This value is about seven times the experimental uncertainty in the entropy at ambient conditions. The table also shows that the largest deviations occur in the second temperature derivative of the Helmholtz energy, the heat capacity. Figure 14 indicates that these deviations dramatically decrease with the number of terms, m, taken in the summation terms of Equation (78).

§(307)

Table 4:
Sensitivity analysis for entropy (S), heat capacity (C_V), temperature derivative of thermal pressure $(\partial P/\partial T)_V$, and isothermal bulk modulus, K. $\sigma\%$ is the relative deviation in % of a calculated property with that resulting from Wallace's theorem, and m denotes the number of terms in Equation (78). The numbers are calculated for an Einstein oscillator with wave number 100 cm^{-1}, Grüneisen parameter 1.03, $V_0 =$ 43.4772 cm^3/mol and an anharmonicity parameter of $-4.92\cdot10^{-5}$ K^{-1} at a constant volume of 43.6584 cm$^3\cdot$mol^{-1}

	$\sigma\%, S$		$\Sigma\%, C_V$		$\sigma\%, (\partial P/\partial T)_V$		$\Sigma\%, K^{vib}$	
m	1000 K	2160 K	1000 K	2160 K	1000 K	2160 K	1000 K	2160 K
1	0.8310	4.5780	5.8980	40.9030	0.8310	4.5780	-1.9230	-11.466
2	-0.0576	-0.7325	-0.5820	-9.3712	-0.0576	-0.7325	0.1834	2.1606
3	0.0037	0.1095	0.0489	1.8215	0.0037	0.1095	-0.0192	-0.5630
4	-0.0002	-0.0157	-0.0037	-0.3210	-0.0002	-0.0157	0.0022	0 0306

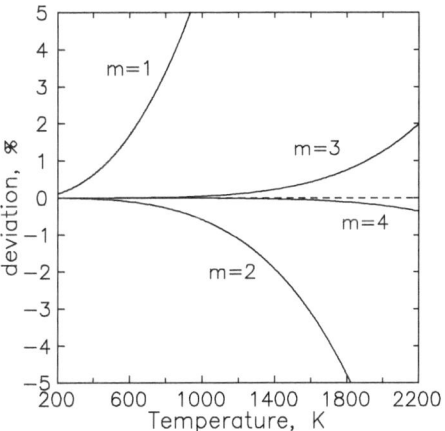

FIG. 14.

Calculated deviations, $100\%*\left(C_V^W - C_V(m)\right)/C_V^W$, of the heat capacity, C_V, for an Einstein oscillator as a function of temperature and 1 bar pressure. C_V^W was calculated from Wallace's theorem when Equation (73) is assumed for the frequency. $C_V(m)$ was calculated with our formalism given by Equation (A3) for m terms in Equation (78). The calculations were performed with input parameters given in Table 3

For $m \geq 2$ the deviations between properties derived from Equation (78) and Equation (71) are smaller than inaccuracies in experimental values.

For completeness we have written the thermodynamic properties derived from the Helmholtz energy, which are crucial in our thermodynamic framework, in the appendix to this section.

§(307)

effects of intrinsic anharmonicity on thermodynamic properties

Through experimental measurements of Raman and infrared spectra by Chopelas (1990, 2000), Wang et al. (1993), Gillet (1991,1997) and Hofmeister (1987) features of the vibrational density of states (VDoS) for forsterite became available. These features comprise the location and density of vibrational modes in the frequency spectrum. We made our method operational as a tool for optimizing experimental data using a least squares method. To that end we combined Equation (78) and Equation (62) to derive the expression for the Helmholtz energy incorporating mode anharmonicity. From the Helmholtz energy all other thermodynamic properties are derived by taking the appropriate derivatives, resulting in analytical expressions for them. The optimization of static lattice properties, the Grüneisen, mode q and anharmonicity parameters are further constrained by isobaric and isothermal experimental data of the vibrational frequencies from Chopelas (1990,2000), Wang et al. (1993) and Gillet et al. (1991,1997), as is shown in Figure 15.

Besides vibrational frequency data, we included thermodynamic and sound velocity data in a least squares optimization and Table 3 gives the resulting model parameters. The cut-off frequencies given in Table 3 refer to zero Kelvin, zero pressure conditions and differ, in terms of wave numbers, by about 1-3 cm^{-1} from the experimental values measured at ambient conditions.

Our thermodynamic analysis prefers the thermal expansivity data of Kajiyoshi (1986). These data appear to be better consistent with experimental data on adiabatic bulk modulus, heat capacity and sound wave velocity relative to other thermal expansivity data given in Figure 16. Intrinsic anharmonicity does not have a large effect on adiabatic bulk modulus, volume and thermal expansivity, 0.3%, 0.01% and 0.04% respectively at 2000 K and 1 bar pressure and therefore its effect is not visible in Figure 16. The effect on Gibbs energy and Helmholtz energy is about 0.13% at these conditions. Figure 17 shows the effect of intrinsic anharmonicity on isobaric heat capacity. From all thermodynamic properties isobaric and isochoric heat capacity are affected most, by about 2.6% and 2.9% respectively.

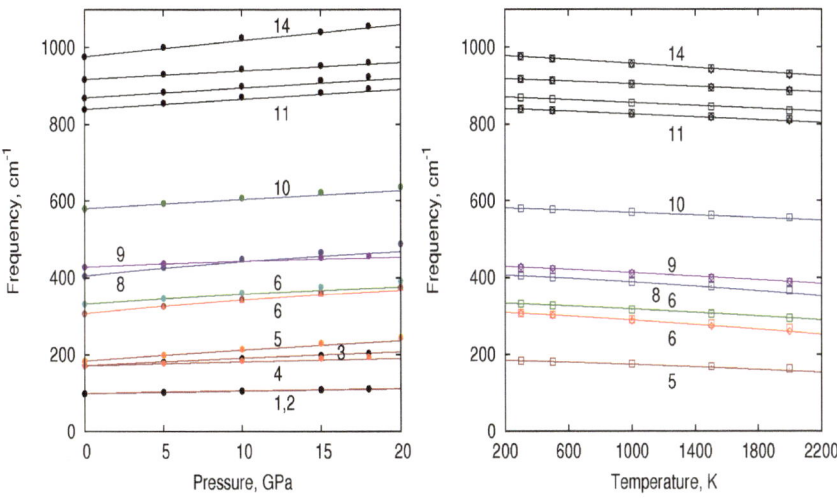

FIG. 15.

Calculated wave numbers for a number of modes of forsterite as function of pressure at 300 K (left frame) and as function of temperature at 1 bar pressure (right frame). The frequencies correspond to experimental peaks at 145, 183,306, 331, 404, 427, 580, 826, 856, 922 and 966 cm^{-1} present in modes 4 to 14 respectively in Table 3. The frequencies at 404, 427 and 580 cm^{-1} are averages of R[SiO$_4$], $v2$ and $v4$ modes respectively. Experimental data are from: left panel: Chopelas (1990) and Wang et al. (1993), right frame: Gillet et al. (1991,1997)

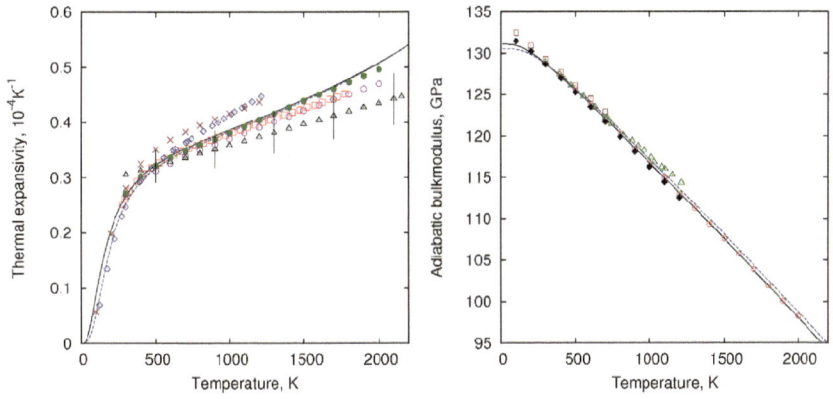

FIG. 16.

Left: thermal expansivity of forsterite calculated with Kieffer's model (solid) and Debye's model (dashed). Experimental data are from: Triangle: Bouhifd et al. (1996); Circle: Gillet et al. (1991); Closed circle: Kajiyoshi (1986); Cross: Matsui and Manghnani (1985); Diamond: Suzuki et al. (1984); Square: Suzuki (1975).
Right: calculated adiabatic bulk modulus calculated with Kieffer's model (solid) and Debye's model (dashed). Experimental data are from: Circle: Isaak et al. (1989); Triangle: Suzuki et al. (1983); Solid diamond: Manghnani & Matsui (1981); Square: Sumino et al. (1977). The effect of anharmonicity is not visible on the scale of the two plots

Heat capacity has a large effect on the *Clapeyron slope* and position of phase boundaries. The Clapeyron slope of phase boundaries between olivine, wadsleyite and ringwoodite significantly affects material transport through the *transition zone* of the Earth, located between 410 and 660 km depth. In § 308 we give an analysis of the phases present in the system $MgO-SiO_2$, based on the work of Jacobs and de Jong (2007), which includes an analysis of the phase boundary between forsterite and wadsleyite. Figure 18 indicates that although the effect of anharmonicity on Gibbs energy is only 0.13%, it has a significant impact on position and slope of the forsterite-wadsleyite phase boundary. Neglecting anharmonicity in forsterite changes this phase boundary by about 2 GPa corresponding to a displacement of the 410 km seismic discontinuity of about 60 km in the mantle of the Earth. This change in pressure is significantly different compared to the experimental uncertainty of about 0.7 GPa in the location of this phase boundary in *P-T* space. Similarly, the 2.6% difference in isobaric heat capacity changes the Clapeyron slope of this boundary from 2.3 MPa/K to 1.6 MPa/K, about a significant factor of two.

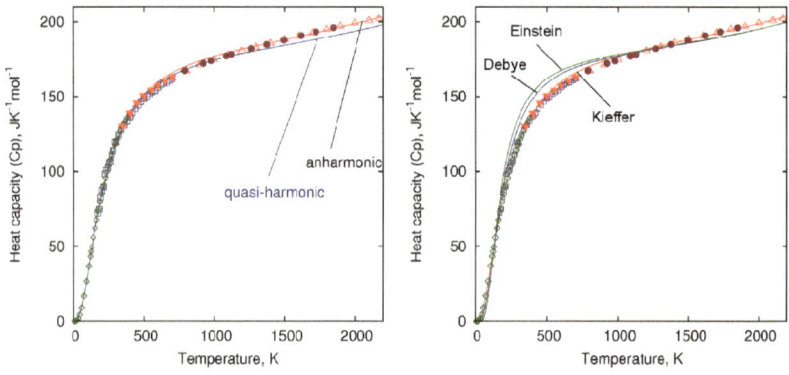

FIG. 17.
Left: isobaric heat capacity of forsterite calculated using Kieffer's model with intrinsic anharmonicity and using it in a quasi-harmonic approximation. Experimental data: Solid circle: Gillet et al. (1991); Triangle: compilation of Barin (1989); Square: Ashida et al. (1987); Diamond: Robie et al. (1982); Inverse solid triangle: Watanabe (1982).
Right: isobaric heat capacity calculated with Kieffer's model including intrinsic anharmonicity, and the quasi-harmonic models of Einstein and Debye

Figure 17 illustrates that the Einstein and Debye models calculated with the values given in Table 2 overestimate entropy at relevant mantle temperatures of about 1500 K. Using these models it is difficult to simultaneously represent one-bar experimental thermophysical data, one-bar transition data, such as between forsterite and wadsleyite, and high-pressure phase diagram data.

§(307)

Substituting Kieffer's model - in the description of Jacobs and de Jong (2007) - for Debye's model would result in a displacement of the forsterite-wadsleyite phase transition by about 3.5 GPa; and because of the overestimation of the entropy the Clapeyron slope increases with more than a factor two.

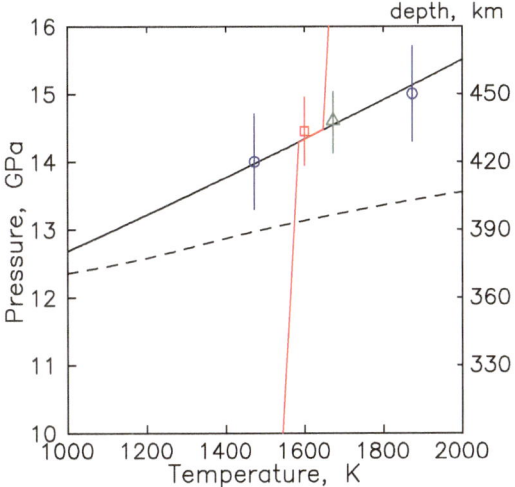

FIG. 18.

Calculated phase boundary between forsterite and wadsleyite. The solid curve was calculated including intrinsic anharmonicity and the dashed one using the quasi-harmonic model for forsterite. The nearly vertical curve represents a *geotherm* with an *adiabatic foot* of 1420 K. Experimental data are from: Square: Fei and Bertka (1999); Triangle: Boehler and Chopelas (1991); Circle: Katsura and Ito (1989)

applying the vibrational model to find a parameterization for perovskite, MgSiO₃

At the end of § 306 we mentioned that we would postpone a thermodynamic description of perovskite based on polynomial parameterizations of one-bar properties. Here we adopt the approach that results from a thermodynamic analysis using a vibrational formalism can be incorporated in the methodology developed in § 306.

Because one-bar thermodynamic data for perovskite are difficult to obtain by experiment, polynomial representations can only be constrained by one-bar measurements in a limited temperature range, typically below about 400 K. The resulting polynomial description is not accurate enough to make the required large extrapolations to conditions prevailing in the *transition zone* and *lower mantle* of the Earth.

§(307)

184

In the case of perovskite the one-bar thermal expansivity and bulk modulus polynomial representations in a thermodynamic analysis must be largely constrained by pressure-volume-temperature data. The one-bar polynomial representation for heat capacity must be constrained by phase diagram data, such as the post-spinel phase transition discussed in § 308. However, the slope and position of this phase transition in the pressure-temperature phase diagram has been under debate for many years (see Fei et al, 2004).

Here we follow a different approach. Jacobs and de Jong (2007) have constructed a thermodynamic description based on lattice vibrations for the system MgO-SiO$_2$, including the perovskite form of MgSiO$_3$. In this description, the thermodynamic properties of perovskite are additionally constrained by the different vibrational motions and their frequency behaviour in pressure-temperature space. Besides that, the formalism warrants that physically unrealistic behaviour of thermodynamic properties does not occur.

In the present approach we have fitted the results of the thermodynamic analysis based on lattice vibrations to extract polynomial functions for the one-bar properties heat capacity, thermal expansivity and isothermal bulk modulus. Additionally values for K' and $(\partial K'/\partial T)_P$ were calculated by the vibrational formalism. The result for our own equation of state is:

$$\frac{C_P^0}{\mathrm{JK^{-1}mol^{-1}}} = 118.1630 + 7.4144 \times 10^{-3} \cdot T + 6.6914 \times 10^3 \cdot T^{-1} - 9.2804 \times 10^6 \cdot T^{-2} +$$
$$+1.1643 \times 10^9 \cdot T^{-3}$$

$$\frac{\alpha^0}{\mathrm{K^{-1}}} = 2.1708 \times 10^{-5} + 3.0872 \times 10^{-9} \cdot T + 3.1938 \times 10^{-3} \cdot T^{-1} - 2.5217 \cdot T^{-2} +$$
$$+3.1996 \times 10^2 \cdot T^{-3}$$

$$K^0(T_0) = 251.69 \, \mathrm{GPa}$$

$$K'_R = 4.3894$$

$$a = +5.2914 \times 10^{-11} \, \mathrm{m^3K^{-1}mol^{-1}} \quad \text{(Equation 306:25)}$$

Heat capacity is represented to within 0.1%, thermal expansivity to within 0.3%, and isothermal bulk modulus to within 0.3% in the temperature range between 300 K and 3000 K. Next we compare the results obtained with the two methods, those obtained with the vibrational method, and those obtained with the polynomial method. This gives us an idea how large is the loss of accuracy when we use results of the vibrational method in the polynomial method.

Figure 19 shows that *isochoric heat capacity* calculated using the vibrational model attains values slightly larger than the Dulong-Petit limit of 124.7 $\mathrm{JK^{-1}mol^{-1}}$ due to the presence of intrinsic anharmonicity. From Raman spectroscopic measurements by Gillet et al. (2000) it became clear that the intrinsic anharmonic effect is small in perovskite relative to forsterite, with anharmonicity parameters of about -5×10^{-6} K^{-1} for the majority of the vibrational modes.

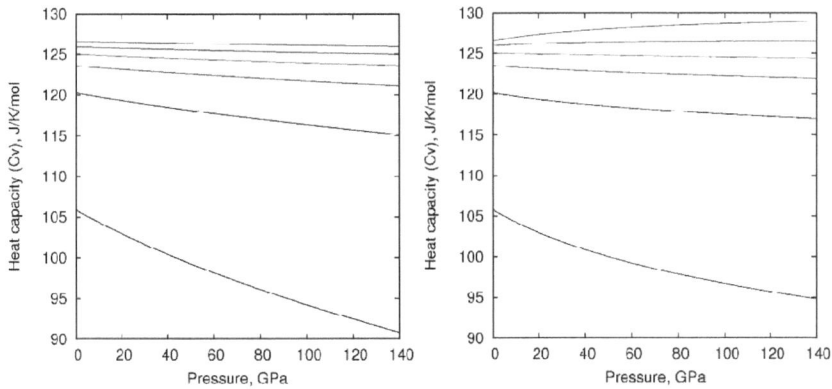

FIG. 19.
Isochoric heat capacity of $MgSiO_3$ (perovskite) as function of pressure at different temperature ranging from 500 K to 3000 K in steps of 500 K. The result of the vibrational model is shown on the left and the result calculated with the polynomial model is shown on the right

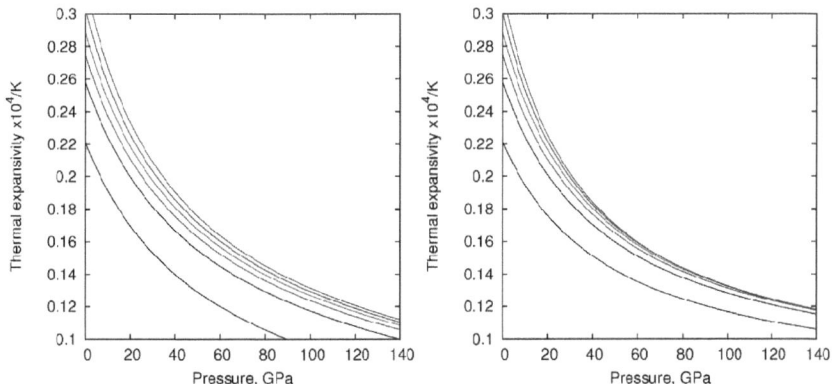

FIG. 20.
Thermal expansivity of $MgSiO_3$ (perovskite) as function of pressure at different temperature ranging from 500 K to 3000 K in steps of 500 K. The result of the vibrational model is shown on the left and the result calculated with the polynomial model is shown on the right

Isochoric heat capacity decreases with pressure and the slope becomes flatter at higher temperature, but the slope will stay negative even if the temperature is twice as high as used in Figure 19. That is not the case for our own equation of state

§(307)

based on polynomial descriptions, which shows an increase of heat capacity with pressure for temperatures above 2000 K. However, the difference between the two methods is fair; about 2.7% at *Core-Mantle-Boundary* (CMB) conditions. This kind of anomalous behaviour is also seen in the thermal expansivity curves depicted in Figure 20. Whereas the thermal expansivity curves calculated with the vibrational model are approximately parallel, they start to intersect when our equation of state is used for temperatures above 2000 K. The resulting values for thermal expansivity calculated by the two methods are quite comparable at CMB conditions, but differ considerably at temperatures below about 500 K. From Figure 19 and 20 we conclude that although physically unrealistic behaviour cannot be completely eliminated in our equation of state, the resulting values are quite comparable in the two models at the *P-T* conditions of the transition zone and at the CMB.

Figure 21 shows *bulk sound velocity* profiles, which are parallel and converging to each other at high pressure by using the vibrational method. When our equation of state is used, they intersect unrealistically at about 95 GPa. Beyond this pressure bulk sound velocity will isobarically increase with temperature. The difference between calculated sound velocities at pressures above 95 GPa in the temperature range between 1000 K and 3000 K is however small. The difference between the two methods at CMB conditions is about 2%, about four times the *tomographic uncertainty* in sound velocities.

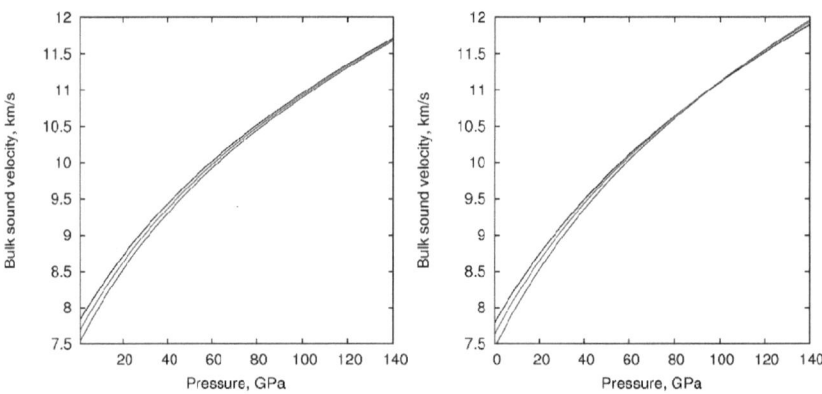

FIG. 21.
Bulk sound velocities of MgSiO$_3$ (perovskite) at 1000 K, 2000 K and 3000 K as function of pressure calculated with the vibrational model (left) and our equation of state (right)

It is also possible to use the results of the vibrational method to derive the High-Temperature-Birch-Murnaghan (HTBM) equation of state, frequently used in geophysics, see Saxena (1993). In addition to the polynomial expressions for heat capacity and thermal expansivity we have polynomial expressions for one-bar isothermal bulk modulus and the pressure derivative of bulk modulus. Fitting the results of the vibrational method in the temperature range between 300 K and 3000 K gives:

$$\frac{K^{\text{o}}}{\text{GPa}} = 252.707 - 2.2463 \times 10^{-2}\,(T - T_0) - 2.6366 \times 10^{-7}\,(T - T_0)^2$$

$$\left(K^{\text{o}}\right)' = 4.3894 + 4.8974 \times 10^{-5} \cdot (T - T_0)\ln\!\left(\frac{T}{T_0}\right)$$

In these expressions T_0 denotes the reference temperature of 298.15 K. The one-bar bulk modulus expression represents the results of the vibrational method to within 0.15% in this temperature range. The expression for the pressure derivative of bulk modulus is of the form frequently used in geophysics, but it shows an unrealistic temperature derivative of zero at the reference temperature and the wrong sign at temperatures below T_0, not present in the vibrational method. Figure 22 illustrates that this expression and our own equation of state represent the temperature trend of the pressure derivative of bulk modulus to a fair extent.

Figure 22 demonstrates that isochoric heat capacity in pressure-temperature space generally shows the same type of behaviour as plotted in Figure 19 but the pressure derivative of heat capacity becomes positive at 3000 K instead of 2500 K in our own equation of state. Figure 23 shows that thermal expansivity is nearly free from physically unrealistic behaviour and that it is about 20% larger than our own equation of state at CMB conditions.

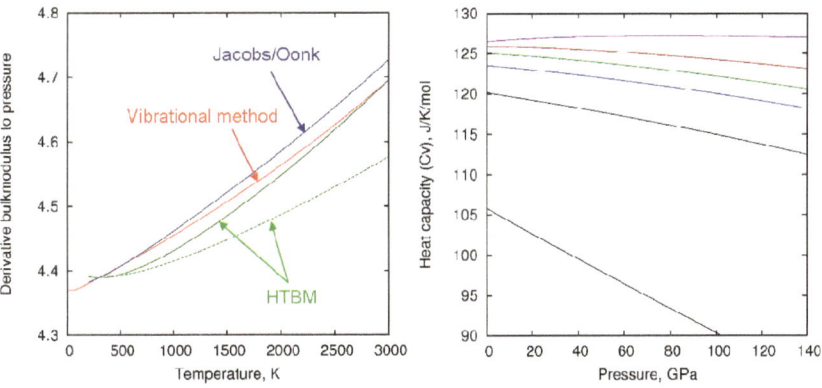

FIG. 22.
Left: Pressure derivative of isothermal bulk modulus of $MgSiO_3$ (perovskite) as function of temperature. The dashed HTBM curve is calculated using a smaller pressure derivative to eliminate the anomalous behaviour in sound velocity.
Right: Isochoric heat capacity as function of pressure at different temperature ranging from 500 K to 3000 K in steps of 500 K

188

Figure 23 also illustrates that bulk sound velocity profiles intersect unrealistically at about 60 GPa, 35 GPa lower than is the case with our own equation of state. At pressures beyond 60 GPa, bulk sound velocity isobarically increases with temperature. This anomalous behaviour can be eliminated by decreasing the value of the coefficient in the expression for the pressure derivative of K' from 4.9×10^{-5} K^{-1} to about 3×10^{-5} K^{-1}. That results in bulk sound velocity profiles converging to each other in the same way as illustrated in Figure 21. In that case the pressure derivative of isochoric heat capacity becomes positive at temperatures larger than 2000 K, just as in our equation of state.

From the calculations above we conclude that our equation of state and the HTBM, both based on polynomial expressions for one-bar properties are not able to represent the vibrational results to a full extent. The two polynomial methods do not produce results free from physically unrealistic behaviour. However, reasonable values for thermodynamic properties can be obtained at transition zone conditions and even at CMB conditions. The Gibbs energy in pressure temperature space of the three methods is quite comparable. Substituting the descriptions based on the two polynomial methods into the description of perovskite in the thermodynamic description of Jacobs and de Jong (2007) results in a post-spinel phase boundary differing by about 0.1 GPa only.

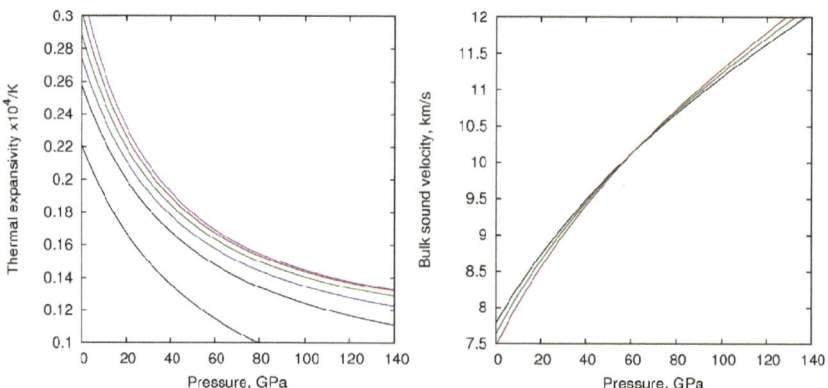

FIG. 23.
Thermal expansivity of $MgSiO_3$ (perovskite) as function of pressure at different temperatures, ranging from 500 K to 3000 K in steps of 500 K calculated with the High-Temperature-Birch-Murnaghan equation of state (left). Bulk sound velocity at 1000 K, 2000 K and 3000 K as function of pressure with the same equation of state (right)

===

A new model has been introduced for solid substances. The model, which is based on lattice vibrations, has four advantages relative to the model used in the preceding section. The first is that static lattice properties at zero Kelvin can be constrained by or compared with ab-initio calculations. Secondly, thermodynamic properties in pressure-temperature space are free from physically unrealistic behaviour. Thirdly, because the model incorporates more physical properties it is more unambiguously constrained. Lastly, the model provides a better means for assessing the quality of data from different sources. The development of the model comes from geophysical requirements.

===

Appendix

thermodynamic properties derived from the Helmholtz energy

This appendix summarizes the vibrational contributions to thermodynamic properties, which are derived from the Helmholtz energy, Equation (78).

In the following I_{j+1} denotes:

$$I_{j+1} = \int_{\infty}^{x} \frac{dx}{x^{j+1}(\exp(x)-1)} \tag{A1}$$

heat capacity C_V

$$C_V = k\sum_{j=0}^{m}\left[\frac{(j+1)(aT)^{j}(j-2aT)}{(1+aT)^{j+3}}\left(\frac{x}{\exp(x)-1} - \ln(1-\exp(-x))\right)\right] + k\sum_{j=0}^{m}\left[\frac{(j+1)(aT)^{j}}{(1+aT)^{j+3}}\frac{x^2\exp(x)}{(\exp(x)-1)^2}\right] \tag{A2}$$

When $m \to \infty$ the heat capacity is given by:

$$C_V = \frac{k}{1+aT}\frac{x^2\exp(x)}{(\exp(x)-1)^2} \tag{A3}$$

thermal pressure

$$P^{vib}(T,V) = \frac{kT\gamma}{V}\left[\frac{x}{\exp(x)-1}\sum_{j=0}^{m}\frac{(j+1)(aT)^{j}}{(1+aT)^{j+1}} + \sum_{j=1}^{n}\left[\frac{j(j+1)(aT)^{j}x^{j+1}}{(1+aT)^{j+1}}I_{j+1}\right]\right] + \frac{1}{2}\frac{kT\gamma}{V}\frac{x}{1+aT} \tag{A4}$$

When $m \to \infty$ the first summations term reduces to $(1+aT)$.

§(307)

temperature derivative of thermal pressure

$$\left(\frac{\partial P^{vib}}{\partial T}\right)_V = \frac{k\gamma}{V}\frac{x^2\exp(x)}{(\exp(x)-1)^2}\sum_{j=0}^{m}\left[\frac{(j+1)(aT)^j}{(1+aT)^{j+2}}\right] \tag{A5}$$

When $m\rightarrow\infty$ the summations term reduces to 1.

isothermal bulk modulus

$$K^{vib} = \frac{kT}{V}\frac{x}{\exp(x)-1}\left\{\left[\gamma^2+\gamma-V\left(\frac{\partial\gamma}{\partial V}\right)_T\right]\sum_{j=0}^{m}\left[\frac{(j+1)(aT)^j}{(1+aT)^{j+1}}\right]+\gamma^2\sum_{j=1}^{m}\left[\frac{j(j+1)(aT)^j}{(1+aT)^{j+1}}\right]\right\}$$

$$-\frac{kT}{V}\frac{\gamma^2 x^2\exp(x)}{(\exp(x)-1)^2}\sum_{j=0}^{m}\left[\frac{(j+1)(aT)^j}{(1+aT)^{j+1}}\right]+\frac{kT}{V}\left[\gamma^2+\gamma-V\left(\frac{\partial\gamma}{\partial V}\right)_T\right]\frac{x}{2(1+aT)}$$

$$+\frac{kT}{V}\sum_{j=1}^{m}\left[\frac{j(j+1)(aT)^j}{(1+aT)^{j+1}}\left[(1+j)\gamma^2+\gamma-V\left(\frac{\partial\gamma}{\partial V}\right)_T\right]x^{j+1}I_{j+1}\right] \tag{A6}$$

The total isothermal bulk modulus is derived from eqn. (10):

$$K = K^{st}(V)+K^{vib}(T,V). \tag{A7}$$

thermal expansivity

Thermal expansivity, α, is calculated from the thermodynamic identity:

$$\left(\frac{\partial P}{\partial T}\right)_V = \left(\frac{\partial P^{vib}}{\partial T}\right)_V = \alpha K \tag{A8}$$

heat capacity C_P

Heat capacity at constant pressure is calculated using the thermodynamic identity:

$$C_P - C_V = \alpha^2 KVT \tag{A9}$$

and Equations (A1), (A7), (A8) and Equations (10), (6), and (A4).

§ 308 THE SYSTEM MgO – SiO₂ UNDER PRESSURE

===

The formalism based on lattice vibrations, in which the vibrational density of states is described with Kieffer's model, is applied to the system MgO-SiO₂ to derive thermodynamic properties of the relevant phases and the phase diagram. The resulting small data base containing all thermodynamic model parameters coupled to the formalism is applied in geodynamic modelling to elucidate material heterogeneity in the transition zone of the Earth.

═══

introduction, the system MgO – SiO₂

In the preceding chapters it has been emphasized that it is of great importance to constrain a thermodynamic model for solid substances not only by experimental data at high pressure, but also by data measured at 1 bar pressure. The reason behind this philosophy is that data obtained at 1 bar pressure do not suffer from uncertainties associated with equations of state of pressure reference materials. It was demonstrated in § 306 that such data are well fitted by methods based on polynomial expressions for thermodynamic properties, such as thermal expansivity, isothermal bulk modulus and isobaric heat capacity. Subsequently it was shown that these methods are cumbersome to use when thermodynamic properties are extrapolated to conditions prevailing in the lower mantle because of, for instance, physical anomalous behaviour occurring in computed thermal expansivity or bulk sound velocity. Therefore these polynomial methods do not perform well, compared to alternative methods based on Mie-Grüneisen-Debye theory (MGD), when applied to the entire range of pressure and temperature covered by the Earth's mantle. Next it was illustrated that due to the simple construction of the vibrational density of states (VDoS) inherent to Debye's model and the application of the quasi-harmonic approximation, experimental data at 1 bar conditions are less well represented by the MGD method relative to the polynomial approach. This means that two methods are available today for making thermodynamic databases, applicable in different pressure ranges: the polynomial method at low pressure and the MGD method at high pressure. Examples of databases based on the polynomial method for multi-component oxide systems are those of Piazonni et al. (2007), Fabrichnaya et al. (2004) and Holland and Powell (1998). Databases based on the MGD formalism have been developed by Stixrude and Lithgow-Bertelloni (2005a,b).

192

Because the use of two different methods is cumbersome in the application of a thermodynamic analysis of thermodynamic properties in a complicated system characterized by phase transitions, a new method was introduced in § 307. This method is based on Kieffer's (1979) more elaborate model for the VDoS and it also includes intrinsic anharmonicity. It was illustrated that this method enables an accurate description of not only 1-bar experimental data but also data obtained at the high pressures occurring in the lower mantle. Such accuracy is important because the representation of phase equilibrium boundaries is sensitive to small variations in the Gibbs energy in pressure-temperature space as was demonstrated for the forsterite-wadsleyite phase boundary in § 307.

   This section deals with the phase behaviour of the solid state in the system MgO–SiO₂ and can be considered as a first step in the application of the *vibrational method* discussed in § 307 to a system characterized by multiple phase transitions. These phase transitions occur in the pressure range up to 30 GPa, and temperatures that range from 1000 K to 2600 K. In this pressure-temperature regime the method will be applied for establishing accurate phase boundaries and thermodynamic properties.

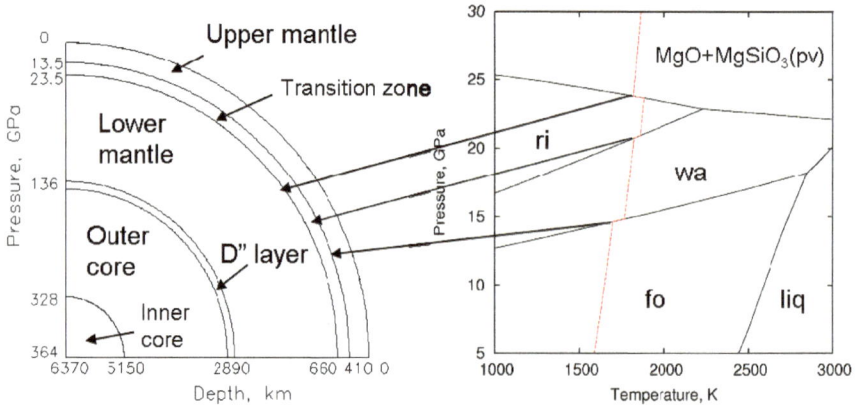

FIG. 1.

Left panel: Section of the Earth showing a heterogeneous distribution of materials in compartments determined by seismic experiments. The core of the Earths consists of a liquid outer core and a solid inner core. Right panel: Seismic discontinuities in the transition zone are related to phase equilibria, here illustrated with a simple backbone system, Mg₂SiO₄. They are close to the transitions between the forsterite, wadsleyite and ringwoodite polymorphs of Mg₂SiO₄. The dashed curve represents an isentrope with a fixed point of 1600 K at 6 GPa determined by a petrological study of Mercier and Carter (1975)

In the pressure-temperature field mentioned here, a region is found which, by geophysicists is referred to as the *transition zone* - a zone in the mantle of the Earth, where phase transitions give rise to discontinuities in *seismic observations*. Figure 1 illustrates that the heterogeneity of Earth's mantle is related to phase transitions, in this case very roughly simulated with the system Mg$_2$SiO$_4$, serving as a backbone system.

In this context the system MgO-SiO$_2$ is a vehicle, a model system to reveal the complexity of the issue, and to demonstrate how discontinuities in *sound velocities* are related to phase transitions. The system at hand also is a first step from another perspective. In the near future the elements Al, Ca, and Fe have to be included in the thermodynamic description to reach a better approximation of the complexity and richness of phase transitions in the Earth. In addition, experimental data on phase transitions and thermodynamic properties for significantly high pressure are still becoming available.

The solid substances that make their appearance are the *end members* MgO and SiO$_2$ and their *compounds* Mg$_2$SiO$_4$ and MgSiO$_3$. Out of these substances MgO is the only one that is not polymorphous. The other three, SiO$_2$ and MgSiO$_3$ in particular, display a rich range of *polymorphism*.

In this section an exhaustive use of lengthy tables is avoided; details can be found in Jacobs and de Jong (2007).

the substances and their forms and thermodynamic properties

The substance MgO (periclase, pc), treated in § 306, has been reanalyzed in terms of the Kieffer model for the VDoS (←307). The output of the new analysis as far as it is relevant for this section is included in Table 1.

As regards to the data in Table 1, an important observation has to be made. In the paper by Jacobs and de Jong (2007) the outcome is presented of two different optimization processes. In one of these processes, the outcome of which is used in this section, the optimization is directed to the data by Shim et al. (2001) for the change from ringwoodite to MgO + perovskite. In the other one, the optimization is directed to the data by Fei et al. (2004) for that change (see below, Figure 1).

The substance, the compound MgSiO$_4$ is trimorphous (←306). The three forms are forsterite (fo), wadsleyite (wa), and ringwoodite (ri). Like in the case of MgO, the forms have been treated in § 306, and re-optimized in terms of Kieffer's model. The results of the latter optimization are included in Table 1.

§(308)

Table 1:

Calculated thermodynamic properties at 298 K and 1 bar pressure in terms of the Kieffer model: enthalpy of formation (MJ mol^{-1}), entropy (J K^{-1} mol^{-1}), heat capacity (J K^{-1} mol^{-1}), volume (cm^3 mol^{-1}), thermal expansivity (10^{-5} K^{-1}), isothermal bulk modulus (GPa), and its pressure derivative

substance	Form	H	S	C_P	V	α	K	K'
MgO	pc	-0.6013	26.97	36.77	11.247	2.977	160.2	4.14
Mg$_2$SiO$_4$	fo	-2.1769	94.98	118.4	43.658	2.671	127.5	4.81
	wa	-2.1468	88.44	116.0	40.289	2.195	171.2	4.28
	ri	-2.1394	82.66	116.8	39.477	1.883	183.2	4.25
MgSiO$_3$	oen	-1.5475	67.63	81.16	31.370	2.996	105.9	6.97
	cen	-1.5475	61.23	80.13	30.452	2.518	115.0	6.28
	mj	-1.5174	60.02	79.40	28.330	1.584	167.5	6.29
	aki	-1.5039	51.37	78.90	26.457	1.705	212.4	5.63
	pv	-1.4513	57.75	82.21	24.464	1.699	252.4	4.39
	ppv	-1.4123	57.73	77.28	24.276	1.760	230.3	4.46
SiO$_2$	st	-0.8735	25.60	40.94	14.02	1.237	291.0	4.20

As far as the polymorphous substance SiO$_2$ is concerned, only its high-pressure form stishovite (st) is involved in the phase equilibria that take place in the transition zone.

It is a general fact that transitions between the forms of a substance are such that on increasing pressure the transition is accompanied with a decrease in volume; and on increasing temperature with an increase in entropy. The volume effect of the transition, or, in other terms, the volume change at the transition (ΔV), and the entropy change (ΔS) together determine the sign and the magnitude of the slope of the equilibrium curve in the PT plane; and so through the *Clapeyron equation* ($\leftarrow$302)

$$dP / dT = \Delta S / \Delta V. \tag{1}$$

In the case of Mg$_2$SiO$_4$ the properties of the forms are such that ΔS and ΔV invariably have the same sign. As a result, in the PT plane the two-phase equilibrium curves have a positive slope: increase in equilibrium temperature goes together with increase in equilibrium pressure. Besides, and for the different forms of a substance, volume and entropy most of the times change with pressure and temperature such that the signs of ΔS and ΔV are not affected. And it means that those signs can be read from the numerical values given in Table 1.

In the case of MgSiO$_3$ ΔS and ΔV do not invariably have the same sign, and it implies that in the PT plane not all of the two-phase equilibrium curves have a positive slope.

From the volume data in Table 1 it follows that a decrease in volume also can be evoked by a chemical reaction. For instance, the molar volume of the ringwoodite form of Mg$_2$SiO$_4$ is greater than the sum of the molar volumes of perovskite and MgO. In other terms, in the PT plane an equilibrium curve may be expected which gives the equilibrium states of the *chemical equilibrium*

$$\text{Mg}_2\text{SiO}_4 \text{ (ri)} = \text{MgSiO}_3 \text{ (pv)} + \text{MgO} .$$

Besides, and because Equation (1) is also valid for this equilibrium case, and moreover assuming that the signs of ΔS and ΔV remain the same as for the numbers in Table 1, it is to be expected that the equilibrium curve for this chemical equilibrium has a negative slope.

phase behaviour of the composition Mg$_2$SiO$_4$

The phase behaviour of Mg$_2$SiO$_4$ for pressures up to about 30 GPa has been plotted in Figure 2. At pressures above about 23 GPa the material decomposes into perovskite and MgO. The calculated phase diagram is shown twice: one time without, and the other time along with the experimental data; see Figure 2.

The thermodynamic analysis of forsterite has been discussed in § 307. Wadsleyite and ringwoodite are quenchable from high pressure conditions and experimental data are available for heat capacity, thermal expansivity, adiabatic bulk modulus and sound wave velocities at 1 bar pressure to temperatures of about 1000 K. The thermodynamic analysis of these two forms is constrained by these data, pressure-volume-temperature data and data on the heat of transition when these phases transform back to forsterite. The analysis using Kieffer's method for the VDoS is additionally constrained by *Raman and infrared spectroscopic data* on vibrational frequencies as function of pressure and in the case of forsterite and wadsleyite also as function of temperature. These data tightly constrain location and slope of the phase boundaries forsterite-wadsleyite and wadsleyite-ringwoodite. Figure 2 shows that the analysis by Jacobs and de Jong (2007) prefers the phase boundary experimented by Katsura and Ito (1989) to within experimental uncertainty.

196

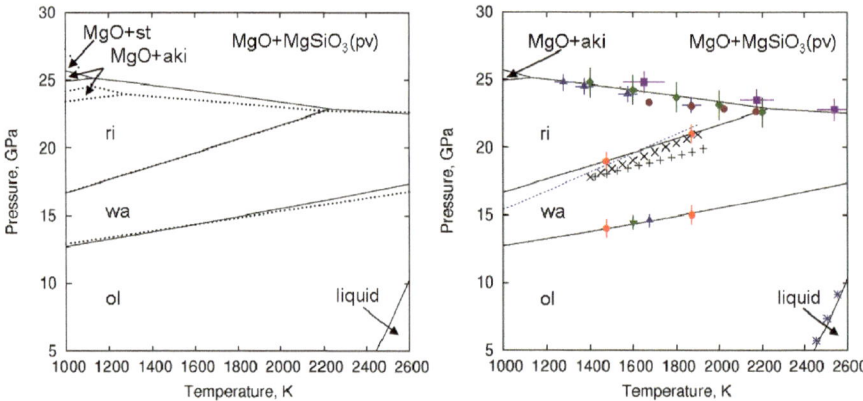

FIG. 2.

Calculated phase diagram of Mg_2SiO_4. Left frame: the solid curves are obtained by directing the thermodynamic analysis towards the post-spinel phase transition data of Shim et al. (2001), whereas the dotted curves are obtained by directing the thermodynamic analysis towards the post-spinel transition data of Fei et al. (2004a). Right frame: calculated curves are obtained by directing the thermodynamic analysis towards the post-spinel data of Shim et al. (2001). Experimental data are from Katsura and Ito (1989) (circle, ol-wa and ol-ri transitions); Ito and Takahashi (1989) (triangle, post-spinel transition); Boehler and Chopelas (1991) (inverse triangle, ol-wa transition); Chopelas (1994) (square, postspinel transition); Fei and Bertka (1999) (triangle, ol-wa transition); Suzuki (2000), dotted curve for the wa-ri transition; Shim et al. (2001) (diamond, post-spinel transition); Gasparik (2003) (cross, wa-ri transition); Fei et al. (2004a) (circle, post-spinel transition, uncertainties are equal to the symbol size); Inoue et al. (2006) (plus, wa-ri transition)

We investigated ringwoodite, Mg_2SiO_4, by the vibrational method because we showed in Jacobs and de Jong (2005a) that thermodynamic properties could not be constrained sufficiently accurate enough using polynomial methods to allow a description for bulk sound velocity of $(Mg_{0.91}Fe_{0.09})_2SiO_4$ to within 0.8%, twice the tomographic accuracy. Figure 2 also illustrates that there is disagreement in location and slope of the wadsleyite-ringwoodite phase boundary, experimented by different investigators. The most recent measurements of Inoue et al. (2006) indicate the smallest slope of 4.1 ± 1.0 $MPa\,K^{-1}$ and was determined using the gold pressure scale of Shim et al. (2002). Differences in slope and position of phase boundaries are likely be caused by reaction kinetics, the neglect of pressure on thermocouple emf, and the use of different pressure scales. In the measurements use is made of a pressure reference material, usually an inert material such as gold or MgO, for which an equation of state is *assumed* to be accurate enough.

§(308)

However, establishing an accurate pressure scale is one of the most complicated and challenging problems with which the geosciences community is faced (see e.g. Fei et al, 2004b for an overview) and work on this issue is still in progress. The phase boundary between MgO+perovskite and ringwoodite is also known as the *post-spinel phase boundary*. The intersection of Inoue et al.'s wadsleyite-ringwoodite phase boundary with the post-spinel phase boundary leads to a triple point located between 2430-2500 K. The temperatures in this range are significantly higher than the value of 2173 K established by Fei et al. (2004a). These investigators refer to a detailed study of Li et al. (2003) to the effect of pressure on thermocouple emf showing that the neglect of this effect results in an underestimation of the temperature by about 100 K at 23 GPa. That still leads to a discrepancy of about 200 K for the triple point when Inoue et al's data are extrapolated to the post-spinel phase boundary. The calculated slope of the wadsleyite-ringwoodite phase boundary in Figure 2, is 5.0±0.3 MPa·K^{-1} and can be reconciled with Inoue et al.'s experimentally established slope. However the position of Inoue's phase boundary is lower than that of Katsura and Ito's by about 1 GPa. Directing the thermodynamic analysis to Inoue et al.'s data resulted in a description of the heat of transition data between the forsterite, wadsleyite and ringwoodite forms, which significantly differs from the experimental data established by Akaogi et al. (1989). Figure 2 shows that the calculated slope of the phase boundary is in between that determined by Suzuki et al. (2000) and Gasparik (2003) and that determined by Inoue et al. (2006); 7.0 and 6.0 MPa·K^{-1} , respectively.

The *post-spinel phase transition* has been the subject of many investigations, resulting in different locations and Clapeyron slopes of the phase boundary. For instance, Ito and Takahashi (1989) determined a Clapeyron slope of −2.8 MPa·K^{-1}, whereas Katsura et al. (2004a) determined a slope of −0.4 MPa·K^{-1}. According to Fei et al. (2004), these differences are attributed to the use of different pressure scales, kinetics of the phase transformation and different experimental techniques. The slope of this phase boundary is an important parameter in determining the style of mantle convection. The smaller the slope, the easier material is transported across the post-spinel phase boundary. By making two thermodynamic analyses, we showed in Jacobs and de Jong (2007) that both datasets for the post-spinel transition, those of Shim et al. (2001) and those of Fei et al. (2004a) can be represented to within experimental uncertainty. Both analyses do not lead to significantly different thermophysical properties relative to the experimental data. Directing the thermodynamic analysis to the post-spinel data of Shim et al. leads to a Clapeyron slope of −2.1±0.2 MPa·K^{-1} , whereas directing the analysis to the post-spinel data of Fei et al. results in a Clapeyron slope of −1.4±0.2 MPa·K^{-1}.

The appearance of the phase fields MgO+stishovite and MgO+akimotoite, at low temperature in Figure 2, is because the phase equilibria in the systems Mg$_2$SiO$_4$ and MgSiO$_3$ are intimately linked together. This will be discussed below.

phase behaviour of the composition MgSiO₃

The phase diagram calculated for the composition MgSiO$_3$ is shown in Figure 3: left-hand side without, and right-hand side along with the experimental data.

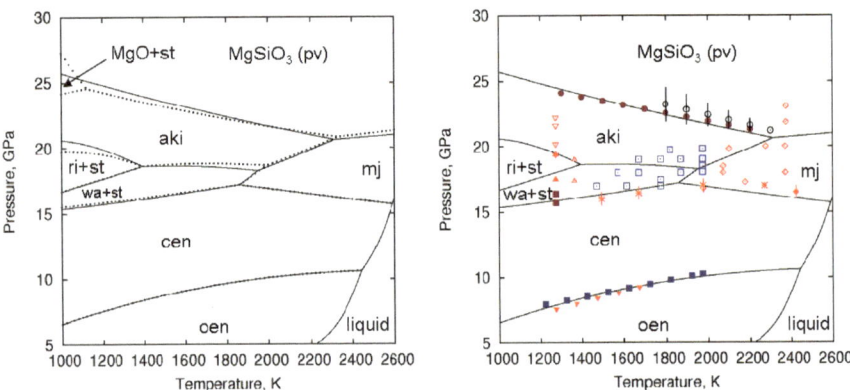

FIG. 3.

Calculated phase diagram of MgSiO$_3$. Left frame: the solid curves are obtained by directing the thermodynamic analysis towards the post-spinel transition data of Shim et al. (2001), whereas the dotted curves are obtained by directing the thermodynamic analysis towards the post-spinel transition data of Fei et al. (2004a). Right frame: calculated curves are obtained by directing the thermodynamic analysis towards the post-spinel transition data of Shim et al. (2001). Experimental data: akimotoite (inverse open triangle, Kanzaki, 1987); wa+st (square, Sawamoto, 1987, and filled square, Kanzaki, 1987); ri+st (filled triangle, Kanzaki, 1987, open triangle, Ito and Navrotsky, 1985); mj (open diamond, Sawamoto, 1987, filled diamond, Presnall and Gasparik, 1990); clino-enstatite (asterisk, Gasparik, 1990, and Presnall and Gasparik, 1990). The phase boundary between akimotoite and perovskite is from: filled circles, combination of Ono et al. (2001) and Hirose et al. (2001) and Fei et al. (2004a) when measured pressures are based on the equation of state for MgO by Speziale et al. (2001); open circle, Chudinovsky and Boehler (2004); Clapeyron slope is −4±2 MPa K^{-1}. The phase boundary between ortho- and clino-enstatite is measured by: square, Gasparik (1990); triangle Kanzaki (1987). The phase boundaries involving liquid are estimated, using results of Gasparik (2003)

The phase equilibria in the MgSiO$_3$ system are intimately linked to the phase equilibria in the system Mg$_2$SiO$_4$ because of the presence of wadsleyite, ringwoodite and perovskite in both systems. Therefore, results of a thermodynamic analysis in one system will affect phase equilibria in the other system.

Figure 3 shows that the transition between akimotoite and perovskite does not change significantly when the Clapeyron slope of the post-spinel phase transition in Figure 2 changes from −2.1 MPa K^{-1} to −1.4 MPa K^{-1}. The calculated Clapeyron slope of the akimotoite-perovskite transition is −3.8 MPa K^{-1} or −3.1 MPa K^{-1} depending whether the thermodynamic analysis is directed to the post-spinel data of Shim et al. (2001) or to those of Fei et al. (2004a). Because this slope is steeper than that of the post-spinel phase transition, the akimotoite-perovskite transition intersects the post-spinel transition in the *P-T* plane at all compositions between that of Mg$_2$SiO$_4$ and MgSiO$_3$ at about 1175 K as demonstrated in Figure 5. The presence of this intersection results in the formation of the phase field MgO+akimotoite. This intersection also results in the formation of the phase field MgO+akimotoite in Figure 2. Because the Clapeyron slope of Fei et al.'s post-spinel phase boundary is less steep, relative to that of Shim et al., the MgO+akimotoite phase field in Figure 2 extends to higher temperatures, about 1280 K. The flattest Clapeyron slope of the post-spinel transition has been measured by Katsura et al. (2004), −0.4 MPa K^{-1}, and keeping the akimotoite-perovskite boundary at the same location results in an extension of the MgO+akimotoite field to about 1700 K. All investigators, whose work has been cited in Figure 2, examined the post-spinel phase transition at temperatures higher than 1273 K.

Because the MgO+akimotoite phase field has not been observed by any of these investigators we anticipate that the slope of the post-spinel phase transition is steeper than about −1.4 MPa K^{-1}. The flatter post-spinel phase boundary experimented by Fei et al. (2004a) has the additional effect that the phase field MgO+stishovite extends to higher temperatures.

sound waves and their velocities

In geophysics, or more precise by *seismology,* the interior of the Earth is studied by elucidating the behaviour of *elastic sound waves* propagating through solid and liquid material. There are two basic types of *sound waves*: *longitudinal sound waves* and *transverse sound waves*. The longitudinal waves propagate through materials with the largest speed; they are the first to appear on the seismogram, and for that reason they are also referred to as *primary waves* or *P-waves*. The transverse waves, also referred to as *shear sound waves*, travel with a smaller speed compared to the longitudinal waves; they appear later on the seismogram (*secondary waves / S-waves*). In *fluids* only longitudinal waves make their appearance.

200

The velocities of the longitudinal and transverse waves are connected through the so-called *bulk sound velocity*. The latter is an auxiliary property, related to the *adiabatic bulk modulus,* and, therefore, directly related to a thermodynamic quantity. The relation between bulk sound velocity, v_B, and adiabatic bulk modulus, K_S, is

$$v_B = \left(\frac{K_S}{\rho}\right)^{1/2} = \left(\frac{K_S \cdot V}{M}\right)^{1/2}, \tag{2}$$

in which ρ stands for *density,* V for *molar volume* and M for the *molar mass* of the material.

Using the well-known thermodynamic relation between the heat *capacities at constant pressure* and at *constant volume* ($\leftarrow$Exc 107:10; Exc 301:28),

$$C_P - C_V = \alpha^2 \, KVT, \tag{3}$$

and the relationship between adiabatic bulk modulus and *isothermal bulk modulus, K,* ($\leftarrow$ Exc 301:28),

$$K_S = K\frac{C_P}{C_V}, \tag{4}$$

the bulk sound velocity is given by

$$v_B = \left(\frac{KC_P V}{M(C_P - \alpha^2 \, KVT)}\right)^{1/2}. \tag{5}$$

In these equations the symbol α is for *cubic expansion coefficient.*

Equation (5) indicates that the calculation of the bulk sound wave velocity only requires *thermodynamic* properties.

The transverse, or shear sound velocity is related to the *shear modulus, G_{sh},* as

$$v_S = \left(\frac{G_{sh}}{\rho}\right)^{1/2} = \left(\frac{G_{sh}V}{M}\right)^{1/2}. \tag{6}$$

The *longitudinal sound wave velocity, v_p,* and the shear sound velocity are related through the equation

$$v_P = \left(v_B^2 + \frac{4}{3}v_S^2\right)^{1/2} = \left(\frac{K_S + \frac{4}{3}G_{sh}}{\rho}\right)^{1/2}. \tag{7}$$

Equations (6) and (7) indicate that the application of a thermodynamic database to seismology requires a formalism which enables the calculation of the relevant *thermodynamic properties and* the *shear moduli* of the materials involved. These requirements are met with the model developed by Susan Kieffer (1979), and which was discussed in the foregoing section (←307).

In Kieffer's theory the longitudinal and transverse sound velocities are related to the maximum *angular frequency* of the three *acoustic modes*. Two of the three modes are transverse and one is longitudinal.

In Table 307:5 for forsterite the two transverse modes are indicated by AC1 and AC2, and the longitudinal one by AC3; the maximum *wave number* of the latter is 172.7 cm^{-1}. To obtain the angular frequency the wave number has to be multiplied by two times π and the velocity of light. The latter is defined as $c = 2.99792458 \cdot 10^8$ $m \, s^{-1}$. This corresponds to an angular frequency of $3.253 \cdot 10^{13}$ s^{-1}.

The relationship between the (directionally averaged) sound velocity, v_i, and the maximum angular velocity, $\omega_{i,max}$, is

$$v_i = \frac{1}{4} (4 \pi Z V / 3 N_A)^{1/3} \omega_{i,max} \, . \, i = 1, 2, 3 \tag{8}$$

In Equation (8), Z denotes the number of 'molecules' per unit cell, V the molar volume, and N_A Avogadro's constant. From the numerical values in Table 307:3, it follows, from Equation (8) for $i = 3$, that the longitudinal sound velocity of forsterite at zero K and zero pressure is about 9 $km \, s^{-1}$.

In terms of Kieffer's model, the maximum wave numbers of the acoustic modes, as well as the upper and lower wave numbers of the *optic modes,* are part of the input data. To derive values for them, they are along with the available thermodynamic data subjected to an optimization process. In that process the differences between the input data and the values calculated for the properties are minimized.

As an example, in Table 2 for clino-enstatite, the optimized properties are given as far as the three acoustic modes (out of 60 normal modes) are concerned. In Figure 4 it is shown how well the experimental sound velocities are represented by the outcome of the optimization process. In Table 2 the symbol γ_0 represents the *Grüneisen parameter*. The Grüneisen parameter is related to the change of frequency v with volume at constant temperature as (←Equation 307:14)

$$\gamma = - (\partial \ln v / \partial \ln V)_T . \tag{9}$$

The symbol q_1 is related to the change of the Grüneisen parameter with volume at constant temperature:

$$q_1 = (\partial \ln \gamma / \partial \ln V)_T . \tag{10}$$

202

The *anharmonicity parameter* (Equation 307:59), symbol a, expresses the change of a given frequency with temperature at constant volume:

$$a = (\partial \ln v / \partial T)_V. \tag{11}$$

Table 2:
Properties at zero Kelvin and zero pressure, resulting from an optimization process, of the acoustic modes of the high-pressure form of clino-enstatite. TA denotes transverse acoustic and LA longitudinal acoustic

motion type	wave number in cm^{-1}	γ_0	$q_{1,0}$	$a \times 10^5$ K^{-1}
TA	0.0-112.7	1.34	3.36	-4.62
TA	0.0-121.7	1.34	3.36	-4.62
LA	0.0-190.4	2.03	0.81	-1.02

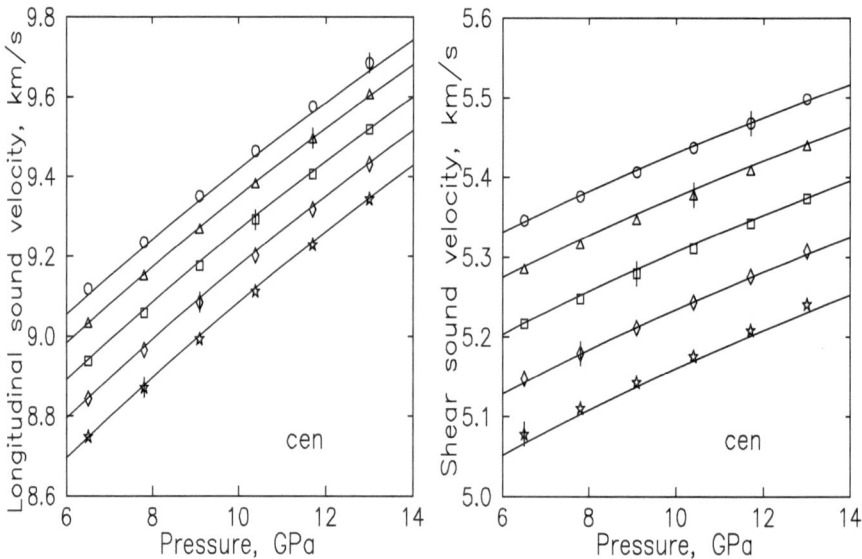

FIG. 4.
Curves represent calculated longitudinal sound velocities (left) and transverse (shear) sound velocities (right) for the high-pressure form of clino-enstatite. The experimental data are from Kung et al. (2005): 298 K (circle); 473 K (triangle); 673 K (square); 873 K (diamond); and 1073 K (star)

seismic observations

Seismological research has revealed that there are two major seismic discontinuities in sound velocity profiles, in the mantle of the Earth. These two discontinuities, located at depths of 410 km and 660 km, are ascribed to phase transitions.

Focusing on the MgO-SiO$_2$ system, the situation is such that the 410-km discontinuity is the result of the change from forsterite to wadsleyite, the 660-km discontinuity being the result of the change to MgO and perovskite.

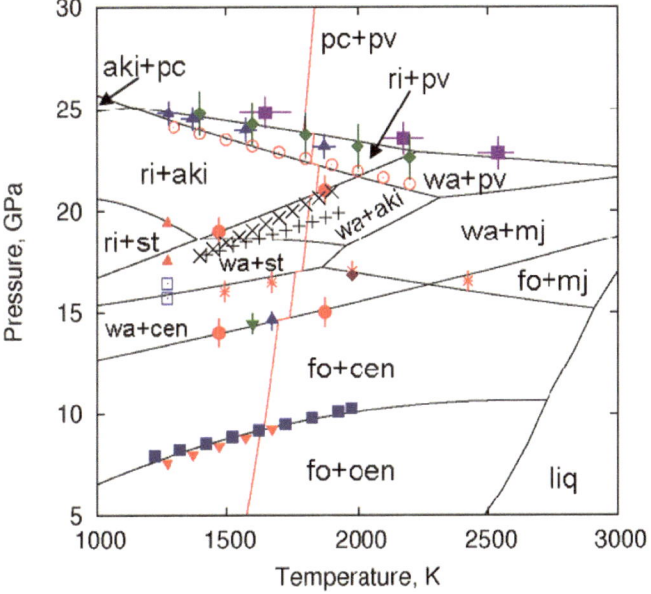

FIG. 5.

Calculated phase behaviour of mixtures of Mg$_2$SiO$_4$ and MgSiO$_3$ with some experimental data. The thermodynamic analysis has been directed to the post-spinel phase transition data of Shim et al. (2001). Experimental data are from Ito and Takahasi (1989), solid triangle; Chopelas et al. (1994), Shim et al. (2001), combined dataset of Ono et al. (2001), Hirose et al. (2001) and Fei et al. (2004), open circle, solid diamond, solid square; Katsura and Ito (1989), solid circle; Gasparik (2003), x; Inoue et al. (2006), +; Boehler and Chopelas (1991), inverse solid triangle; Fei and Bertka (1999), solid triangle; Gasparik (1990), solid square; Kanzaki (1991), inverse solid square; Presnall and Gasparik (1990), star; Kanzaki (1987), open square and solid triangle; Sawamoto (1987), solid diamond. The heavy solid curve represents the adiabatic path that is in line with the petrological study by Mercier and Carter (1975)

In Figure 5, which in fact is a superposition of Figures 2 and 3, it is shown how, for a *mechanical mixture* of Mg$_2$SiO$_4$ and MgSiO$_3$, the various forms of the two substances make their appearance as a function of pressure and temperature.

The heavy curve in Figure 5 is a so-called *adiabatic path;* along this path the entropy of the material is constant. And it is assumed that the path, as it is shown, is representative of the temperature of the material as a function of depth.

In spite of the fact that Figure 5 is just a first step in delineating the phase behaviour in the transition zone, the presence of elements other than Mg, Si, and O being ignored, its major phase transformations are in harmony with the seismic discontinuities at 410 and 660 km.

The figure also shows that between the two major discontinuities, a number of additional phase transformations are present. This statement finds expression in a number of weaker discontinuities observed in recent seismic studies, such as those by Deuss and Woodhouse (2001), Deuss et al. (2006), and Cao et al. (2010). Figure 6 illustrates that the phase transitions in Figure 5 are visible in the sound wave velocities. The figure shows that sound velocities are roughly commensurate with the seismically derived PREM (*Preliminary Reference Earth Model*, Dziewonsky and Anderson, 1981). The density profile, in particular, illustrates that the thermodynamic model lacks iron.

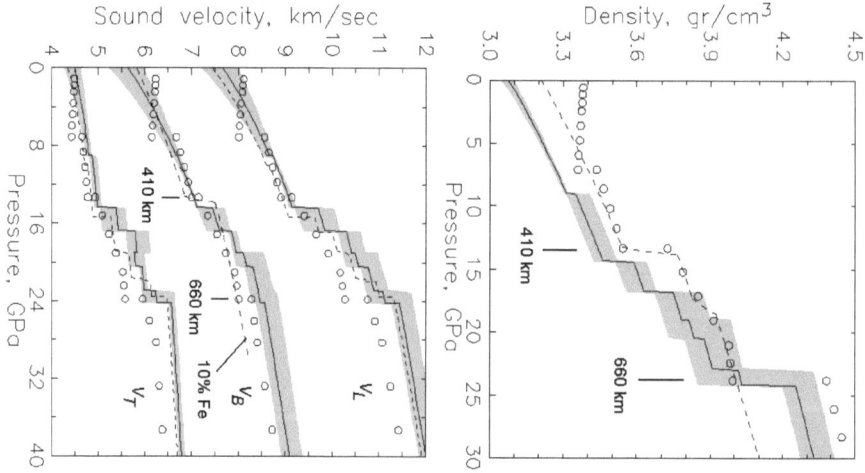

FIG. 6.

Left: Calculated longitudinal (V_L), Transverse (V_T) and bulk sound wave velocities (V_B) along adiabats using the thermodynamic analysis by Jacobs and de Jong (2007). The solid curves were calculated for a mixture of 60 vol% Mg_2SiO_4 and 40 vol% $MgSiO_3$, along adiabats with a foot temperature of 1420 K, whereas the dashed ones were calculated with a foot temperature of 1700 K. The grey fields were calculated for compositions ranging from Mg_2SiO_4 to $MgSiO_3$. The dashed bulk sound velocity was calculated for a $(Mg_{0.9}Fe_{0.1})_2SiO_4$ mixture (foot temperature 1420 K). Right: corresponding density profile. Dashed curve corresponds to a $(Mg_{0.9}Fe_{0.1})_2SiO_4$ mixture. Circles are from taken from PREM

geodynamic models, mantle convection

The preceding paragraphs might give the impression that the Earth is a static system in which phase boundaries have fixed depth-locations. However on a large time-scale of the order of millions of years this is not the case; the Earth appears as a dynamic system. Hot plumes rise from the core-mantle-boundary, cold slabs descent from the lithosphere through the transition zone, tectonic plates move. The elucidation of dynamic processes in the Earth is a challenging field in geophysics, where many disciplines come together, not only equilibrium thermodynamics, but also for example seismology, tectonophysics, petrology, rockphysics, and geochemistry.

We have applied our thermodynamic description of MgO-SiO$_2$ in a mantle convection model (Jacobs and van den Berg, 2011, van den Berg et al., 2011). This model describes the dynamic processes in the mantle of the Earth in terms of material convection, i.e. by means of heat, mass and momentum transfer. In this model we were especially interested in the small-scale heterogeneity in the transition zone located between about 400 km and 660 km depth in the Earth and how it is expressed in high resolution seismic imaging. Of course our thermodynamic description is not complete because important elements such as iron, aluminum and calcium are not included in it. However it results in the most significant phase boundaries in the transition zone and the locations of them are not unrealistic.

The effect of the deficiency of for instance iron in our thermodynamic description is that density is too low as described above, and that the olivine–wadsleyite and wadsleyite-ringwoodite transitions are about 1 GPa too high. The location of the post-spinel phase boundary is less affected. Another effect is that incorporating other elements in the thermodynamic description will split-up phase boundaries in multi-phase regions, resulting is less sharp velocity jumps across the transitions.

Although the MgO-SiO$_2$ system does not fully represent the real Earth, we consider it useful to employ it as an exploratory system to investigate the effects of heterogeneity in the Earth on properties resulting from a convection process, which in principle can be revealed by seismic methods.

Compared to other thermodynamic methods, our thermodynamic description is unique in the sense that it includes a realistic VDoS and anharmonicity in the lattice dynamic model of substances, thereby coupling the microscopic world to the macroscopic one. The use of microscopic data such as Raman and infrared spectroscopic data in the model has been employed to put tighter constraints on phase boundaries and thermodynamic properties.

At this place we briefly demonstrate some of the outcomes of the application of our thermodynamic description of the MgO-SiO$_2$ system to mantle convection modeling.

A mantle convection calculation consists of solving the equations for mass, momentum and energy conservation. The energy equation includes the effect of adiabatic heating, thermal conductivity, heating by viscous dissipation and radiogenic heating. These equations were solved using finite element methods on a two-dimensional quarter circle cylindrical domain with a prescribed temperature contrast of 3500 K between the top and bottom of the mantle. The thickness of the Earth's mantle is about 3000 km.

Thermodynamic properties needed in the convection calculation were pre-calculated using the thermodynamic descriptions of the phases in the system MgO-SiO$_2$. Because the Helmholtz energy in the formalism presented in §307 is trivially connected to the Gibbs energy, it is straightforward to incorporate the formalism in a Gibbs energy minimization program to determine which phases are in equilibrium with each other at given pressure and temperature conditions.

 The resulting properties of the equilibrium assemblages were stored in a table for about 4×10^6 P-T points. Because the convection calculation uses about 10^7 evaluations of thermodynamic properties per integration time step, the method of preparing a large table with thermodynamic properties appeared to be computationally quite efficient. The full details of the mantle convection model fall outside the scope of this text and are described by Jacobs and van den Berg (2011).

A result to illustrate the effects of heterogeneity in the transition zone is given in Figure 7. It illustrates a snapshot in time after simulating convection in the mantle of the Earth of about 10^9 year (1 Gyr). Figure 7a shows a snapshot of the temperature distribution characterized by two colliding cold downwellings descending from the surface towards the core-mantle boundary simulating subducting slabs, and Figure 7b shows a zoom-in of this field.

The interaction of material with a phase boundary is determined by the density contrast between the phases above and below the transition and by the Clapeyron slope. Figure 7a and 7b show that these downwellings pass almost unhindered through the endothermic post spinel phase boundary entering the lower mantle, and this is mainly due to the small value of the Clapeyron slope, -1.9 MPa K^{-1}. The smoothness of the temperature distribution does not give many clues about the occurrence of phase transitions in the transition zone. Figure 7c shows the distribution of phases for this snapshot in time of the transition zone and Figure 7d shows that they are well visible in the lateral variation of the transverse sound velocity. Figure 7c shows that along the radial profiles marked by 39° and 35°, the forsterite+clinoenstatite phase field is strongly distorted towards shallower depths. This is related to the positive Clapeyron slope between that field and the forsterite-orthoenstatite field. Because the Clapeyron slope of the phase boundary is positive, a cold downwelling consisting of forsterite and othoenstatite has a lower equilibrium pressure, when it forms equilibrium with forsterite and clinoenstatite, compared to the hotter environment surrounding this downwelling. That results in a displacement of the phase boundary to shallower depths.

The Clapeyron slope between forsterite+clinoenstatite and wadsleyite+clinoenstatite is also positive but smaller, resulting in a smaller uplift of the wadsleyite+clinoenstatite phase field. The explanation of the strong heterogeneity in this snapshot can be illustrated with the help of the phase diagram. Figure 8 shows the same three radial temperature profiles in the Earth, so-called geotherms at 35°, 37° and 39°. The temperature profiles through the two cold downwellings 35° and 39°, differ strongly from adiabatic (isentropic) behaviour. Additionally these two profiles differ strongly from each other.

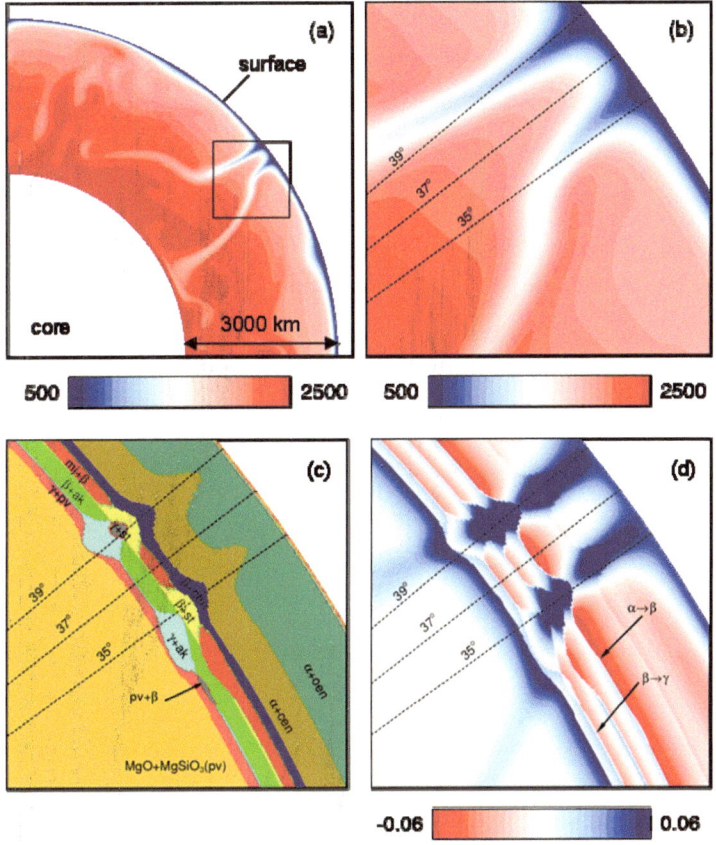

FIG. 7.
Snapshot of a mantle convection calculation after about 1 Gyr. (a) temperature snapshot with a box marking two cold downwellings. (b) Zoom-in of the box given in (a) with thin dashed lines marking three radials. (c) Phase field partitioning. (d) Perturbation of shear velocity δv_s. In these frames α denotes forsterite, β wadsleyite and γ ringwoodite

208

This is in contrast with the behavior of the temperature profile between the two downwellings located at 37° in Figure 8, which resembles an adiabat with a *foot temperature* (apparent temperature at the surface) of 1700 K. The three radial temperature profiles address different phase fields in the phase diagram and that results in a strong lateral variation of the phase distribution in the transition zone as depicted in Figure 7c. Figure 7d illustrates that although the geotherms indicate a smooth behavior in pressure-temperature space this is not the case with shear wave velocity. From Figure 7 and 8 we conclude that seismic shear wave velocity depends not only on pressure and temperature, but that it is also strongly related to the phase regions through which the shear waves pass. Figure 7c and 7d illustrate that, individual pockets of mineral phase assemblages are clearly visible in the velocity of shear waves. Stated in another way, seismic visibility of the mineral phases in the transition zone would put tighter constraints on the local temperature in that zone.

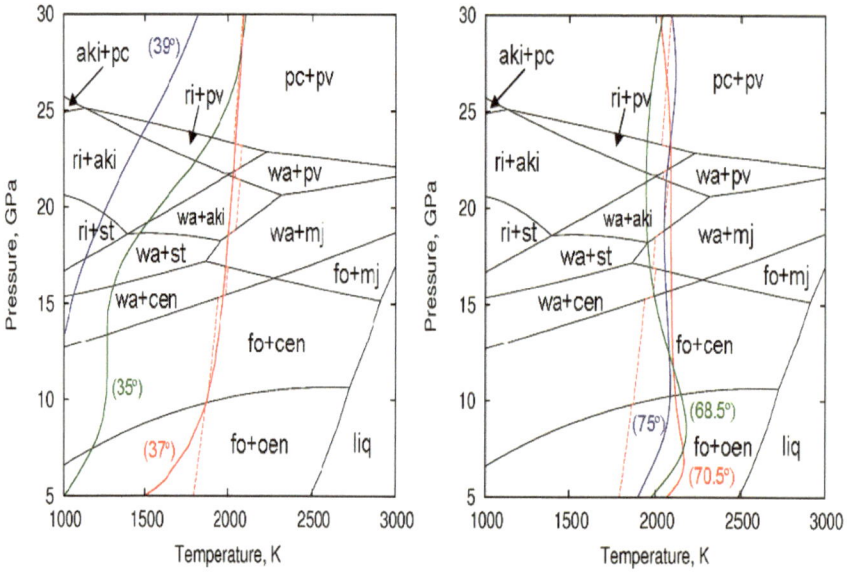

FIG. 8.

Phase diagram of Mg_2SiO_4 and $MgSiO_3$ mixtures. Left: curves labeled (35°), (37°) and (39°) are radial geotherms with the same labels as in Figure 7. Right: curves labeled (68.5°), (70.5°) and (75°) are radial geotherms with the same labels as in Figure 9. The dashed curve is a calculated isentrope for a mixture of 0.56 mol Mg_2SiO_4 and 0.44 mol $MgSiO_3$ and a foot temperature of 1700 K

Figure 9 shows the transition zone above a hot upwelling of material in the mantle. Figure 9c shows that the transition zone is less disturbed as in Figure 7. That also applies to the courses of the calculated geotherms illustrated in the right frame of Figure 8, which show smaller excursions in temperature than those plotted in the left frame of Figure 8.

Heterogeneity is observed as a variation in the thickness of the phase fields. This is clearly visible in the phase field of wadsleyite+akimotoite. The right frame of Figure 8 illustrates that depending on the course of the *geotherm* a different thickness of this field is probed. Figure 9d shows that shear wave velocity changes are moderate compare to those in Figure 7d. The wadsleyite-ringwoodite transition responsible for the 520 seismic discontinuity appears in two different settings. Firstly, it appears in colder regions as the transition between the phase fields wadsleyite+akimotoite and ringwoodite+akimotoite and it is more prominently present in cold downwellings such as illustrated in Figure 9c. Secondly, it appears in warmer areas as the transition between wadsleyite+perovskite and ringwoodite+perovskite. This behaviour might be related to lateral variations in seismic observations of the 520 km discontinuity observed by Deuss (2009) and Cao et al. (2010).

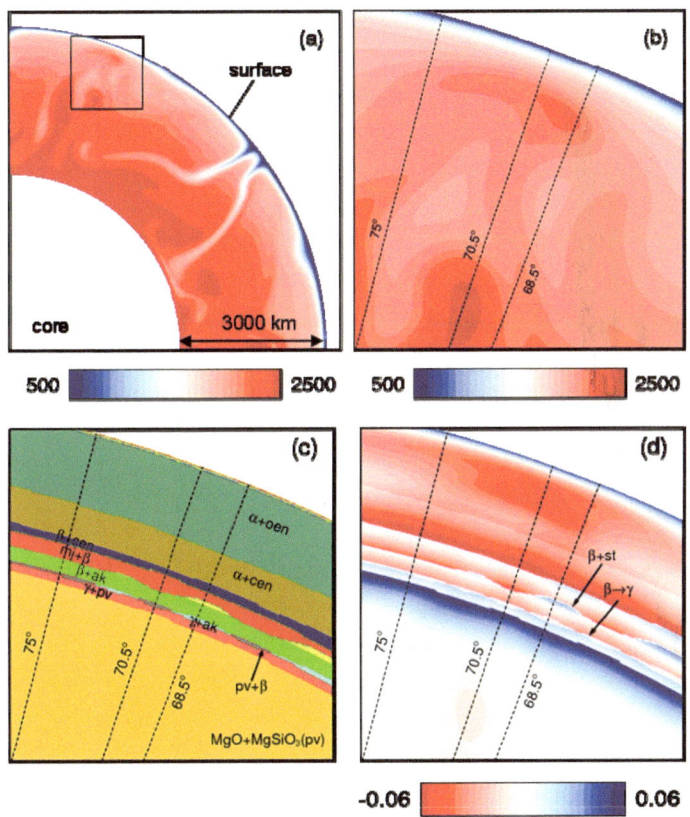

FIG. 9.

Snapshot of a mantle convection calculation after about 1 Gyr. (a) temperature snapshot with a box marking a hot upwelling. (b) Zoom-in of the box given in (a). (c) Phase field partitioning. (d) Perturbation of shear velocity as defined in Figure 7. In these frames α denotes forsterite, β wadsleyite and γ ringwoodite

§(308)

Elucidating the seismic velocity structure in the transition zone is a hot topic of recent research as is apparent from the works of Cao et al. (2011,2010), Deuss (2009), Deuss et al. (2006), Lebedev et al. (2002), Deuss and Woodhouse (2001) and Shearer et al. (1999). In studies such as those of Cao et al., and Deuss et al. reflectivity-time traces were obtained as a measure of the radial distribution of velocity contrast by investigating SS precursors originating from underside wave reflections of the transition zone. In Jacobs and van den Berg (2011) we derived reflectivity-time traces from our mantle convection modeling in an analogous way as in seismic studies. Details of these calculations fall outside the scope of this text and can be found in that study. An important outcome of these simulations is that we came to the conclusion that the major phase transitions in the $MgO-SiO_2$ system are visible in these reflectivity traces derived from the shear velocity distribution and that these are in principle observable by seismic studies. It should be stressed however that our calculations were carried out using equilibrium thermodynamics and apart from the simplified mineralogy the effects of phase transition kinetics were neglected. However the coupling between thermodynamic models and geodynamic models as demonstrated above will be helpful in the interpretation of seismic results.

The thermodynamic formalism based on lattice vibrations has been applied to the system $MgO-SiO_2$ and it has been used in mantle convection modeling to elucidate the structure of the transition zone. The convection simulations indicate a heterogeneous phase distribution in the transition zone accompanied with a heterogeneous shear wave velocity structure.

EXERCISES

1. *the phase diagram of Mg$_2$SiO$_4$ from a zeroth point of view*

As some kind of zeroth order approximation, the Gibbs energy of a given form of a given substance, at given T and high pressure P, can be represented by the formula

$$G(T,P) = H^o - T S^o + P V^o,$$

in which the superscript ^{0} stands for the standard conditions of 298 K and 1 bar. As an example, for forsterite, using the data in Table 1, the formula reads

$$G(T,P) = \{- 2176900 - 94.98\ (T/\text{K}) + 43658\ (P/\text{GPa})\}\ \text{J}\,\text{mol}^{-1}.$$

- In this approximation and using the data in Table 1, construct the PT phase diagram for the composition Mg$_2$SiO$_{4,}$ for $1000 \leq (T/\text{K}) \leq 2600$; and $10 \leq (P/\text{GPa}) \leq 30$.
- Calculate the coordinates of the invariant point.

2. *partial derivatives of volume in terms of bulk modulus and its derivative*

In terms of the bulk modulus, the first partial derivative of volume with respect to pressure is given by $(\partial V/\partial P)_T = - (V/K)$.

- Set up the equation for the second partial derivative of volume with respect to pressure in terms of bulk modulus and its derivative with respect to pressure at constant temperature.

3. *an exercise on extrapolation*

The data presented in Table 1 allow the calculation of Gibbs energy, enthalpy, entropy and volume in a modest range of temperature and pressure around 298 K and 1 bar.

The formula for the calculation / extrapolation of the Gibbs energy at / to temperature T from data valid for the reference temperature Θ is (from Equation 109:13):

$$G(T) = H(\Theta) - T S(\Theta) - C_P(\Theta)\ [T \ln(T/\Theta) - T + \Theta]\ . \tag{A}$$

For the extrapolation of volume to pressure P from data valid for the reference pressure Π the information in Table 1 suggests the use of a *Taylor series*:

$$V(P) = V(\Pi) + V'(\Pi)(P - \Pi) + \tfrac{1}{2} V''(\Pi)(P - \Pi)^2. \tag{B}$$

V' and V'', which are the first and the second partial differential coefficients of volume with respect to pressure, are related to the bulk modulus (K) and its partial derivative with respect to pressure (K''); see Exc 2.

From Equation (A) and through Equation (B), the formula for the calculation/extrapolation of the Gibbs energy at $(T = T$ and $P = P)$ from data valid for $(T = \Theta$ and $P = \Pi)$ is obtained as

$$G(T,P) = H(\Theta) - T S(\Theta) - C_P(\Theta) [T \ln(T/\Theta) - T + \Theta]$$

$$+ V(T)(P - \Pi) + \tfrac{1}{2} V'(T)(P - \Pi)^2 + 1/6\, V''(T)(P - \Pi)^3. \tag{C}$$

- Use the data in Table 1 to calculate the Gibbs energy of forsterite for $(T = 308$ K and $P = 250$ bar). To appreciate the influence of the individual terms in Equation (C), calculate/give the value of each of them in $J\,mol^{-1}$ to three decimal places. How many decimal places are justified, when it comes to the *difference* between the calculated Gibbs energy and the one at 298 K and 1 bar?

It goes without saying that the above equations are inadequate for extrapolations over large distances. The following task may be useful to get an idea of the implications. It's about forsterite's molar volume and its representation by Equation (B). Forsterite's molar volume at 298 K as a function of pressure is shown in Figure 306:18. At 298 K, and for $\Pi = 0$ GPa, the Taylor series, Equation (B), takes the numerical form of

$$V(P) = \{43.658 - 0.3424\,(P/GPa) + 7.802 \times 10^{-3}\,(P/GPa)^2\}\ cm^3\,mol^{-1}.$$

- From Figure 306:18 read the value of forsterite's molar volume for $5 \leq P/GPa \leq 35$ in steps of 5 GPa from the computed curve. Next, for each of the seven pressures calculate the molar volume generated by the above numerical equation, hereafter referred to as Equation (D), and determine the sum of the squares of the differences between value read from the figure and value generated. Finally, repeat the task after having modified Equation (D) to Equation (E), which is

$$V(P) = \{43.658 - 0.3424\,(P/GPa) + 3.901 \times 10^{-3}\,(P/GPa)^2\}\ cm^3\,mol^{-1}.$$

4. *the Grüneisen parameter derived from sound velocities*

Equation (9) for the Grüneisen parameter, can be transformed to

$$\gamma = (K / v) \, (\partial v / \partial P)_T \, ;$$

see § 307, Equation (26).

- From the experimental data in Figure 4 for the longitudinal sound velocity in clino-enstatite, make an educated guess of the change of the sound velocity with pressure for the conditions zero K and zero pressure. From the result, derive the value of the Grüneisen parameter at zero K and zero pressure.

5. *a classical thermodynamic relationship*

In Equation (307:66), which is

$$a_i = \left(\frac{\partial \ln v_i}{\partial T} \right)_V = \left(\frac{\partial \ln v_i}{\partial T} \right)_P + \alpha \, K \left(\frac{\partial \ln v_i}{\partial P} \right)_T \, ,$$

the first equality sign pertains to the definition of the *anharmonicity parameter*. The expression after the second equality sign follows from the definition "using a classical thermodynamic relationship".

- Write down the derivation of the expression after the second equality sign.

6. *properties of forsterite at 1800 K and 1 bar*

From the information scattered in § 306, derive, for the forsterite form of Mg$_2$SiO$_4$ at 1800 K and 1 bar, the numerical values of the molar heat capacities C_P and C_V; the cubic expansion coefficient and its temperature coefficient; the adiabatic bulk modulus and its temperature coefficient; and through Equations (306:17 and 11) the numerical value of the molar volume.

Next, calculate the numerical values of the isothermal bulk modulus and its temperature coefficient.

NB. The outcome is needed for some of the exercises that follow.

§(308)

7. *thermal expansivity and intuition*

Intuitively one is inclined to say that the *cubic expansion coefficient* of a material will diminish when the pressure on it is increased. To verify the statement, and to quantify the effect, the following actions are suggested.

- By applying the *cross-differentiation identity* to the mixed second derivatives of volume, derive the equation that relates the pressure derivative of the expansion coefficient to the isothermal bulk modulus and the temperature derivative of the latter.

- For forsterite, at 1800 K and 1 bar, determine the numerical value of $(\partial\alpha/\partial P)_T$, and express the effect in terms of percentage per GPa ($\leftarrow$ Exc 6).

8. *heat capacity at constant pressure under pressure*

As a consequence of the *cross-differentiation identity*, the *change of C_P with pressure* is related to the second derivative of volume with respect to temperature ($\leftarrow$ Exc 107:6). The relationship is

$$(\partial C_P/\partial P)_T = -T\,(\partial^2 V/\partial T^2) \Rightarrow -TV\,[\alpha^2 + (\partial\alpha/\partial T)_P].$$

- Verify the change represented by the sign $\Rightarrow$.

- For forsterite, at 1800 K and 1 bar, determine the numerical value of $(\partial C_P/\partial P)_T$, and express the effect in terms of percentage per GPa ($\leftarrow$ Exc 6).

9. *the PX phase diagram of Mg_2SiO_4 + Fe_2SiO_4*

The phase diagram which is shown is a copy of Figure 306:8. For $T = 1673$ K it represents the phase relationships in the system Mg_2SiO_4 + Fe_2SiO_4. The diagram is an expression of the readiness of the two components to form *mixed crystals* in all proportions.

- Estimate the *metastable transition pressure* for the change from the olivine to the ringwoodite form in Mg_2SiO_4.

- From the *initial slopes* of the two-phase equilibrium curves at the right-hand side of the diagram, calculate the change in molar volume for the olivine to ringwoodite transition in Fe_2SiO_4.

§(308)

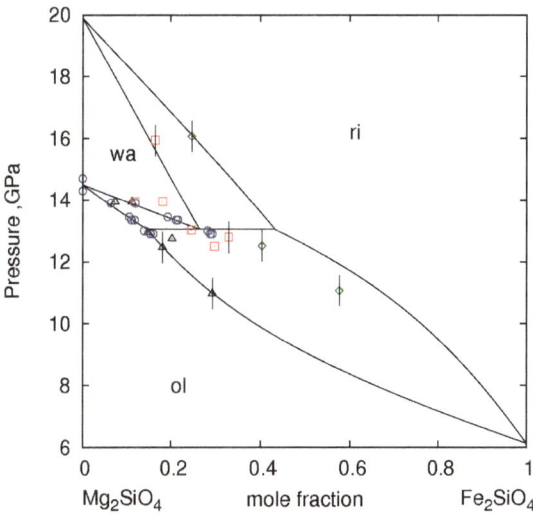

10. *the ideal PX loop*

A fragment of § 210 on the isobaric (*TX*) equilibrium between liquid and vapour, in a system where liquid and vapour are ideal mixtures of the components A and B, is as follows:

"After substitution of the recipes for the potentials into the equilibrium equations, the following relations are obtained.

$$(1 - X^{liq}) = (1 - X^{vap})\exp(\Delta G_A^* / RT) ; \tag{9}$$

$$X^{liq} = X^{vap}\exp(\Delta G_B^* / RT) . \tag{10}$$

By addition of the last two equations, X^{liq} is eliminated and X^{vap} is obtained as a function of temperature:

$$X^{vap}(T) = \frac{1 - \exp(\Delta G_A^* / RT)}{\exp(\Delta G_B^* / RT) - \exp(\Delta G_A^* / RT)} , \tag{11}$$

which is, in other words, the formula for the vaporus."

NB. In this fragment the operator Δ stands for property in vapour minus same property in liquid.

- Replacing the superscript *vap* by the phase symbol β ,and *liq* by α, what other changes are there to be made to the fragment to make it applicable to the isothermal equilibrium between two ideal mixed forms α and β, such that α is the low-pressure and β the high-pressure form?

§(308)

- Calculate the ideal PX diagram at $T = 1673$ K for a system, whose components have the following transition properties: $P_A{}^o = 15.838$ GPa; $\Delta V_A{}^* = -1.513$ cm^{3}mol^{-1}; $P_B{}^o = 6.133$ GPa; and $\Delta V_B{}^* = -2.055$ cm^{3}mol^{-1}. NB. These properties are the same as the real, read optimized properties in the system $\{(1-X)\ Mg(SiO_4)_{0.5}(A) + X\ Fe(SiO_4)_{0.5}(B)\}$. At this stage, ignore the changes of $\Delta V_A{}^*$ and $\Delta V_B{}^*$ with pressure.

For $T = 1673$ K and $P = 12$ GPa, the optimized values of $\Delta G_A{}^*/RT$ and $\Delta G_B{}^*/RT$ in the system $\{(1-X)\ Mg(SiO_4)_{0.5}(A) + X\ Fe(SiO_4)_{0.5}(B)\}$ are 0.436661 and minus 0.791403, respectively.

- Calculate the mole-fraction values generated by Equations (11) and (10).

11. *mixed crystals in the system* $Mg(SiO_4)_{0.5} + Fe(SiO_4)_{0.5}$

Mixed crystals generally have excess properties that are well represented by the $AB\Theta$ model ($\leftarrow$ 304). Their excess enthalpies generally are positive, and that is also true for their excess Gibbs energies and excess entropies.

This exercise is about the excess Gibbs energies of the olivine and ringwoodite mixed crystals in the system $Mg(SiO_4)_{0.5} + Fe(SiO_4)_{0.5}$. The expectation that the two mixed forms have positive excess Gibbs energies is supported by the findings of Exc 10, in the sense that the real two-phase region in the PX diagram is wider than the ideal one ($\rightarrow$ Exc 12). Moreover, because the real one of the two regions is more to the right of the PX diagram than the ideal one, the excess Gibbs energy is larger in the ringwoodite form (β) than it is in the olivine form (α).

For an explorative assessment of the magnitudes of the excess Gibbs energies of the two forms at 1673 K and 12 GPa, the Gibbs functions of the two mixed states are given the form

$$G(X) = (1-X)\ G_A{}^* + X\ G_B{}^* + RT\ \{(1-X)\ \ln(1-X) + X\ \ln X\} + g\ X(1-X),$$

in which the last member is the most simple expression for the excess Gibbs energy.

It is instructive to carry out the assessment in two steps. In the first step the difference between the two parameters of the excess Gibbs energy is determined by means of an *equal-G analysis* ($\leftarrow$ Equations 301:15-16). In the second step the individual parameters g^α and g^β act as the two unknowns in the two equations based on the equality of *chemical potentials* ($\leftarrow$ Equations 304:1-5).

§(308)

From Exc 10, the data at hand are the following: $\Delta G_A{}^*/RT = 0.436661$; $\Delta G_B{}^*/RT = -0.791403$; compositions read from the phase diagram $X^\alpha = 0.22$; $X^\beta = 0.57$; equilibrium compositions if the system were ideal $X^\alpha = 0.227$; $X^\beta = 0.500$.

- Make an educated guess of the mole fraction of the point of intersection of the two Gibbs functions G^α and G^β. Next solve $\Delta g\ (= g\beta - g^\alpha)$ from the equation $(1-X)\Delta G_A{}^* + X\Delta G_B{}^* + \Delta g\,X(1-X) = 0$. Give the result as an integer in kJ mol^{-1}.

- Solve g^α and g^β from the two equations contained in Equation (304:1): $N = N\left[\mu_A^I = \mu_A^{II};\ \mu_B^I = \mu_B^{II}\right]$. Give the result as integers in kJ mol^{-1}.

12. *the usefulness of GX sketches*

The two *GX* diagrams shown here are from the intermezzo in § 205 about the usefulness of *linear contributions*. For the purpose of this exercise, the curves shown are assumed to pertain to ideal mixing.

- First sketch a reproduction of one of the two *GX* diagrams. Next, extend the sketch drawing by two other curves, which, for unchanged properties of the pure components, represent the Gibbs energies of α and β mixtures having positive excess Gibbs energies. Observe that the distance between the points of contact of the common tangent is greater for the non-ideal case than it is for the ideal case.

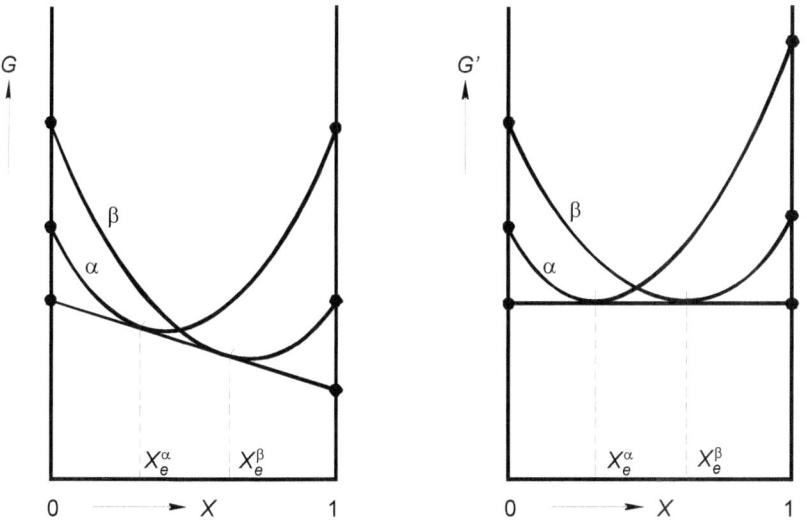

FIG. 2. By the addition of equal linear contributions to the two Gibbs functions, the points of contact can be made to coincide with the minima

§(308)

218

13. *estimating phase behaviour*

The *PT* phase diagram for Fe_2SiO_4, printed below, along with the *PT* phase diagram for Mg_2SiO_4 (Figure 1), and the *TX* phase diagram, printed above (Exc 9) for the combination of the two substances, can be used for making estimates of *TX* and *PX* phase diagrams for the binary $Mg_2SiO_4 + Fe_2SiO_4$.

- Make a sketch drawing of the *TX* phase diagram you expect for $P = 14$ GPa, and up to temperatures where the substances are liquid. Assume that the substances do not dissociate.

- Do the same for the *PX* phase diagram at the normal melting temperature of Mg_2SiO_4; for $0 \leq (P/GPa) \leq 15$.

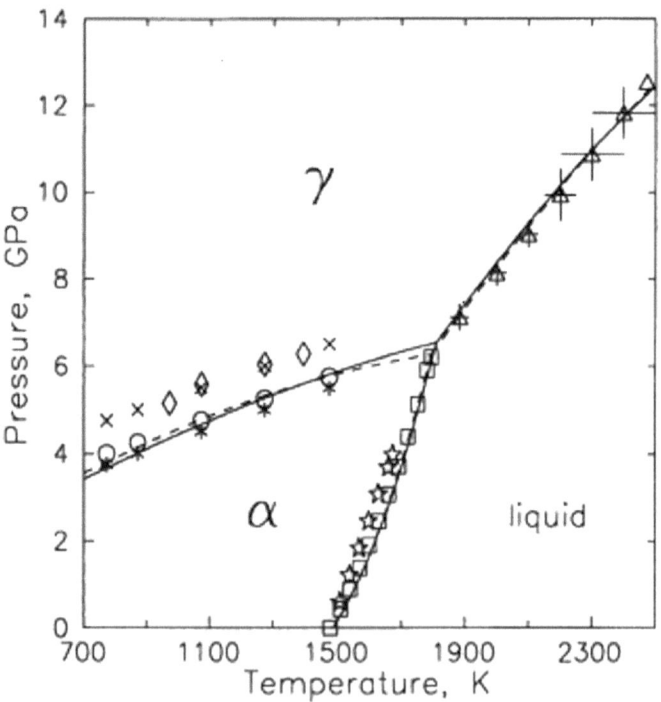

SOLUTIONS OF EXERCISES

301   *central role of Gibbs energy*

301:1   *characteristic*

Numerical values in SI units: m.p = 2131.3; ΔS = 7.95; ΔH = 16943;
ΔC_P = −11.47.

301:2   *a coexistence problem*

1) f = M[T,P] − N[2 conditions] = 0 ; this excludes the first option;
2) by eliminating ln$P$ from the two equilibrium equations in the ideal-gas approximation, the invariant point should follow from [60000 + 16 (T/K) = 0]; this excludes the second option.

301:3   *trigonometric excess Gibbs energy – spinodal and binodal*

$RT_{spin}(X)$ = (4000 J mol^{-1})$\pi^2X(1-X)$sin$2\pi X$;
limits at zero K are X = 0 (obviously) and X = 0.75;
at 0.75 T_c the binodal mole fractions are about 0.11 and 0.52

301:4   *from experimental equal-G curve to phase diagram*

$\Delta\Omega$ = 1342 J mol^{-1}; X^{sol} = 0.635; X^{liq} = 0.435;
see also van der Linde et al. 2002.

301:5   *an elliptical stability field*

The property ΔG^*_A , which invariably is negative, passes through a maximum, when plotted against temperature. The maximum is reached for a temperature which is about 'in the middle of the ellipse'. The property $\Delta\Omega$ is positive.

301:6   *a metastable form stabilized by mixing*

The stable diagram has three three-phase equilibria:
α_I + liquid + β at left; eutectic;
liquid + β + α_{II} at right; peritectic;
α_I + β + α_{II} (below the metastable eutectic α_I + liquid + α_{II}); eutectoid.

301:7   *the equal-G curve and the region of demixing*

For the left-hand case, where the EGC has a minimum, the stable phase diagram has a eutectic three-phase equilibrium. The eutectic three-phase line is just above the EGC minimum and below the intersection of EGC and BIN.
For the right-hand case, the EGC having a maximum and a minimum, the stable phase diagram has a peritectic three-phase equilibrium. The three-phase line is below the left-hand intersection of EGC and BIN.

301:8 *crossed isodimorphism*

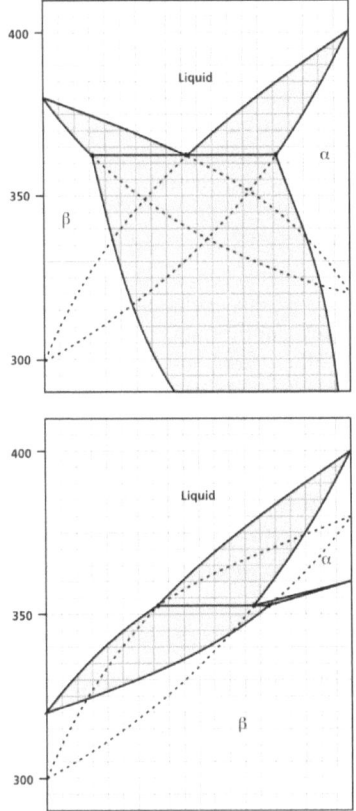

301:9 *activities and Gibbs-Duhem*

For this case, the Gibbs-Duhem equation comes down to
$X_A \, d(\ln a_A) + X_B \, d(\ln a_B) = 0$,
from which it follows that, when one of the activities decreases, the other has to increase; in other words, if one of the activities has a minimum for given X, the other must have a maximum for the same X. The G curve is not convex over the whole X range – typically of a miscibility gap. See also Oonk (1981) pp 94-95.

301:10 *statements about solubilities*

Yes for the first statement, provided that the solubility is expressed in mole fraction. A metastable form has a higher solubility than the stable form; the solubility is dependent of the choice of solvent; the ratio, indeed, is independent of that choice.

301:11   *a different analysis of the region of demixing in AgCl + NaCl*

From the special condition two equations are obtained, which are
(1) $RT' \ln[X'/(1-X')] + g_I(T=T')(1-2X') = 0$
(2) $RT' \ln[Y'/(1-Y')] + g_I(T=T')(1-2Y') = 0$
Equation (A) is (1) minus (2).
The nine function values, Equation (A), yield the result
$g_I = [9100 - 2.8(T/K)]$ J mol^{-1}.
The 18 function values generated by (1) and (2) yield virtually the same result.

301:12   *partial pressure versus concentration*

$[A] = 100 \, X_A \cdot P/ \, RT = 100 \, P_A / RT$
Using the symbol K_p for the equilibrium constant in partial pressures, the equations become
$\ln K_c = \ln K_p + 2 \ln(RT/100) = -\Delta G^\circ/ RT + 2 \ln(RT/100)$
$\mathrm{d} \ln K_c / \mathrm{d}T = \Delta H^\circ/ (RT)^2 + 2/T$

301:13   *reminiscence of James Boswell*

In the state where there are equal amounts of the three substances and the pressure indicated by the manometer is 3 bar, the partial pressures of all three substances are 1 bar. Accordingly, the equilibrium constant in terms of partial pressures (K_p) has the value of 1 bar^{-1}. The equilibrium temperature, as a result, is the solution of
$\Delta G^\circ = \Delta H^\circ - T\Delta S^\circ = -195274 + 185.48T = 0$.
The solution is $T = 1053$ K. The numerical value of the equilibrium constant in terms of concentration is $K_c = 87.55$ dm^3 mol^{-1}.
The equilibrium state at $T = 1103$ K has mole fractions $X_{SO3} = 0.25$; $X_{SO2} = 0.39$; $X_{O2} = 0.36$; and pressure $P = 3.25$ bar.
The volume of the vessel is $V = 87.55$ dm^3.
The values of the equilibrium constants at 1000 K are $K_p = 3.2477$ bar^{-1} and $K_c = 270.03$ dm^3 mol^{-1}. After the addition of one mole of O_2, the equilibrium state has $P = 3.61$ bar, $n_{SO3} = 1.404$; $n_{SO2} = 0.596$; $n_{O2} = 1.798$.

301:14   *vapour pressures over liquid mixtures of methanol and methyl acetate*

The maximum is at about $X = 0.67$ and $P = 73.22$ kPa; these values involve the g_I value of 2842 J mol^{-1} shown in the table below.
The quadruplets that follow give (1) the experimental liquid composition; (2) component A's partial excess Gibbs energy; (3) component B's partial excess Gibbs energy; (4) the integral excess Gibbs energy:
[0.0392; 6; 2594; 107] [0.0805; 34; 2342; 220] [0.1184; 60; 2152; 308] [0.1715; 107; 1883; 411] [0.2330; 180; 1612; 514] [0.3810; 438; 1054; 673] [0.4830; 686; 746; 715] [0.6200; 1096; 410; 671] [0.7060; 1414; 247; 590] [0.9240; 2392; 24; 204].

The ensemble of quadruplets involves the g_1 and g_2 values shown in the table below.
The quadruplets that follow give (1) the experimental liquid composition; (2) the value of Q; (3) X_{EGC}; and (4) the value of the excess Gibbs energy:
[0.0392; 2.902; 0.0795;211] [0.085; 2.048; 0.1486; 372] [0.1184; 1.572; 0.2039; 476] [0.1715; 1.145; 0.2704; 570] [0.2330; 0.847; 0.3369; 642] [0.3810; 0.520; 0.4662; 710] [0.4830; 0.451; 0.5411; 711] [0.6200; 0.477; 0.6369; 659] [0.7070; 0.566; 0.7000; 597] [0.9240; 1.278; 0.9029; 254].
The g_1 and g_2 are shown in the table below.

	g_1 / J mol^{-1}	g_2 / J mol^{-1}	Δ_P / kPa	Δ_X
maximum	2842		0.106	0.0040
partial props	2879	25	0.300	0.0055
EGC fraction	2877	34	0.299	0.0055

One can observe that the results of the second and third rounds of calculation, although they resemble one another, do not show an improvement with respect of the outcome of the simplistic calculation in the first round.

In more detail, and concentrating on pressure: in all three cases there is a systematic deviation between calculated and experimental pressures. For the last two cases, nine of the ten calculated pressures are higher than the experimental ones.

The origin of the lack of agreement between experimental and calculated equilibrium states is the subject of the two following exercises. Is it a question of methodology, or rather a lack of compatibility between thermodynamic description and the nature of the experimental data?

See also Figurski et al. (1997)

301:15 *a consistency test*

The rule is not fully satisfied: the area at the negative side is about 1.06 times the area at the positive side.

301:16 *assessment of methodology*

	g_1 / J mol^{-1}	g_2 / J mol^{-1}	Δ_P / hPa	Δ_X
maximum	2728		2.00	0.0028
partial props	2750	100	0.00	0.0000
EGC fraction	2779	150	1.55	0.0016

In the first task the calculated pressures at the A side of the system are somewhat too low, and at the B side somewhat too high. This is an expression of the need to introduce a positive g_2. The partial-properties method gives a one-to-one reproduction of the data set. The result of the EGC-fraction method corresponds to an overestimation of the asymmetry of the excess property. Clearly, it is difficult to locate the exact position of the EGC. Notwithstanding these remarks, the G^E values provided by the three approaches are the same within a few percent.

In order to fully comply with the thermodynamic description, the experimental two-phase region, Exc 14, should be somewhat broader. As a matter of fact, the phase diagram for the same system at the same temperature measured by V.Bekarek (1968) Collect. Czech Chem. Commun. 33:2608 (figures 546; 547; 549 in Ohe 1989) has a broader two-phase region than the one in Exc 14.

301:17 *dimolybdenum nitride*

The Gibbs energy change of the formation reaction changes from a negative (spontaneous formation at 298.15 K; 1 bar)) to a positive value (spontaneous decomposition at 1400 K; 1 bar). If the property were to change with temperature in a linear manner, it would become zero at 1001 K.

The values of $\Delta_f S^o$: -89.81 J K^{-1}mol^{-1} at 298.15 K and -66.43 J K^{-1}mol^{-1} at 1400 K. The heat capacity property is

$\Delta_f C_p{}^o = a + bT = (0.548 + 0.020438\ T / K)$ J K^{-1}mol^{-1}. The Gibbs energy of formation changes sign between 961 K and 962 K.

Ignoring deviation from ideal-gas behaviour, and assuming that the vessel does not expand, the pressure at 1400 K should be at least 211 bar, which comes down to 45 bar at 298.15 K.

301:18 *the Helmholtz energy in its capacity of equilibrium arbiter*

A	PV	G	A
0.20	3991	-8504	-12495
0.25	4157	-8498	-12655
0.30	4324	-8413	-12737
0.35	4490	-8261	-12751
0.40	4656	-8049	-12705
0.45	4822	-7781	-12603
0.50	4989	-7460	-12499

This time, i.e. at constant T and V, the Helmholtz energy is the equilibrium arbiter (←Exc 108:11)

301:19 *the Wilson equation*

$\mu^E_A / RT = -\ln(X_A + \Lambda_{AB} X_B) + X_B \{\Lambda_{AB} / (X_A + \Lambda_{AB} X_B) - \Lambda_{BA} / (X_B + \Lambda_{BA} X_A)\}$

$\mu^E_B / RT = -\ln(X_B + \Lambda_{BA} X_A) - X_A \{\Lambda_{AB} / (X_A + \Lambda_{AB} X_B) - \Lambda_{BA} / (X_B + \Lambda_{BA} X_A)\}$

For $X = 0$, the expression of the derivative of μ^E_A / RT with respect to X takes the form

$(\Lambda_{AB} - 1) + \Lambda_{AB} - \Lambda_{BA}^2 / \Lambda_{BA}^2 = 0$,

and it means that Raoult's law, read, the condition $(\partial \mu^E_A / \partial X_B)_{X_B \rightarrow 0} = 0$, is satisfied. The condition is not satisfied if Λ_{BA} is put equal to zero: the Wilson parameters should not be put equal to zero.

Observe that if both of the two parameters were put equal to zero, the excess Gibbs energy would be equal to minus the ideal Gibbs energy of mixing – and the complete Gibbs energy of mixing would be zero for each value of X.

Unlike e.g. the expression

$G^E / RT = X(1-X) \{g_1 / RT + g_2 / RT(1 - 2X)\}$

the Wilson equation is not applicable to systems with limited miscibility. The second derivative of the complete Gibbs energy of mixing is positive for every X, no matter the values of Λ_{AB} and Λ_{BA}.

$\Lambda_{BA\downarrow}$

$\Lambda_{AB\rightarrow}$	0.01	0.2	0.4	0.6	0.8	1.0	1.2	1.4	1.6	1.8	2.0
0.01	**0.683**	*0.15*	*0.27*	*0.38*	*0.49*	*0.59*	*0.70*	*0.81*	*0.95*	*1.11*	*1.34*
0.2	0.597	**0.511**	*0.11*	*0.20*	*0.29*	*0.37*	*0.45*	*0.55*	*0.66*	*0.85*	*1.26*
0.4	0.520	0.434	**0.357**	*0.08*	*0.16*	*0.22*	*0.30*	*0.39*	*0.54*	*1.90*	*-0.01*
0.6	0.453	0.367	0.290	**0.223**	*0.07*	*0.13*	*0.19*	*0.32*	*0.13*	*0.24*	*0.28*
0.8	0.394	0.308	0.231	0.164	**0.105**	*0.06*	*0.14*	*0.14*	*0.18*	*0.22*	*0.25*
1.0	0.342	0.255	0.178	0.112	0.053	**0.000**	*0.05*	*0.09*	*0.13*	*0.19*	*0.22*
1.2	0.294	0.208	0.131	0.064	0.005	−.048	**−.095**	*0.04*	*0.08*	*0.11*	*0.15*
1.4	0.250	0.164	0.087	0.020	−.038	−.091	−.139	**−.182**	*0.04*	*0.07*	*0.10*
1.6	0.210	0.124	0.047	−.020	−.079	−.131	−.179	−.222	**−.262**	*0.03*	*0.06*
1.8	0.173	0.087	0.010	−.057	−.116	−.168	−.216	−.259	−.299	**−.336**	*0.03*
2.0	0.139	0.053	−.024	−.091	−.150	−.203	−.250	−.294	−.334	−.371	**−.405**

For given values of its parameters, the two most important characteristics of the Wilson equation are its numerical value at the equimolar composition and its degree of asymmetry with respect to that composition. These characteristics are displayed in the table on p 224 for a number of parameter pairs. Along the diagonal of the table with the bold numbers the values of the two parameters are the same, and it means that the Wilson function is symmetrical. Below the table's diagonal, function values are given for the equimolar composition; these values do not change if the values of the two parameters are interchanged. As regards the asymmetry of the function, the numbers in italics above the diagonal represent (function value for $X = 0.25$ – function value for $X = 0.75$) / function value for $X = 0.75$. The same numbers are obtained when the Wilson parameters are interchanged, and at the same time component A (B) is made to change into component B(A).

As regards the excess Gibbs energy of liquid mixtures of water (A) and methanol (B), almost quantitative agreement is obtained for $\Lambda_{AB} = 1.406$ and $\Lambda_{BA} = 0.257$.

301:20   *Marius Ramirez's view from the arc*

The arc has $b = 0.000449$ K^{-1}; $h = 0.0366$; and $T_{max} = 317$ K. Difference in heat capacity amounts to -120 $J K^{-1} mol^{-1}$. Heat of vaporization at 298.15 K is 80.6 $kJ mol^{-1}$.

301:21   *the Rackett equation*

$\alpha = -(2/7)(\ln A_1)[1 - (T/T_c)]^{-(5/7)}/T_c$

Obviously, and also in view of the validity range, the information is on liquids under their own equilibrium vapour pressure. The influence of pressure, therefore, is hidden in the equation. Strictly speaking,

$(1/V)dV/dT = \alpha - \kappa \, dP/dT$,

where the symbol κ is for isothermal compressibility, and dP/dT for the slope of the liquid+vapour equilibrium curve.

$V = 89.416$ $cm^3 mol^{-1}$; $\alpha = 1.14 \times 10^{-3}$ K^{-1}; $P_c \approx RT_c/A_2 = 4.90 \times 10^6$ Pa = 49 bar.

301:22   *the Simon equation*

	A / bar	b	Δ_P / bar	Δ_T / K
Babb	4266	4.44	34	0.26
footsteps	4048	4.60	38	0.31

NB. In the original paper (Simon and Glatzel 1929) the relation between T and P was given as $\log(a + p) = c \cdot \log T + b$; and three data pairs were used to evaluate the values of the system-dependent constants a, b, and c.

For sodium molybdate the change in molar volume on melting is 6.1 $cm^3 mol^{-1}$.

301:23 *an exploratory calculation*

e.g. for kyanite:

$G^{III}(T,P) = [-2443881 - 84.467 \{(T / K) - 298.15\} + 44090 (P / GPa)]$ J·mol⁻¹.

e.g. for (I + II):

$(T / K) = 1085 - 882 (P / GPa)$.

Triple point coordinates: $P = 0.405$ GPa; $T = 728$ K; the coordinates given by Tonkov (1992) are 0.55 GPa and 893 K (taken from Richardson et al 1969) NB. One can read in Tonkov that the experimental triple point, reported in seven independent investigations, varies from about 0.2 GPa; 693 K to 0.9 GPa; 663 K !
In the phase diagram (pressure axis horizontal and temperature axis vertical) the sillimanite field is at the upper side; the andalusite field is at the left-hand side; and the kyanite field at the right-hand side.

301:24 *fluid carbon dioxide*

$\Delta G = 21.5$ kJ·mol⁻¹ $= RT \ln 176$; *fugacity coefficient* $= f / P = 0.88$; Angus gives $f / P = 0.88210$.

301:25 *compounds in the role of component*

(1) $\{(1 - X)$ mole of $A_2BC + X$ mole of $AB_2C\}$: A's mole fraction in the liquid mixture $x(A) = \{2(1 - X) + X\}/4 = (2 - X)/4$;
$x(B) = (1 + X)/4$; $x(C) = 1/4$;
For left-hand liquidus: $A_2BC = 2A + B + C$;
$K = \{x(A)\}^2 x(B)x(C) = (2 - X)^2 (1 + X)/64$; $K_o = 4/64$.
$R \ln (K/K_o) = R \ln [(2 - X)^2 (1 + X)/4] = - \Delta H_{A2BC} [(1/T) - (1/T_o)]$.
For reasons of symmetry the eutectic point is at $X = 0.5$. Apparently it is assumed that heat capacities may be ignored: the calculated eutectic temperature is 1479.1 K. At $X = 0$ and $X = 1$ the two liquidi have a horizontal start.

(2)   For the second task, $x(A) = (3 - 2X)/(4 - 2X)$; $x(B) = 1/(4 - 2X)$;
Left-hand liquidus :
$R \ln [256(3 - 2X)^3/27(4 - 2X)^4] = - \Delta H_{A3B} [(1/T) - (1/T_{oA3B})]$;
right-hand liquidus :
$R \ln [4(3 - 2X)/(4 - 2X)^2] = - \Delta H_{A3B} [(1/T) - (1/T_{oAB})]$.
Eutectic point at $X = 0.812$; $T = 1468.9$ K.
Again horizontal start of liquidi.

(3)   By the change in melting point to 1400K, AB has become an *incongruently melting compound*. And because it is not so wise to take an incongruently melting end-member, it is recommendable to change the definition of the system; e.g. to $\{(1-X) A + X B\}$.

The liquidus for the compound A_3B is given by

$R \ln [(256/27)(1 - X)^3 X] = - \Delta H_{A3B} [(1/T) - (1/T_{oA3B})]$;

and the liquidus for AB by

$R \ln [4(1 - X) X] = - \Delta H_{AB} [(1/T) - (1/T_{oAB})]$.

Crystallization starts at 1437.5 K: A_3B is formed. At the three-phase equilibrium temperature, about 1394 K, the last part of the liquid 'reacts' with the last part of the precipitated A_3B to yield AB (that is to say, if during crystallization the precipitated solid is not removed).

301:26 *Professor Geus's nickel coin*

The calculated pressure, for $\alpha = 0$, is 548 Torr; in line with Horstmann's data.

$f = M [T, P, \alpha] - N [\mu_{NH4Cl} = \mu_{NH3} + \mu_{HCl} ; \mu_{NH3} = 0.5 \mu_{N2} + 1.5 \mu_{H2}] = 3 - 2 = 1$.

$X_{HCl} = 1 / (2 + \alpha)$; $X_{NH3} = (1 - \alpha) / (2 + \alpha)$; $X_{N2} = 0.5 \alpha / (2 + \alpha)$; $X_{H2} = 1.5 \alpha / (2 + \alpha)$.

At 600 K the value of $\alpha = 0.9332$; and $P = 4.14$ bar.

NB ($\leftarrow$Exc 007:4; Exc 111:7)

301:27 *montroydite and dephlogisticated air*

Two changes and their corresponding equilibria make their appearance. First, the change from mercuric oxide to gaseous mercury and oxygen; and second, the change from liquid mercury to gaseous mercury. The two equilibrium conditions are:

$$\mu_{HgO} = \mu_{Hg}^{vap} + 0.5\mu_{O2}^{vap}, \tag{1}$$

and

$$\mu_{Hg}^{liq} = \mu_{Hg}^{vap} \quad , \text{respectively.} \tag{2}$$

After substitution of the function recipes for the chemical potentials, and with allowance for the approximations made, Equation (1) reads:

$$G^*_{HgO} (T) = G^o_{Hg} (T) + RT \ln (X_{Hg} P) + 0.5 [G^o_{O2} (T) + RT \ln (X_{O2} P)],$$

which is equivalent to

$$\Delta_f G^o_{HgO} (T) - RT \ln (X_{Hg} P) - 0.5 \, RT \ln (X_{O2} P) = 0 . \tag{1a}$$

Analogously, Equation (2) becomes

$$\Delta_{liq}^{vap} G^o_{Hg} (T) - RT \ln (X_{Hg} P) = 0. \tag{2a}$$

From the Gibbs energies of formation of HgO: mercury's boiling point is about 630 K. Next, from the heat-of-formation data it follows that the heat of vaporization is about 59 kJ·mol^{-1}. The entropy of vaporization at standard pressure is 59000 J·mol^{-1}/ (630 K) $\approx$ 94 J·K^{-1}·mol^{-1} (which is 11.3 times the gas constant; NB, this value complies well with Trouton's rule ($\leftarrow$004)). As a result, the standard change in Equation (2a) is approximated by

$$\Delta_{liq}^{vap} G^o_{Hg} (T) = \{59000 - 94 \, (T/K)\} \, J·mol^{-1}. \tag{2b}$$

From the formation data it follows that, for temperatures above 630 K, the standard change in Equation (1a) is approximated by

$$\Delta_f G^o_{HgO}(T) = \{-144000 + 192\,(T/K)\}\ \text{J mol}^{-1}. \tag{1b}$$

In the test tube the air in the tube, having a pressure of 1 atm, say 1 bar, has to be pushed away (the situation corresponds to a cylinder+piston experiment with an external pressure of 1 bar). The temperature at which this happens is the solution of Equation (1a) for $P = 1$ bar; $X_{Hg} = 2/3$; and $X_{O2} = 1/3$. The solution is $T = 720$ K.

For the equilibrium (solid HgO + liquid Hg + gaseous mixture of Hg and O_2) the system formulation is

$$f = \text{M}\,[T, P, X_{O2}] - \text{N}\,[\text{Equation (1); Equation (2)}] = 3 - 2 = 1.$$

The system has one degree of freedom (is monovariant). NB, the number of components in the sense of the Phase Rule is two: with HgO and Hg all three phases can be prepared in arbitrary amounts (←007).

For $T = 700$ K and using Equations (1b) and (2b), the solution - of this case of two equations with two unknowns - is $P = 3.2$ bar; $X_{O2} = 0.0011$.

301:28 *the adiabatic bulk modulus*

$dV = \alpha V\, dT - (V/K)\, dP$
$(\partial V/\partial P)_S = \alpha V\,(\partial T/\partial P)_S - (V/K)$ (A)

$dS = (\partial S/\partial T)_P\, dT + (\partial S/\partial P)_T\, dP = (C_P/T)\, dT - (\partial V/\partial T)_P\, dP =$
$(C_P/T)\, dT - \alpha V\, dP$ (B)

from (B): $(\partial T/\partial P)_S = \alpha TV/C_P$; to be inserted in (A), and subsequently combined with $C_p - C_V = \alpha^2 K\, V\, T$

302   *slopes of curves in phase diagrams*

302:1   *the isothermal counterparts*

The total differential of $\Delta_e\mu_A$, in terms of the variables X^α, X^β, and P, leads to

$$X^\alpha\,(\partial^2 G^\alpha/\partial X^2)\, dX^\alpha - X^\beta\,(\partial^2 G^\beta/\partial X^2)\, dX^\beta + \Delta_e V_A\, dP;$$

and the total differential of $\Delta_e\mu_B$ to

$$-(1-X^\alpha)\,(\partial^2 G^\alpha/\partial X^2)\, dX^\alpha + (1-X^\beta)\,(\partial^2 G^\beta/\partial X^2)\, dX^\beta + \Delta_e V_B\, dP,$$

realizing that the chemical potentials of A and B in α (β) have nothing to do with X^β (X^α), and making use of the relationships between molar Gibbs energy (the integral molar property) and chemical potentials (the two partial molar properties) ←203.

After substitution of the two expressions in the two equations in N', Equation (6), the two solutions, i.e. the 'partial forms' of Equations (15a, and b), are readily found.

302:2 *a variant of the linear contribution property*

Suppose that the entropy of pure component B in form α deviates from its absolute entropy by the amount of C. If so, and because of the fact that ΔS^{*}_{B} is an experimental reality, the entropy of B in form β will also deviate by that amount.
The entropy of α as a function of X, because of $S^{\alpha}(X) = (1-X) S^{*\alpha}_{A} + X S^{*\alpha}_{B}$, etc., will deviate by the amount of CX; same for $S^{\beta}(X)$.
However, The Equations (13a, and b) remain unaffected: the second derivative in the denominator part does not change; and in the numerator parts the changes cancel one another. For the case of Equation (13a): $\partial S^{\alpha}/\partial X$ changes into $\partial S^{\alpha}/\partial X + C$, and $\Delta_{e}S/\Delta_{e}X$ into $[S^{\beta} + CX^{\beta} - (S^{\alpha} + CX^{\alpha})] / (X^{\beta} - X^{\alpha}) = \Delta_{e}S/\Delta_{e}X + C$.
Note that this property has been applied in Figure 3 – the entropies of pure A and B in the low-temperature form are on the same level.

302:3 *exceptional circumstances*

In Figure 3 the line through A and C should become steeper than the line through A and B. This is favoured by a small difference in entropy between the forms β and α; by a great difference in composition of the α and β phases; and for X_{e}^{α} approaching zero.

302:4 *negligible solid solubility*

This time the total differential of $\Delta_{e}\mu_{A}$ (see above 302:1) is given by

$$- X^{liq} (\partial^2 G^{liq}/\partial X^2) \, dX^{liq} - \Delta_{e}S_{A} \, dT .$$

For the numerical example:

$$\Delta_{e}S_{A} = \Delta S_{A}^{*} - R \ln X_{A}^{liq} - 4 R (X^{liq})^2 = 9.693 R;$$

$$(\partial^2 G^{liq}/\partial X^2) = RT/X(1-X) + 4000 R (1- T / 500 \text{ K}) = 79.6 R.$$

d $X^{liq} /dT = - 0.2435$ K^{-1}
Very flat liquidus, indicating the proximity of a critical point; see next Exc.

302:5 *a simple eutectic system*

See preceding exc, and § 212.
There is a symmetrical region of demixing, having a lower critical point at 1000 K.
At 1025 K the compositions of the coexistent liquid phases are at 0.37 and 0.63; at 1050 K the compositions are 0.32 ande 0.68; and at 1100K the compositions are 0.26 and 0.74.
The left-hand liquidus is at 980.1 K for $X = 0.5$.
The eutectic point is at $X = 0.934$; $T = 894.6$ K.

302:6 *the metastable region of demixing in copper + cobalt*

From the diagram for $X^{liq} = 0.5$: $dX^{liq}/dT = (50$ K$)^{-1}$; $t = 1390$ °C; $X^{sol} = 0.96$.
The second derivative of the liquid's G : $(\partial^2 G/\partial X^2) = \{RT/X(1-X)\} - 2\omega$, which, for $X=0.5$, and with $\omega = 2RT_{c}$, reduces to $4R (T - T_{c})$.

As for the numerator part of the Van der Waals equation, the entropy diagram, Figure 3, can be made such (see Exc 2) that, at $X=0.5$, $(\partial S^{liq}/\partial X)=0$; accordingly,
$\Delta_e S/R = - LN(0.5) +1.04 + 0.04(1.15-1.04) + LN(0.96) = 1.57$.
As a result $(1/50\ K) = [1.57R\ /\ 0.46]\ /\ 4R\ (T-T_c)$; and $(T-T_c) = 43$ K.
The critical temperature of the metastable ROD is at about 1620 K.

302:7 *'statistics' of stationary points*

Horizontal point of inflexion, at $T = 407.4$ K, for $T^o{}_B = 466.67$ K;
minimum for $200 < (T^o{}_B\ /\ K) < 466.67$ K;
minimum plus maximum for $400 < (T^o{}_B\ /\ K) < 466.67$ K.

302:8 *from van der Waals to van 't Hoff*

For ideal behaviour and $X \rightarrow 0$:
the numerator parts reduce to $\Delta S^*{}_A$;
in the denominator parts the second derivative, which for ideal-mixing behaviour is given by $RT\ /\ X(1-X)$, reduces to $RT\ /\ X$, for α and for β;
after subtraction of the two equalities, the α and β mole fractions disappear at the right-hand side, because of $(a-b)\ /\ (a-b) = 1$.

302:9 *the idealized liquid + vapour equilibrium*

All $\Delta_e V$ quantities are given by $V^{vap} = RT/P$; and the second derivatives by $RT/X(1-X)$. After replacing dP/P by $d\ln P$, the desired result is obtained, making use of the property $(a^2-b^2) = (a+b) \cdot (a-b)$.

302:10 *idealized equilibrium and initial slopes*

The expressions for B's chemical potentials are
$\mu_B{}^{vap} = G_B{}^o + RT \ln X^{vap} + RT \ln P$;
$\mu_B{}^{liq} = G_B{}^* + RT \ln X^{liq} + \mu_B{}^E$.
Realizing 1) that $G_B{}^* = G_B{}^o + RT \ln P_B{}^o$; 2) that, for $X \rightarrow 0$, pressure P reaches $P_A{}^o$; and 3) that B's excess chemical potential, for $X \rightarrow 0$, is given by $\partial G^E\ /\ \partial X$; and after division by RT, the following expression/equation is obtained:

$(\Delta_e \mu_B) = \ln\ (P_A{}^o/\ P_B{}^o) + \ln X^{vap} - \ln X^{liq} - (\partial G^E\ /\ \partial X)_{X \rightarrow 0}\ /RT = 0$
or, $X^{liq} = X^{vap}\ (P_A{}^o/\ P_B{}^o)\ \exp[- (\partial G^E\ /\ \partial X)_{X \rightarrow 0}\ /RT]$.

Next, $dX^{liq}/d\ln P = (dX^{vap}/d\ln P) \times (P_A{}^o/\ P_B{}^o)\ \exp[-(\partial G^E\ /\ \partial X)_{X \rightarrow 0}\ /RT]$,
which, after substitution in $(dX^{liq}\ /d\ln P)_{X \rightarrow 0} = (dX^{vap}\ /d\ln P)_{X \rightarrow 0} -1$, gives the expression for $(dX^{vap}\ /d\ln P)_{X \rightarrow 0}$.

NB Appreciate the parallel between this case and its isobaric counterpart (last part of the section, under the heading *'the equal-G curve'*), and the role of the EGC, which, for the case at hand is given by ($\leftarrow$212)

$$\ln P_{EGC}(X) = (1 - X)\ln P_A^o + X \ln P_B^o + \frac{G^{E\ liq}(X)}{RT} \ .$$

See also Exc 13.

302:11 *the azeotropic equilibrium*

$$G^{vap}(T,P,X) = (1-X)\, G_A^{\,o}\,(T) + X\, G_B^{\,o}\,(T) + RT\,\mathrm{LN}(X) + RT\,\ln P$$

$$G^{liq}(T,X) = (1-X)\, G_A^{\,*}\,(T) + X\, G_B^{\,*}\,(T) + RT\,\mathrm{LN}(X) + \omega\, X(1-X)$$

$\partial F_1 / \partial X = F_2 = 0$ $\partial F_2 / \partial X = 2\omega$

$\partial F_1 / \partial T = -\,\Delta S$ $\partial F_2 / \partial T = -\,\partial \Delta S / \partial X = \Delta S_A^{\,o} - \Delta S_B^{\,o}$

$\partial F_1 / \partial \ln P = RT$ $\partial F_2 / \partial \ln P = 0$

$dF_1 = -\,\Delta S\, dT + RT\, d\ln P$

$dF_2 = 2\omega\, dX - (\partial \Delta S / \partial X)\, dT$

$dT / d\ln P = RT / \Delta S;$

$dX / d\ln P = (RT / \Delta S) \times (\partial \Delta S / \partial X) / 2\omega$

302:12 *the eutectic point under pressure*

$$dX/dP = [\Delta_e S_B \times \Delta_e V_A - \Delta_e S_A \times \Delta_e V_B] / \Delta_e S \times (\partial^2 G^{liq} / \partial X^2);$$

$$dT/dP = \Delta_e V / \Delta_e S.$$

$dX/dP = -\,9.08 \times 10^{-10}\ \mathrm{Pa^{-1}};\ dT/dP = 1.162 \times 10^{-7}\ \mathrm{K\cdot Pa^{-1}}.$

302:13 *initial slopes of liquidus and solidus*

For $X \to 0$ and $T = T^o_A$, the initial slope (E) of the EGC follows from Equation (27). Next, define ε by $\varepsilon = \exp(E \Delta S_A^* / RT^o_A)$. Then the initial slopes (dT/dX) of liquidus *(L)* and solidus (S) are given by $L = (\varepsilon - 1)\, RT^o_A / \Delta S_A^*$, and $S = L / \varepsilon$.
For A = AgCl: $E = 178$ K; $\varepsilon = 1.64$; $L = 230$ K; $S = 140$ K.
For A = NaCl: $E = -290$ K; $\varepsilon = 0.4268$; $L = -195$ K; $S = -457$ K.

Calculated mole fractions:
$T = 800$ K, $X^{liq} = 0.174$, $X^{sol} = 0.398$; $T = 900$ K, 0.386, 0.682; $T = 1000$ K, 0.683, 0.860.
NB. Within a few K, the calculated liquidus is in agreement with the experimental one (Sinistri et al. 1972). The calculated solidus, on the other hand, is, on the average, some 50 K higher than the experimental one.

302:14 *the 'opening angle' in TX phase diagrams*

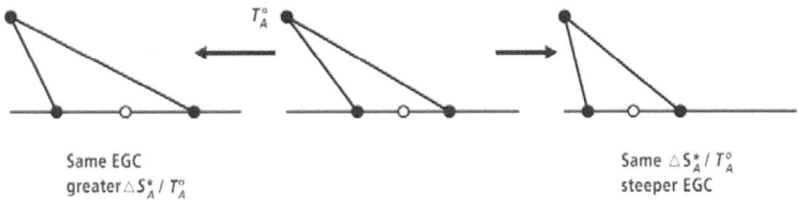

Same EGC   Same $\Delta S_A^* / T^o_A$
greater $\Delta S_A^* / T^o_A$ steeper EGC

302:15 *a rule of thumb for heteroazeotropes*

The slopes of the two vapori both can be calculated by means of Equation (16b). The second derivative in the denominator part is the same for the two. For the left-hand vaporus the X_e^α in the numerator part of Equation (16b) is the composition of the liquid phase L_I. For the right-hand vaporus the X_e^α is the composition of L_{II}. If the assumption is made that the numerator part of (16b) has the same value for $X^\alpha = X^{liql}$ $=X_L$ and $X^\alpha = X^{liqll} =X_R$, then the ratio of the two slopes is exactly given by expression (A).

In experimental reality the numerator part of Equation (16b) is a rather complex function: both $\Delta_e S_A$ and $\Delta_e S_B$ depend not only on X^α, but also on X^β. The circumstances that make that, for liquid+vapour equilibria, the difference in numerical value, between the numerator part for $X^\alpha = X^{liql}$ and the one for $X^\alpha = X^{liqll}$, is relatively small are: 1) the entropies of mixing are small with respect to the entropy of vaporization; 2) the entropy of vaporization does not change much from substance to substance (Trouton's rule, according to which the entropy of vaporization of substances at their normal boiling point is equal to $11R$; →004).

For system (1) the values of the numerator parts of Equation (16b) are:

87 J K^{-1}mol^{-1} for $X^{liq} = 0.2008$ and 101 J K^{-1}mol^{-1} for $X^{liq} = 0.7992$.

For system (2), of which the components have the same entropy of vaporization, those values are:

84 J K^{-1}mol^{-1} for $X^{liq} = 0.1961$ and 87 J K^{-1}mol^{-1} for $X^{liq} = 0.8039$.

The values for the open places in the table are $L_L = -5$ K; $V_L = -15$ K; $V_R = 60$ K; $L_R = 25$ K.

Unlike the two vapori, which emanate from the same composition, the two liquidi at the three-phase equilibrium emanate from different compositions. And it means that for the two liquidi the second derivative - in the denominator part of Equation (16a) - not necessarily has the same value.

For system (3) the numerical values of the second derivatives are:

11.56 kJ mol^{-1} for L_I and 5.04 kJ mol^{-1} for L_{II}. The numerical values of the numerator part of Equation (16a) are 86 J mol^{-1} and 84 J mol^{-1}, respectively.

The systems (1) and (2) have regions of demixing that are symmetrical with respect to the equimolar composition: their liquid phases in equilibrium have equal values for the second derivative of the Gibbs energy with respect to mole fraction.

303   *retrograde equilibrium curves*

303:1 *two similar arcs*

The heat capacity is given by $C_p = 8 h \cdot T_m / b^2$, where T_m is the temperature at which the arc is at its maximum.

$T_m \approx 1420$ K; $h = 0.73$ kJ mol^{-1} ; $b = 481$ K → $C_p \approx 36$ J K^{-1}mol^{-1}.

Numerical values given by Barin (1989) are 33.890 for 1184 K, and 37.907 for 1665 K.

303:3 *the gamma-loop in iron + chromium and its EGC analysis*

From the numerical values given for the high-temperature transition it follows that the low-temperature transition is at 1185 K, where $\Delta S^*_A = -0.0672 R$.

The gamma-loop (the EGC) of the arithmetical example has a maximum at about 1840 K; a minimum at about 1040 K; and at about 1425 K its greatest extension, X_{EGC} = 0.22.

For the initial slope of the EGC, see Equations (302:26, 27).

For the phase diagram, Figure 7, $\Delta G^*_B \approx -950\ R$ K, and

$\Delta \Omega\ (T) \approx [1435\ (1 - T / 4800\ \text{K})]\ R$ K.

For more theory, see Wada (1961).

303:4   *retrograde solubility*

700	0.349	0.001765	6.34	0.625	4.71	+	−
750	0.283	0.001910	6.26	0.927	5.62	+	−
800	0.221	0.001914	6.26	1.259	6.91	−	−
850	0.162	0.001745	6.35	1.645	8.93	−	−

303:5   *Breusov's retrograde liquidi*

The left-hand liquidus is given by

$R \ln X_A = - \Delta G^*_A (T) / T$.

The change of $\Delta G/T$ with T, is given by minus $\Delta H/T^2$; and with

$\Delta H^*_A = a_A + b_A T$

the formula $R \ln X_A = - a_A (T^{-1} - T^{o\ -1}_A) + b_A \ln(T/T^o_A)$ is obtained – the log analogue of the formula to be derived.

The derivative of X_A with respect to temperature is given by

$R\ (dX_A/dT) = (X_A/T)\ (b_A + a_A/T)$.

Retrograde behaviour requires $dX_A/dT = 0$, say at T_t, so that $T_t = -\ a_A/b_A$, or $(a_A + b_A T_t) = 0$. The constants a_A and b_A must have opposite signs. Moreover, the condtion $(a_A + b_A T^o_A) > 0$ has to be satisfied. Putting these things together we have: $a_A < 0$; $b_A > 0$; and $T_t < T^o_A$.

The two liquidi of the hypothetical system do not intersect. The A liquidus has its greatest extension at 50 K, where $(1-X_A) = 0.135$, and it intersects the axis $X_A = 1$ at $T \approx 28$ K. The B liquidus has a point of inflexion at $T = 15$ K, and $X_B = 0.183$ (←Exc 210:4).

303:6   *the elliptical stability field once more*

With $\Delta G^*_A = \Delta G^*_B = \Delta G^*$,

the EGC is given by $\Delta G^* + \Delta \Omega X (1 - X) = 0$.

$X_{EGC} = 0.5 \pm 0.5\ [1 + 4\ \Delta G^*/ \Delta \Omega]^{\ 0.5}$

Outcome of Exc 301:5:

1) ΔG^* is invariably negative, passing through a maximum in the vicinity of 400 K;
2) $\Delta \Omega$ is positive.

At the minimum and the maximum of the elliptic EGC, $X_{EGC} = 0.5$; so that $\Delta G^* = -\ 0.25 \Delta \Omega$.

There are two leading equations:

(I) $\quad \Delta S^*(T) = \Delta S^*(\Theta) + \Delta C_p^* \ln(T/\Theta)$

(II) $\quad \Delta G^*(T) = \Delta G^*(\Theta) - \Delta S^*(\Theta)(T - \Theta) - \Delta C_p^*[T \ln(T/\Theta) - T + \Theta]$.

Considerations and consequences:

a)   At 500 K and 300 K the values of ΔG^* are the same: (II) provides a relation between $\Delta S^*(300\ K)$ and ΔC_p^*.
That relation is $\Delta S^*(300\ K) = -0.2771\ \Delta C_p^*$.

b)   At the maximum of ΔG^*, the property ΔS^* is equal to zero: (I) gives the temperature at which ΔG^* is at its maximum.
$T = 396$ K.

As a result, and after the choice of 500 K for the maximum and 300 K for the minimum, the system comes down to the choice of two constants; these are $\Delta\Omega$ and ΔC_p^*.

Numerical example: $\Delta\Omega = 1000$ J mol^{-1} and $\Delta C_p^* = 5$ J K^{-1}mol^{-1}.
At $T = 396$ K. the value of ΔG^* is -186.7 J mol^{-1}; the X_{EGC} values are 0.248 and 0.752.
The individual Ω values of $\Omega^\alpha = 4500$ J mol^{-1}, $\Omega^\beta = 5500$ J mol^{-1} involve regions of demixing whose critical temperatures are 271 K and 331 K, respectively. And it means that the β region of demixing interferes with the [$\alpha + \beta$] equilibrium, giving rise to a three-phase [$\beta^I + \alpha + \beta^{II}$] equilibrium.
The three-phase equilibrium has $T = 307$ K, along with 0.275; 0.500; and 0.725 for the X values of the coexisting phases.

304   phase-equilibrium research in practice

304:1   prediction by analogy

The table displays the given molar volumes in two decimal places; and the estimated ones in one decimal place. The phase symbol α is for the NaCl type of structure, and the symbol β for the CsCl type of structure.

	α	β
CsF	36.91	28.4
CsCl	50.8	42.22
CsBr	57.4	47.93
CsI	68.4	57.61

The eutectic in the system CsF + CsI.

Taking ideal liquid mixing, the coordinates are calculated as $X_{eut} = 0.545$ and $T_{eut} = 754$ K.
Making allowance for deviation from ideal liquid mixing in $AB\Theta$ terms,
($A = -6480$ J mol^{-1}; $B = 0$; $\Theta = \infty$), the coordinates are calculated as
$X_{eut} = 0.525$ and $T_{eut} = 701$ K.
The experimental eutectic temperature is 704 K (for this system and the other ones, see also the compilation by Sangster and Pelton 1987).

The systems CsF + CsCl; CsCl + CsBr; CsCl + CsI.

For the systems, for which CsCl is one of the components, the numerical values for mismatch and the A(in kJ·mol^{-1}) and B constants of the $AB\Theta$ model (Θ invariably being 2565 K) are as shown.

	$m(\alpha)$	$A(\alpha)$	$B(\alpha)$	$m(\beta)$	$A(\beta)$	$B(\beta)$
CsF+CsCl	0.38	17.1	0.15	0.49	26.9	0.20
CsCl+CsBr	0.13	2.9	0.05	0.14	3.2	0.06
CsCl+CsI	0.35	14.8	0.14	0.36	15.5	0.14

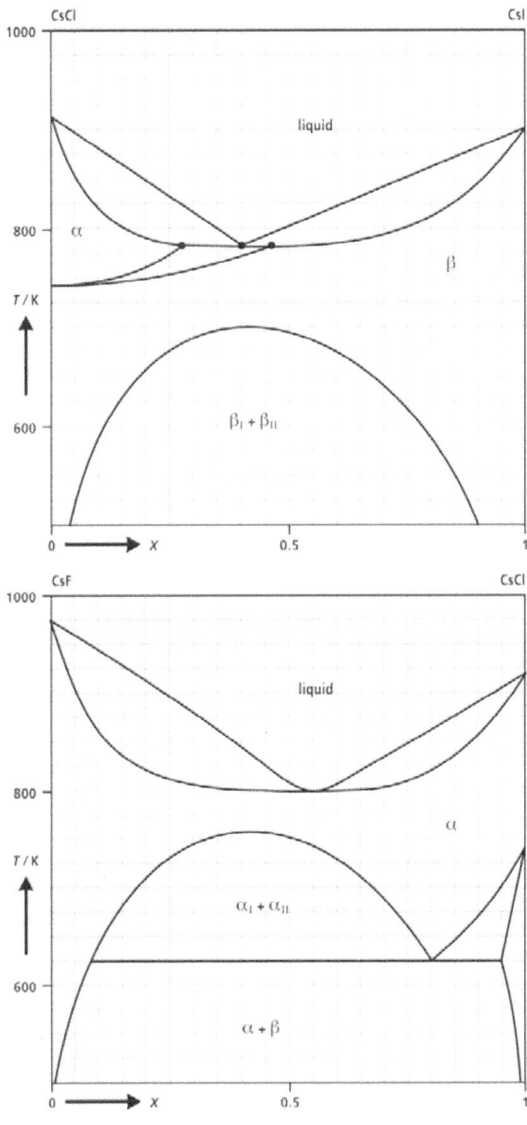

The phase diagrams of the systems CsCl + CsI and CsF + CsCl, which are shown, have been calculated using the thermochemical data given, ignoring any heat capacities, and treating the liquid state as an ideal mixture.

To avoid the appearance of the metastable forms of CsF and CsI, their entropies of melting were made equal to the entropies of melting of the stable form. The melting points of the metastable forms have arbitrarily been set at 875 K for α-CsI and 950 K for β-CsF.

By making allowance for the liquid's deviation from ideal mixing behaviour, the minimum of the melting loop moves to a lower temperature; however, without changing the subsolidus nature of the phase diagram.

As regards the system CsCl + CsBr, the upper part of its phase diagram is like the upper part in CsCl + CsI, the region of demixing in the lower part being absent.

an observation

The heats of melting published by Dworkin and Bredig (1960) for CsF and CsBr are found again in Barin's (1989) compilation. D & B's data for CsCl and CsI, on the other hand, differ from the ones in Barin, the latter being 15899 and 25606 $J \cdot mol^{-1}$, respectively. Barin's heat of melting for CsCl corresponds to an entropy of melting of 2.08 R - against (2.88 ± 0.22) times R for the whole family of the alkali halides having the NaCl type of structure at their melting point.

See also Oonk & Calvet (2007) Exc 110:9, which is based on the data by Clark (1959) and Dworkin and Bredig (1960).

304:2   *has minimal Gibbs energy been reached ?*

The mismatch m = 0.2119; and it yields A = 6303 $J \cdot mol^{-1}$ and B = 0.085. The value of 0.085 for B implies a critical point whose mole fraction is 0.439. Next, the critical temperature follows from the spinodal equation as T_c = 335 K. To reproduce the experimental critical temperature A's value must be set at 6550 $J \cdot mol^{-1}$.

The TX diagram - with experimental data points and calculated binodal - indicates that, the position of the ROD on the temperature scale is defined by the 'family description' with fair precision; and, next, that the true equilibrium states probably have not been reached. NB.The second part of the conclusion finds support in the fact that a thermodynamic analysis of the experimental data triplets in terms of Equations (5a,b) would imply a negative compensation temperature (see Exc 3).

304:3   *RbBr + RbI as an isolated system*

g_1 = { 2533 + 9.2232 (T/K)} $J \cdot mol^{-1}$;
g_2 = { −812 + 4.891 (T/K)} $J \cdot mol^{-1}$

A = 2533 $J \cdot mol^{-1}$; B = 0.10 ; Θ = −275 K
X_c = 0.43 ; T_c = 358.5 K

In the TX diagram the experimental data are shown, along with the calculated bimodal for the printed $AB\Theta$ values. The disagreement between the experimental data and the calculated result is given by Δ_x = 0.020.

For the binodal corresponding to the above g_1 and g_2 functions, the disagreement, obviously, is somewhat smaller: Δ_X = 0.017.

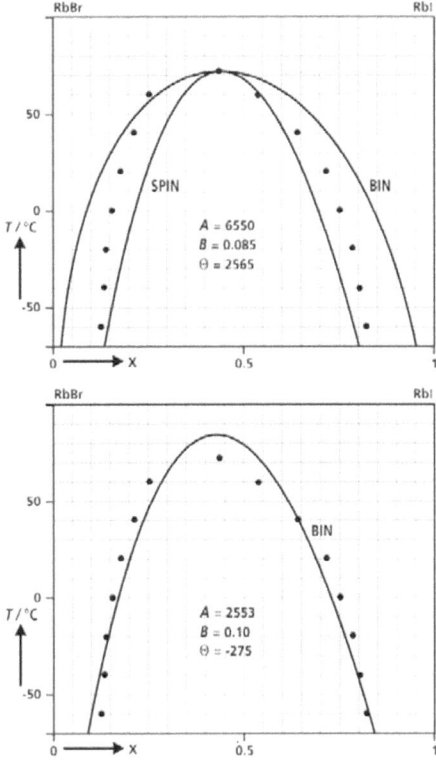

304:4 *crossed isotrimorphism*

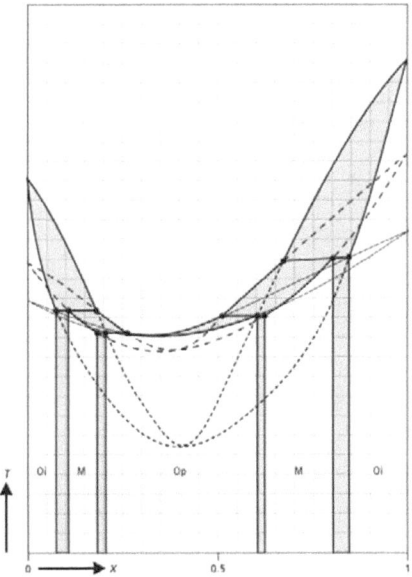

304:5 *the isothermal transition in C19 + C21*

The isothermal transition takes place at the EGC temperature. The interpolated part of the EGC has a minimum at about 283 K; and $T \approx 285$ K at the equimolar composition.

$$\Delta G^E(X) = X(1\text{-}X)\{-2300 - 600(1-2X)\} \; J\,mol^{-1}.$$

In view of Equation (15), the equimolar excess Gibbs energy of RI is 243 J mol⁻¹; the same property of Mdci as a result is 243 +575 = 818 J mol⁻¹.

304:6 *time for a paradox?*

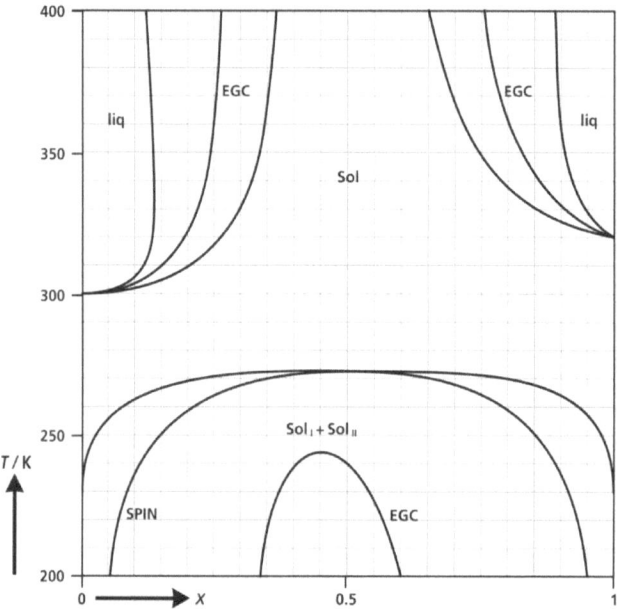

304:7 *the AnΘ model*

The spinodal temperature, which is the quotient of the second derivatives of the enthalpy and entropy of mixing ($\leftarrow$212), is given by

$$T(X) = a \, / \, (- R/X(1 - X) + a), \text{ with } a =$$
$$A \, (n{+}1)^{(n+1)} \, n^{-n} \, (1 - X)^{(n-2)} \, (n^2 X + nX - 2n).$$

Component A's excess chemical potential is given by

$$\mu_A^{\,E} = G_A^{\,*}(T) + RT \ln(1- X) + A \, (1- T/\Theta) \, (n{+}1)^{(n+1)} \, n^{-n} \, n \, X^2 (1- X)^{(n-1)} \, .$$

Component A's liquidus temperature is the quotient of A's change in partial enthalpy and A's change in partial entropy on melting. The former is given by

$\Delta H_A^* + A\,(n+1)^{(n+1)}\,n^{-n}\,n\,X^2\,(1-X)^{(n-1)}$;

and the latter by

$\Delta S_A^* - R\ln(1-X) + (A/\Theta)\,(n+1)^{(n+1)}\,n^{-n}\,n\,X^2\,(1-X)^{(n-1)}$.

304:8   *Kracek and the heat of fusion of cristobalite*

The best agreement, having $\Delta_T = 7.5$ K, is obtained for 6.6 kJ mol^{-1}.

NB If the assessment is carried out for $n = 8.1$ and $A = 1060$ J mol^{-1} (from the critical temperature of the metastable region of demixing, 1275 K, according to Moriya et al. 1967), a value of about 10 kJ mol^{-1} is obtained, along with $\Delta_T = 6.0$ K.

304:9   *an inconvenient result*

The results presented here, have been obtained for the dummy set composed of the following eleven $\{X;\,(T/K)\}$ pairs:

{0; 1996}; {0.015; 1975}; {0.126; 1950}; {0.163; 1925}; {0.182; 1900}; {0.198; 1875}; {0.213; 1850}; {0.223; 1825}; {0.232; 1800}; {0.239; 1775}; {0.248; 1750}.

a)   For 8.92 kJ mol^{-1}, and excluding, for the least-squares calculation, the A value calculated for {0.015; 1975}, the result is $A = 927$ J mol^{-1}; $\Theta =$ minus 1622 K; $\Delta_T = 5.2$ K.

b)   Similarly, for two times 8.92 kJ mol^{-1}, the result is $A =$ minus 4006 J mol^{-1}; $\Theta = 1335$ K; $\Delta_T = 2.7$ K.

a)   The critical point, at the SiO$_2$ side, is at $\{X = 0.084;\,T = 2627$ K$\}$.

b)   The spinodal consists of two branches: at the SiO$_2$ side there is a lower   (!) critical point at $\{X = 0.084;\,T = 1929$ K$\}$; at the BaO side an upper   critical   point at $\{X = 0.403\,;\,T = 701$ K$\}$.

NB   The spinodals for the two solutions a) and b) have in common that their extrema are at the same X, namely $X = 0.0843$ and $X = 0.4032$. The $An\Theta$ model is such that the X coordinate of a critical point is determined by n's value.

The principal conclusion from this exercise is that excellent agreement between an experimental curve in a phase diagram and its calculated reproduction, on its own, is no guarantee that the adopted model is physically meaningful.

304:10   *the system d-limonene + l-carvone by adiabatic calorimetry*

First heating run: glass transition at 152 K with a jump of ≈ 91 J K^{-1} mole^{-1}; nucleation of carvone at 158 K followed by crystallization of carvone starting around 179 K. At the arrest at 185 K, 68 mole% of the sample is crystalline; 32 mole% is liquid, containing 6 mole% of carvone.

Second heating run: glass transition at 131 K with a jump of ≈ 27 J K^{-1} mol^{-1}; nucleation of limonene at 153 K followed by crystallization of limonene starting around 165 K; eutectic event at 196.38 K, at the end of which 33.4 mole% of the sample is liquid and 66.6 mole% crystalline carvone; further melting until melting is complete at 237.5 K.

304:11 *the system l-limonene + l-carvone by DSC*

first option

The observations diagram, constructed out of the DSC information and the melting point of limonene, suggests a cigar-type of phase diagram for a system whose ΔG^E property is small (the diagram looks like a diagram for a fully ideal system), say virtually zero.

The heat effects, when plotted as a function of temperature, reveal a large negative ΔH^E property, having a value of about -7600 J mol^{-1} at the equimolar composition. And to give rise to a ΔG^E, which is zero are about 220 K, it must be 'compensated' by a ΔS^E property of $-7600 / 220 = -34.5$ J K^{-1} mol^{-1}.

If the liquid state is taken as an ideal mixture (which is not unreasonable), then the solid state's A and Θ values, that follow from the adopted interpretation, are $A = 30.4$ kJ mol^{-1}, and $\Theta = 220$ K. The value of Θ complies well with the empirical relationship expressed by Equation (9).

For details, see Calvet et al. 1996.

second option

The lowest temperature the samples were exposed to was 153 K. As follows from Figure 13, the isotherm $T = 153$ K is above the nucleation curve for limonene; and, up to the equimolar composition, also above the nucleation curve for carvone. In other terms, only samples that have $X \geq 0.5$ will show the phenomenon of crystallization; and from these samples carvone is the only component which crystallizes.

Next, and taking one mole of the sample whose overall composition is (0.2 mol limonene + 0.8 mol carvone): from the heat effects in Table 4 it follows that only 6.2 / 11.3 = 0.55 mole of carvone had crystallized. Therefore, at the onset of the melting phenomenon the sample consisted of 0.55 mol solid carvone and 0.45 mole of a liquid mixture, containing 0.20 mol limonene and 0.25 mol carvone, and as a result having $X = 0.56$.

From the liquidus curve - i.e. the end of melting registered by DSC as a function of overall sample composition - it can be read that a liquid mixture, having $X = 0.56$, is in equilibrium with solid carvone at 228-229 K. In other terms, a sample having these characteristics, will, when heated, start to produce more liquid at about 229 K; in agreement with the onset temperature displayed in Table 4.

discussion

The idea of dual crystallization not only provides a full interpretation of the observations made by adiabatic calorimetry on the system *d*-limonene + *l*-carvone, it also gives a conclusive interpretation of the observations made by DSC on the systems *l*-limonene + *l*-carvone and *d*-limonene + *l*-carvone. The second option seems to be the one to be preferred; it would exclude the formation of mixed crystals. In addition, if limonene and carvone would form mixed crystals, it would be unlikely that the *l* + *l* combination and the *d* + *l* combination are giving rise to the same phase diagram. Last but not least, there is the X-ray experiment described by Gallis (1998). Liquid mixtures of *l*-limonene and *l*-carvone having $X = 0.6$ and $X = 0.8$ were cooled to 183 K; next heated to 218 K, at which temperature X-ray powder diffraction patterns were measured. The diffraction patterns were identical to those obtained by Sañé et al. (1997) for pure *l*-carvone under the same experimental circumstances.

304:12 *Würflinger's lower loop*

The columns in the table:
(1) EGC mole fraction
(2) EGC temperature
(3) minus ΔG^E at EGC
(4) rotator excess Gibbs energy at EGC
(5) normal solid excess Gibbs energy at EGC
(6) normal solid excess Gibbs energy at 285 K
(7) Gibbs energy of mixing in normal solid
(8) Gibbs energy of mixing in normal solid, assuming ideal mixing in rotator and absence of excess entropy in normal solid

(1)	(2)	(3)	(4)	(5)	(6)	(7)	(8)
0.00	294.9	000	000	0000	0000	000	000
0.05	287.2	398	043	0441	0471	+ 001	− 072
0.10	284.1	581	090	0671	0654	− 116	− 189
0.15	282.7	676	132	0808	0758	− 244	− 326
0.20	282.1	732	168	0900	0831	− 355	− 454
0.25	281.9	774	198	0972	0893	− 440	− 559
0.30	281.9	802	222	1024	0941	− 507	− 646
0.40	282.9	807	247	1054	0994	− 601	− 788
0.50	285.0	757	243	1000	1000	− 643	− 885
0.60	287.7	676	215	0891	0965	− 630	− 919
0.70	291.1	559	168	0727	0880	− 568	− 889
0.80	295.3	399	110	0509	0721	− 464	− 787
0.90	300.0	206	050	0256	0448	− 322	− 564
1.00	305.0	000	000	0000	0000	000	000

The *G*-curves implied in (7) and (8) have concave parts at the side of the axis $X=0$. This fact excludes the existence of a continuous series of mixed crystals. This can be clearly shown after the addition of a *linear contribution* of $2000X$ to the numbers in (7) and (8).
For each of the two options, the points of contact of the double tangent line are $X = 0.00$ and $X = 0.22$.

304:13 *a delicate technique for the determination of cooling curves*

The opening angle (see Exc 302:14) of the solid-liquid loop at the dibromo side of the experimental *TX* diagram is much too small (about 50 % of the value it should have). The value, which follows from the initial slope of the liquidus along with the heat of melting, is $dT^{sol}/dX = -103$ K.
The EGC position for the temperature of the second liquidus point is $X_{EGC} \approx 0.58$. With this position the interaction parameter is $\Omega^{sol} = 1380$ J mol^{-1}.
From the equality of B's chemical potentials in liquid and solid and their recipes, the equation to be solved for given T and X^{liq} is

$$\Delta G^*_B + RT \ln X^{liq} - RT \ln X^{sol} - 1380\,(1{-}X^{sol})^2 = 0.$$

The solution of the equation is $X^{sol} = 0.685$. $\rightarrow$ p 242

Unfortunately, the statement in italics is false. The delicacy of the technique is in the unparalleled accuracy of the liquidus temperatures.

305   *ternary systems*

305:1   *ternary azeotropes*

The main assumptions are: 1) binary excess Gibbs energies are symmetrical with respect to the equimolar composition, as expressed by $\Omega X(1-X)$, and their change with temperature may be ignored; 2) the pure components have equal entropies of vaporization; in accordance with Trouton's rule ($\leftarrow$004; however, without using the magnitude of $11R$)

acetone + chloroform + methanol
With $\Theta_{AC} = 20.0$ K; $\Theta_{AM} = -15.8$ K; and $\Theta_{CM} = -40.8$ K, the coordinates of the saddle point are calculated as $X = 0.149$; $Y = 0.437$; and $56.45°C$ for the azeotropic temperature. The figures in the Handbook correspond to $X = 0.24$; $Y = 0.44$; and $57.5°C$.

benzene+chloroform+methanol
There is no ternary stationary point.

A+methanol+benzene
The coordinates of the stationary point, which is a saddle point, are $X = 0.2741$; $Y = 0.1480$; $t = 58.328°C$.

305:2   *solubility of mixed crystals*

The system formulation is $f = M\,[X, m_A, m_B\,] - N\,[\mu_A{}^{sol} = \mu_A{}^{liq}; \mu_B{}^{sol} = \mu_B{}^{liq}] = 3 - 2 = 1$.
The equilibrium system is fixed by the choice of X.

$$m_A(X) = m_A{}^o\,(1-X)\,\exp(\mu_A{}^E / RT)$$

$$m_B(X) = m_B{}^o\,X\,\exp(\mu_B{}^E / RT)$$

$$\mu_A{}^E = X^2\,[g_1 + g_2\,(3 - 4X)]$$

$$\mu_B{}^E = (1-X)^2\,[g_1 + g_2\,(1 - 4X)$$

X	.1	.2	.3	.4	.5	.6	.7	.8	.9
m_A	.550	.514	.487	.461	.430	.390	.332	.251	.141
m_B	.119	.178	.211	.234	.254	.276	.300	.329	.362

The rectangular diagram has a maximum at $X = 0.305$; $Y = 0.698$. The maximum is the common maximum of the so-called *solutus* and the *solidus*. For given Y, the X value of the saturated solution can be read from the solutus and the X value of the mixed-crystalline phase in equilibrium with the solution from the solidus.
In the rectangular diagram the field below the solutus is the single-phase field for the unsaturated solutions, corresponding to the field in the triangular diagram above the *solubility curve.*
How about the field in the rectangular diagram above the solidus?

305:3 *isothermal section with binary and ternary compound*

$$\ln (K / K_o) = \ln [27(1-X-Y) X Y] = - (\Delta H^o / R) (1/T - 1/T_o) = - 2500 \text{ K} / T + 2.5$$

The quadratic equation in Y has solutions for $0.158 < X < 0.547$; for $X = 0.5$, the roots of the equation are 0.170 and 0.330.

For the binary compound AC the K / K_o is given by $4 \times (1-X-Y) \times Y$, which is for the given point $4 \times 0.2 \times 0.8 = 0.64$. The maximum extension of its liquidus inside the triangle is at $X = 0.2$; $Y = 0.4$.

B's liquidus is the straight line $X = 0.5$ (in an ideal mixture, the chemical potential of a component B is a function of its own mole fraction only; ←204).

The triangular phase diagram is symmetrical with respect to the axis {compound AC – compound ABC – vertex substance B}. At the C side of the symmetry axis there are two three-phase triangles: one for compound AC + liquid (of composition $X = 0.175$; $Y = 0.512$) + compound ABC; the other for compound ABC + liquid ($X = 0.5$; $Y = 0.330$) + solid B.

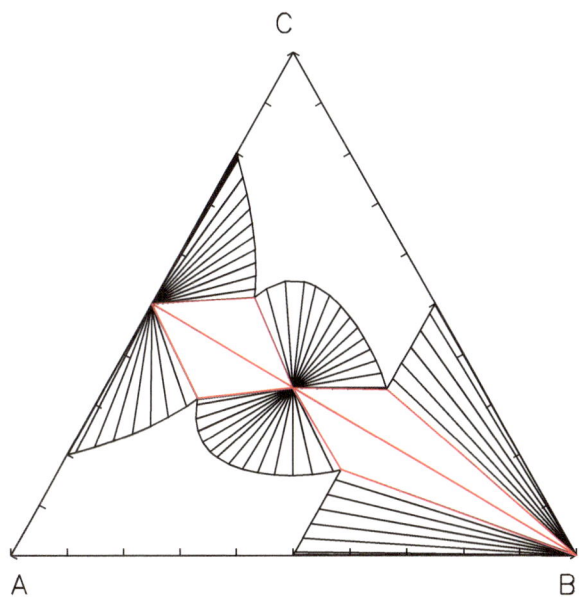

305:4 *an amusing faux pas*

For every point on the solidus there has to be a companion point on the liquidus. In Figure 2 this obvious rule is violated at the right-hand side of the ternary two-phase field. See also Exc 003:2.

305:5 *the idealized isobaric ternary liquid+vapour equilibrium*

Under isobaric circumstances the system formulation is

$$f = M \, [T, X^{liq}, Y^{liq}, X^{vap}, Y^{vap}] - N \, [\mu_A{}^{liq}{=}\mu_A{}^{vap}; \; \mu_B{}^{liq}{=}\mu_B{}^{vap}; \; \mu_C{}^{liq}{=}\mu_C{}^{vap}] = 5 - 3 = 2$$

$$\exp [i] \equiv \exp (\Delta G_i{}^* / RT) = \exp [\Delta S_i{}^*(T{-}T_i{}^o) / RT], \; \text{for } i = A; B; C.$$

$$(1{-}X^{vap}{-}Y^{vap}) = (1{-}X^{liq}{-}Y^{liq}) \cdot \exp [A] \qquad (1)$$
$$X^{vap} = X^{liq} \cdot \exp [B] \qquad (2)$$
$$Y^{vap} = Y^{liq} \cdot \exp [C] \qquad (3)$$

From sum of Equations (1), (2) and (3):

$$X^{liq} = \{1{-}Y^{liq} \cdot (\exp [C] - \exp [A]) - \exp [A]\} / (\exp [B] - \exp [A]) \qquad (4)$$

Independent variables T and Y^{liq}; the dependent variables: X^{liq} from (4); next X^{vap} from (2), and Y^{vap} from (3).

From (4) it follows that
$dX^{liq}/dY^{liq} = - (\exp [C] - \exp [A]) / (\exp [B] - \exp [A])$. For given T this is constant:
the liquidus is a straight line.
As regards the vaporus: see the last part of next Exc, here below.

305:6 *no more than Antoine constants*

substance	nbp / K	ΔH / kJ·mol^{-1}
A = n-hexane	343	30.6
B = n-heptane	371	33.6
C = n-octane	398	36.6

The formula for the heat of vaporization is $\Delta H = B \cdot RT^2 \cdot K^{-1} / [(T / K) - C]^2$.
NB see also solution Exc 110:5, where, besides, the unit K is omitted.

The equilibrium states are calculated as indicated in Exc 5:

state	Y^{liq}	X^{liq}	X^{vap}	Y^{vap}
(1)	0.45	0.536	0.694	0.266
(2)	0.55	0.391	0.506	0.326
(3)	0.65	0.246	0.318	0.385
(4)	0.75	0.100	0.130	0.444

To calculate the boiling point of the equimolar mixture: for $Y^{liq} = 1/3$, T is varied until $X^{liq} = 0.333$. Answer: $T = 363.7$ K.
In order to calculate the dew-point temperature, the four equations in Exc 5 have to be modified first; and such that Y^{vap} instead of Y^{liq} becomes the second independent variable. After that, to calculate the dew point: for $Y^{vap} = 1/3$, T is varied until $X^{vap} = 0.333$. Answer: $T = 377.8$ K.

305:7   *linear contributions at work*

Along the perpendicular bisector the Gibbs function is clearly not convex over the whole range. The points of contact of the double tangent are at $Y = 0.24$ and $Y = 0.78$. This does suggest that the upper and lower boundaries of the two-phase region are very much the same as the ones shown in the phase diagram of the system {methylcyclohexane (A) + n-hexane (B) + methanol (C)} at 30 °C.

First round. Starting values $A = 2752$ J mol^{-1}; $B = -738$ J mol^{-1}. For $X = 0.2$, local minimum at $Y = 0.220$ with $G = -1062.80$ J mol^{-1}; the other local minimum at $X \approx 0.055$; $Y \approx 0.795$ with $G = -1069.63$ J mol^{-1}.

Second round. Adapted value $B = -726$ J mol^{-1}. For $X = 0.2$, local minimum at $Y \approx 0.2172$ with $G = -1060.177$ J mol^{-1}; the other local minimum at $X \approx 0.0555$; $Y \approx 0.7900$ with $G = -1060.154$ J mol^{-1}.

305:8   *the incidence of symmetry*

To four decimal places the X values of the coexisting phases for $Y = 0.5$ are $X^I = 0.1696$ and $X^{II} = 0.3304$; and the coordinates of the lower and upper critical points are ($X = 0.2971$; $Y = 0.4058$) and ($X = 0.2029$; $Y = 0.5942$), respectively.

308   *the system MgO – SiO$_2$ under pressure*

308:1   *the phase diagram of Mg$_2$SiO$_4$ from a zeroth point of view*

The resemblance between the constructed phase diagram and Figure 1 is remarkable. One of the differences is that the constructed diagram, in contrast to Figure 308:1, has a positive slope for the two-phase equilibrium line for the change from wadsleyite to MgO + perovskite.

The coordinates of the calculated invariant point are 1818 K and 22.05 GPa.

308:2   *partial derivatives of volume in terms of bulk modulus and its derivative*

$$(\partial^2 V/\partial P^2)_{TT} = (V/K^2)(K' + 1)$$

308:3   *an exercise on extrapolation*

$G(308$ K; 250 bar$) = -2205086.219$ J mol^{-1};
$G(298$ K; 1 bar$) = -2205204.040$ J mol^{-1}.
The difference is 117.821 J mol^{-1}; it is related to $dG = -SdT + VdP$ and for that matter the number of significant figures is five (the entropy's value is in the vicinity of 100 where four significant figures change to five, that is to say if the number of decimal places remains equal to two): 117.82 J mol^{-1}.

P/GPa	Exp V/cm^3·mol^{-1}	Calc V, Eq (D)	Calc V, Eq (E)
05	42.13	42.14	42.04
10	40.80	41.01	40.62
15	39.73	40.28	39.40
20	38.76	39.93	38.37
25	37.91	39.97	37.54
30	37.13	40.41	36.90
35	36.43	41.23	36.45

Sum of squares: for (D) 39.8; for (E) 0.5

308:4 *the Grüneisen parameter derived from sound velocities*

The change of the sound velocity with pressure is about 110 m·s^{-1}·GPa^{-1}. This change comes down to a value for γ_0 of about 1.9 (against 2.03 in Table A1):
Differentiating Equation (8) with respect to pressure and changing from angular frequency to normal frequency gives:

$$\left(\frac{\partial V_P}{\partial P}\right)_T = \frac{\pi}{6}\left(\frac{4\pi ZV}{3N_A}\right)^{1/3}\left\{3\left(\frac{\partial v}{\partial P}\right)_T - \frac{v}{K}\right\}$$

Use Table 1 in the appendix to insert values for molar volume, number of molecules in the primitive cell, frequency and bulk modulus at zero K and zero pressure.
110×10^{-9} m·s^{-1}·Pa^{-1} = 4.948×10^{-10} ($3\times\Delta v/\Delta P - 5.712\times10^{12}/123.6\times10^{9}$);
$\gamma_0 = (K / v)(\Delta v / \Delta P)_T = 1.94$

308:5 *a classical thermodynamic relationship*

Mathematically we have for the change of a quantity Q with volume
$(\partial Q/\partial T)_V = (\partial Q/\partial T)_P + (\partial P/\partial T)_V (\partial Q/\partial P)_T$. Next, the differential coefficient $(\partial P/\partial T)_V$ is related to the pressure and temperature partial derivatives of V, read to the cubic expansion coefficient α and the bulk modulus K, through the *"minus one identity"* ($\leftarrow$107, Equation (24); Exc 107:8):
$(\partial P/\partial T)_V (\partial V/\partial P)_T (\partial T/\partial V)_P = -1$.
This brings the classical thermodynamic relationship to
$(\partial Q/\partial T)_V = (\partial Q/\partial T)_P + \alpha K(\partial Q/\partial P)_T$.

308:6 *properties of forsterite at 1800 K and 1 bar*

C_P = 195 J·K^{-1}·mol^{-1}; C_V = 178 J·K^{-1}·mol^{-1}; α = 4.7 x 10^{-5} K^{-1}; $(\partial\alpha/\partial T)_P$ = 1.19 x 10^{-8} K^{-2}; K_S = 1.02 x 10^{11} Pa ; $(\partial K_S/\partial T)_P$ = $-$ 2.31 x 10^{7} Pa·K^{-1}; V = 46.33 x 10^{-6} m^3·mol^{-1}; K = 9.2 x 10^{10} Pa ; $(\partial K/\partial T)_P$ = $-$ 2.08 x 10^{7} Pa·K^{-1}.

308:7 *thermal expansivity and intuition*

$(\partial\alpha/\partial P)_T = K^{-2}(\partial K/\partial T)_P$; for forsterite at 1800 K and 0 GPa:
$(\partial\alpha/\partial P)_T = -$ 2.46 x 10^{-6} K^{-1}·GPa^{-1}, which comes down to a decrease of 5 % per GPa.

308:8 *heat capacity at constant pressure under pressure*

The effect is a decrease of 0.6 % per GPa.

308:9 *the PX phase diagram of* $Mg_2SiO_4 + Fe_2SiO_4$

The temperature for which the diagram is valid is 1673 K.

The metastable transition pressure is about 15.8 GPa.

$dX^{ri}/dP - dX^{ol}/dP = [0.056 - 0.200 = -0.144]$ GPa^{-1} $= -0.144$ x 10^{-9} Pa$^{-1} = \Delta V^*/RT$.
The change in molar volume $\Delta V^* = -2.00$ cm^3 mol^{-1}.

NB! Equation 302:24 has been derived for a system in which N_{Av} particles of two different types that are distributed in a random manner over N_{Av} lattice sites. In other words: the result is valid for the system defined as $\{(1 - X)$ mole of Mg(SiO$_4$)$_{0.5}$ + X mole of Fe(SiO$_4$)$_{0.5}\}$ ($\leftarrow$ Exc 304:8; *Van 't Hoff factor*). Referred, therefore, to (Mg$_2$SiO$_4$ and) Fe$_2$SiO$_4$ the change in molar volume is minus 4.00 cm^3 mol^{-1}. The optimized values of the molar volumes and the transition pressure at 1673 K are $V^{*ol} =$ 46.1574 cm^3 mol^{-1}; $V^{*ri} = 42.0476$ cm^3 mol^{-1}; and $P^{ol \rightarrow ri} = 6.133$ GPa.

308:10 *the ideal PX loop*

Above Equation (11): "as a function of pressure"; in Equation (11); in left-hand side: (P); below Equation (11): "the formula for the equilibrium composition of the β phase as a function of pressure".
If restriction is made to 12 GPa the calculated compositions are: for the second task X^{α} = 0.198; X^{β} = 0.472; and for the third task X^{α} = 0.227; X^{β} = 0.500.
The values that can be read from the phase diagram ($\leftarrow$ Exc 9) are X^{α} = 0.22; X^{β} = 0.57.

308:11 *mixed crystals in the system* $Mg(SiO_4)_{0.5} + Fe(SiO_4)_{0.5}$

If the equal-G equation is solved for the ideal case, it is found that the fraction $f = (X_{EGC} - X^{\alpha})/(X^{\beta} - X^{\alpha})$ has the value of 0.4725. Using the same f for the real case, X_{EGC} is found as 0.385, and subsequently $\Delta g = 2$ kJ mol^{-1}.
The outcome of the analysis in terms of chemical potentials is $g^{\alpha} = 6$ kJ mol^{-1} and $g^{\beta} = 8$ kJ mol^{-1}.

308:12 *the usefulness of GX sketches*

308:13 *estimating phase behaviour*

See phase diagrams next page.

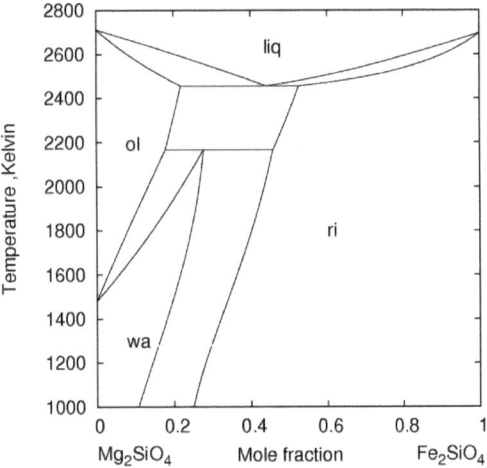

Calculated phase diagram for $P = 14$ GPa

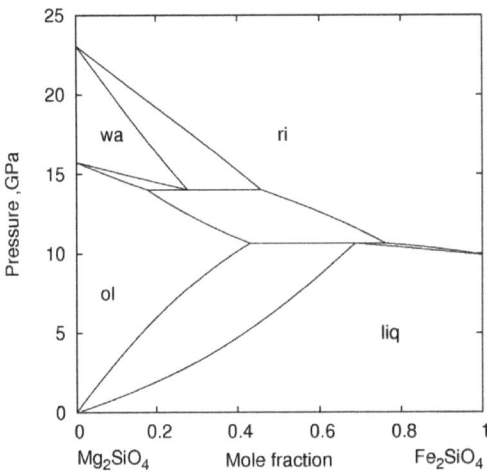

Calculated phase diagram for $T = 2171$ K, which is the melting temperature of forsterite.

REFERENCES

1819 Petit AT, Dulong PL (1819) Recherches sur quelques points importants de la théorie de la chaleur. Annales de Chimie et de Physique 10:395-413

1869 Horstmann A (1869) Dampfspannung und Verdampfungswärme des Salmiaks. Ber. D. Chem. Geselsch. 2:137-140

1881 Konowalow D (1881) Ueber die Dampfspannungen der Flüssigkeitsgemische. Annalen der Physik und Chemie 250:34-52; 219-226

1894 Le Chatelier H (1894) Sur la fusibilité des mélanges salins isomorphes. Compt. rend. 118:350-352

1903 Ruff O, Plato W (1903) Regelmässigkeiten in der Zusammensetzung der niedrigst schmelzenden Mischungen anorganischer Salzpaare. Ber. D. Chem. Geselsch. 36:2357-2370

1906 Einstein A (1906) Die Plancksche Theorie der Strahlung und die Theorie der spezifischen Wärme. Ann. Phys. 327 (1):180-190

1907 Kurnakow N, Żemcżużny S (1907) Isomorphismus der Natrium- und Kaliumverbindungen. Z. anorg. Chem. 52:186-201

1908 van Laar JJ (1908) Die Schmelz- oder Erstarrungskurven bei binären Systemen, wenn die feste Phase ein Gemisch (amorphe feste Lösung oder Mischkristalle) der beiden Komponenten ist. Z. physik. Chem. 63:216-253; 64:257-297

1909 Żemcżużny S, Rambach F (1909) Schmelzen der Alkalichloride. Z. anorg. Chem. 65:403-428

1912 Debye P (1912) Zur Theorie der spezifischen Wärmen. Ann. Phys. 344 (14):789-839

1914 Bowen NL, Andersen O (1914) The binary system: $MgO-SiO_2$. Am. J. Sci. 37:487-500

1914 Bridgman PW (1914) Change of phase under pressure. I. The phase diagram of eleven substances with special reference to the melting curve, Phys. Rev. 3:126-141

1918 Nacken R (1918) Über die Grenzen der Mischkristallbildung zwischen Kaliumchlorid und Natriumchlorid. Sitzungsber. Preuss. Akad. Wiss. Phys. Math. Kl 192

1924 Eastman ED, Williams AM, Young TF (1924) The specific heats of magnesium, calcium, zinc, aluminium and silver at high temperatures. J. Am. Chem. Soc. 46:1178-1183

1924 Honda K, Okubo Y (1924) On the measurements of the coefficients of thermal expansion for aluminium alloys and alloys of nickel-iron and cobalt-iron. Sci. Rep. Tohoku Imp. Univ. 13:101-107

1926 Grüneisen E (1926) Zustand des festen Körpers. Handbuch der Physik 10:1-59

1927 Greig JW (1927) Immiscibility in silicate melts. Am. J. Sci. 5th Ser. 13:1-44, 133-153

1929 Morse PM (1929) Diatomic molecules according to the wave mechanics. II. Vibrational levels. Phys. Rev. 34:57-64

1929 Simon FE, Glatzel G (1929) Bemerkungen zur Schmelzdruckkurve. Z. anorg. allg. Chem. 178:309-316

1930 Kracek FC (1930) The cristobalite liquidus in the alkali oxide-silica systems and the heat of fusion of cristobalite. J. Am. Chem. Soc. 52:1436-1442

1937 Kok JA, Keesom WH (1937) Measurement of the atomic heat of aluminium from 1.1 to 20°K. Physica 4:835-842

1937 Murnaghan FD (1937) Finite deformation of an elastic solid. Am. J. Math. 59:235-239

1937 Quinney H, Taylor GI (1937) The emission of the latent energy due to previous cold working when a metal is heated. Proc. R. Soc. Lond. 163A:157-181

1939 Avramescu A (1939) Temperaturabhängigkeit der wahren spezifischen Wärme von Leitungskupfer und Leitungsaluminuim bis zum Schmelzpunkt. Z. Tech. Physik 20:213-217

1940 Reinders W, de Minjer CH (1940) Vapour-liquid equilibria in ternary systems.II. The system acetone-chloroform-benzene. Rec. Trav. Chim. Pays-Bas 59:369-391

1941 Giauque WF, Meads PF (1941) The heat capacities and entropies of aluminum and copper from 15 to 300°K. J. Am. Chem. Soc. 63:1897-1901

1941 Wilson AJC (1941) The thermal expansion of aluminium from 0° to 650°C. Proc. Phys. Soc. 53:235-244

1943 Kelley KK (1943) Specific heats at low temperatures of magnesium orthosilicate and magnesium metasilicate. J. Am. Chem. Soc. 65:339-341

1944 Fürth R (1944) On the equation of state for solids. Proc. Roy. Soc. Lond. 183:87-110

1944 Murnaghan FD (1944) The compressibility of media under extreme pressures. Proc. Nat. Acad. Sci. U.S.A. 30:244-247

1948 Campbell AN, Prodan LA (1948) An apparatus for refined thermal analysis exemplified by a study of the system p-dichlorobenzene – p-dibromobenzene – p-chlorobromobenzene. J. Am. Chem. Soc. 70:553-561

1948 Meijering JL (1948) Retrograde solubility curves, especially in alloy solid solutions. Philips Res. Rep. 3:281-302

1949 Haase R (1949) Zur Thermodynamik flüssiger Dreistoffgemische. Z.Naturforsch. A 4:342-352

1950 Haase R (1950) Verdampfungsgleichgewichte von Mehrstoffgemischen VII: Ternäre azeotrope Punkte. Z. physik. Chem. 195:362-385

1950 Uchida S, Ogawa S, Yamagushi M (1950) Studies in distillation. Japan Sci. Rev. Eng. Sci. 1:41-49

1951 Ol'shanskii YaI (1951) Equilibria of two immiscible liquids in systems of alkaline earth silicates. Dokl. Akad. Nauk SSSR 76:93-96

1952 Andon RJL, Cox JD (1952) Phase relationships in the pyridine series. Part I. The miscibility of some pyridine homologues with water. J. Chem. Soc. 4601-4606

1952 Birch F (1952) Elasticity and constitution of the Earth's interior. J. Geophys. Res. 57:227-286

1952 Wheeler LP (1952) Josiah Willard Gibbs: The history of a great mind. Revised Ed. Yale University Press, New Haven

1953 Bunk AJH, Tichelaar GW (1953) Investigations in the system NaCl-KCl. Proc. K. Nedl. Akad. Wet. Ser. B. 56:375-384

1953 Orr RL (1953) High temperature heat content of magnesium orthosilicate and ferrous orthosilicate. J. Am. Chem. Soc. 75:528-529

1953 Pochapsky TE (1953) Heat capacity and resistance measurements for aluminium and lead wires. Acta. Metall. 1:747-751

1953 Thurmond CD, Struthers JD (1953) Equilibrium thermochemistry of solid and liquid alloys of germanium and of silicon. II. The retrograde solid solubilities of Sb in Ge, Cu in Ge and Cu in Si. J. Phys. Chem. 57:831-835

1954 Barrett WT, Wallace WE (1954) Sodium chloride-potassium chloride solid solutions. I. Heats of formation, lattice spacings, densities, Schottky defects, and mutual solubilities. J. Am. Chem. Soc. 76:366-369

1954 Keane A (1954) An investigation of finite strain in an isotropic material subjected to hydrostatic pressure and its seismological implications. Aust. J. Phys. 7:322-333

1955 Harker RI, Tuttle OF (1955) Studies in the system CaO-MgO-CO$_2$; Part 1, The thermal dissociation of calcite, dolomite and magnesite. Am. J. Sci. 253:209-224

1956   Haase R (1956) Thermodynamik der Mischphasen. Springer-Verlag. Berlin

1956   Kortüm G, Haug P (1956) Thermodynamische Untersuchungen am System 2,4-Lutidin – Wasser mit geschlossener Mischungslücke. Z. Eleektrochem. 60:355-362

1957   Mazee WM (1957) Thermal analysis of normal alkanes. Anal. Chim. Acta 17:97-106

1957   Strelkov PG, Novikova SI (1957) Silica dilatometer for low temperatures. 1. Thermal expansion of copper and aluminium. Prib. Tekh. Eksp. 5:105-110

1958   Nakagawa Y (1958) Liquid immiscibility in copper-iron and copper-cobalt systems in the supercooled state. Acta Metallurgica 6:704-711

1958   Walling JF, Halsey GD (1958) Solid solution argon-krypton from 70 to 96 °K. J. Phys. Chem. 62:752-753

1959   Clark SP (1959) Effect of pressure on the melting points of eight alkali halides. J. Chem. Phys. 31:1526-1531

1959   Phillips NE (1959) Heat capacity of aluminum between 0.1 °K and 4.0 °K. Phys. Rev. 114:676-685

1960   Dworkin AS, Bredig MA (1960) The heat of fusion of the alkali metal halides. J. Phys. Chem. 64:269-272

1960   Kohler F (1960) Zur Berechnung der thermodynamischen Daten eines ternären Systems aus den zugehörigen binären Systemen. Monatsh. Chem. 91:738-740

1961   Fraser DB, Hollis-Hallett AC (1961) The coefficient of linear expansion and Grüneisen gamma of Cu, Ag, Au, Fe, Ni, and Al from 4 to 300 K. In: Proceedings of 7th international conference on low temperature physics, pp 689-692

1961   Toropov NA, Bondar IA (1961) Silicates of the rare earth elements communication 6. Phase diagrams of the binary systems Sm_2O_3 – SiO_2 and Yb_2O_3 – SiO_2, and comparison of these silicates with other rare earth silicates which have been studied. Bull. Acad. Sci. USSR. Div. Chem. Sci. (Engl. Transl.) 10:1278-1285

1961   Wada T (1961) Thermodynamic studies on the α-γ transformation of iron alloys. Sci. Rep. Res. Tohuko Univ. Ser. A 13:215-224

1962   Nagata I (1962) Vapor-liquid equilibrium at atmospheric pressure for the ternary system methyl acetate-chloroform-benzene. J. Chem. Eng. Data 7:360-366

1962   Nagata I (1962) Isobaric vapor-liquid equilibria for the ternary system chloroform-methanol-ethyl acetate. J. Chem. Eng. Data 7:367-373

1962   Toop GW, Samis CS (1962) Activity of ions in silicate melts. TMS-AIME 224:878-887

1963   Babb SE (1963) Parameters in the Simon equation relating pressure and melting temperature. Rev. Modern Phys. 35:400-413

1963   McGlashan ML (1963) Deviations from Raoult's Law. J. Chem. Education 40:516-518

1963   Nicklow RM, Young RA (1963) Thermal expansion of silver chloride. Phys. Rev. 129:1936-1943

1964   Kamm GN, Alers GA (1964) Low temperature elastic moduli of aluminium. J. Appl. Phys. 35:327-330

1964   Nguyen-Ba-Chanh (1964) Équilibres des systèmes binaires d'halogénures de sodium et potassium à l'état solide. J. Chim. Phys. 61:1428-1433

1964   Wilson GM (1964) Vapor-liquid equilibrium. XI. A new expression for the excess free energy of mixing. J. Am. Chem. Soc. 86:127-130

1965   Fraser DB, Hollis-Hallett AC (1965) The coefficient of thermal expansion of various cubic metals below 100 K. Can. J. Phys. 43:193-219

1965   Kleppa OJ, Meschel SV (1965) Heats of formation of solid solutions in the systems (Na-Ag)Cl and (Na-Ag)Br. J. Phys. Chem. 69:3531-3537

1965   Malesiński W (1965) Azeotropy and other theoretical problems of vapour-liquid equilibrium. Wiley, London

1965   Schlaudt CM, Roy DM (1965) Crystalline solution in the system MgO-Mg2SiO4-MgAl2O4. J. Amer. Ceram. Soc. 48:248-251

1966   Born M, Huang K (1966), Dynamical theory of crystal lattices. Oxford University Press, Oxford, pp 42-51

1966   Breusov ON (1966) Some types of equilibrium diagrams for binary systems. Russ. J. Phys. Chem. Engl. Transl. 40:107-109

1966   Fischer WA, Lorenz K, Fabritius H, Hoffmann A, Kalwa G (1966) Phase transformations in pure iron alloys with a magnetic balance. Arch. Eisenhüttenwes. 37:79-86

1966   Gopal ESR (1966) Specific heats at low temperatures. Plenum Press, New York

1966   Luova P, Tannila O (1966) Miscibility gaps in the systems KBr-KI and NaCl-KCl. Suom. Kemistil. B39:220-224

1966   Pistorius CWFT (1966) Phase diagrams of sodium tungstate and sodium molybdate to 45 kbar. J. Chem. Phys. 44:4532-4537

1966   Sellers CM, Maak F (1966) The thermodynamic properties of solid Au-Ni alloys at 775° to 935°C. Trans. Metall. Soc. *AIME* 236:457-464

1967 Colinet C (1967) Estimation des grandeurs thermodynamiques des alliages ternaires. Thesis. Université de Grenoble, France

1967 Lupis CHP, Elliot JF (1967) Prediction of enthalpy and entropy interaction coefficients by the central atoms theory. Acta Metall. 15:265-276

1967 Moriya Y, Warrington DH, Douglas RW (1967) A study of metastable liquid-liquid immiscibility in some binary and ternary alkali silicate glasses. Phys. Chem. Glasses 8:19-25

1967 Rawson H (1967) Inorganic glass-forming systems. Academic Press, London

1967 Serafimov LA (1967) The azeotropic rule and the classification of multicomponent mixtures. I. The principles of the classification of multicomponent mixtures. Russ. J. Phys. Chem. 41:1599-1601

1968 Ahtee M, Koski H (1968) Solubility gap in the system rubidium bromide – rubidium iodide. Ann. Acad. Sci. Fenn. Ser.A6 297:8pp

1968 Berg WT (1968) Heat capacity of aluminum between 2.7 and 20°K. Phys. Rev. 167:583-586

1968 Leadbetter AJ (1968) Anharmonic effects in the thermodynamic properties of solids. I. An adiabatic calorimeter for the temperature range 25-500°C: the heat capacities of Al2O3, Al and Pb. J. Phys. C. 1:1481-1488

1968 Seward TP, Uhlmann DR, Turnbull D (1968) Phase separation in the system BaO – SiO₂. J. Amer. Ceram. Soc. 51:278-285

1969 Althaus E (1969) Evidence that the reaction of kyanite to form sillimanite is at least bivariant. Amer. J. Sci. 267:273-277

1969 Charles RJ (1969) The origin of immiscibility in silicate solutions. Physics Chem. Glasses 10:169-178

1969 Gehrlich D, Fisher ES (1969) The high temperature elastic moduli of aluminium. J. Phys. Chem. Solids 30:1197-1205

1969 Ho PS, Ruoff AL (1969) Pressure dependence of the elastic constants for aluminum from 77° to 300°K. J. Appl. Phys. 40:3151-3156

1969 Kumazawa M, Anderson OL (1969) Elastic moduli, pressure   derivatives, and temperature derivatives of single-crystal olivine and single- crystal   forsterite. J. Geophys. Res. 74:5961-5972

1969 Nagata I (1969) Vapor-liquid equilibrium data for the binary systems methanol-benzene and methyl acetate-methanol. J. Chem. Eng. Data 14:418-420

1969 Richardson SW, Gilbert MC, Bell PM (1969) Experimental determination of kyanite-andalusite and andalusite-sillimanite equilibria; the aluminum silicate triple point. Amer. J. Sci. 267:259-272

1970 Pathak PD, Vasavada NG (1970) Thermal expansion and the law of corresponding states. J. Phys. C. 3:L44-L48

1970 Rackett HG (1970) Equation of state for saturated liquids. J. Chem. Eng. Data 15:514-517

1970 Vaidya SN, Kennedy GC (1970) Compressibility of 18 metals to 45 kbar. J. Phys. Chem. Solids 31:2329-2345

1971 Decker DL (1971) High-pressure equation of state for NaCl and CsCl. J. Appl. Phys. 42:3239-3244

1971 Fately WG, McDevitt NT, Bently FF (1971) Infrared and Raman selection rules for lattice vibrations: the correlation method. Appl. Spectroscopy 25: 155-174

1971 Oonk HAJ, Sprenkels A (1971) Isothermal binary liquid-vapour equilibria and the EGC method. I. General aspects. Z. physik. Chem. Neue Folge 75:225-233

1971 Oonk HAJ, Brouwer N, de Kruif CG, Sprenkels A (1971) Isothermal binary liquid-vapour equilibria and the EGC method. II. The interpretation of experimental phase diagrams. Z. physik. Chem. Neue Folge 75:234-241

1971 Serafimov LA, Zharov VT, Timofeyev VS (1971) Rectification of multicomponent mixtures, I. Topological analysis of liquid-vapour phase equilibrium diagrams. Acta Chim. Acad. Sci. Hung. 69:383-396

1972 Ferguson J, Currie KL (1972) Silica immiscibility in the ancient "basalts" of the Barberton Mountain Land Transvaal. Nature Physical Science 235:86-89

1972 Sinistri C, Riccardi R, Margheritis C, Tittarelli P (1972) Thermodynamic properties of solid systems AgCl+NaCl and AgBr+NaBr from miscibility gap measurements. Z. Naturforsch. 27a:149-154

1972 Wallace DC (1972) Thermodynamics of crystals. Wiley, New York

1972 Zeman L, Patterson D (1972) Pressure effects in polymer solution phase equilibriums. II. Systems showing upper and low critical solution temperatures. J. Phys.Chem. 76:1214-1219

1973 Grover R, Getting IC, Kennedy GC (1973) Simple compressibility relation for solids. Phys. Rev. (B) 7:567-571

1973 Kirchner G, Nishizawa T, Uhrenius B (1973) Distribution of chromium between ferrite and austenite and the thermodynamics of the α/γ equilibrium in the iron-chromium and iron-manganese systems. Metall. Trans. 4:167-174

1973 Oonk HAJ (1973) Equal-G Surfaces for heterogeneous equilibria in multicomponent systems. Z. physik. Chem. Neue Folge 84:67-83

1973 Richter PW, Pistorius CWFT (1973) The effect of pressure on the melting point of acetone. Z. physik. Chem. Neue Folge 85:82-85

1973 Wagner W (1973) New vapour pressure measurements for argon and nitrogen and a new method for establishing rational vapour pressure equations. Cryogenics 13:470-482

1973 Würflinger A, Schneider GM (1973) Differential Thermal Analysis under high Pressures II: Investigation of the rotational transition of several n-alkanes. Ber. Bundesges. Phys. Chem. 77:121-128

1974 Figurski G (1974) Die Bestimmung des thermodynamischen Verhaltens binärer Essigsäureester / Alkohol Systeme und binärer Gemische cyclischer Kohlenwasserstoffe aus Messungen des Dampf-Flüssigkeitsgleichgewichts unter besonderer Berücksichtigung der Fehler möglichkeiten. Thesis, University Rostock

1974 Ito H, Kawada K, Akimoto S (1974) Thermal expansion of stishovite. Phys. Earth Planet. Inter. 8:277-281

1975 Mercier JC, Carter NL (1975) Pyroxene geotherms. J. Geophys. Res. 80:3349-3362

1975 Mizukami S, Ohtani A, Kawai N, Ito E (1975) High-pressure X-Ray diffraction studies on β- and γ-Mg_2SiO_4. Phys. Earth Planet. Inter. 10:177-182

1975 Muggianu YM, Gambino M, Bros JP (1975) Enthalpies de formation des alliages liquides bismuth-étain-gallium à 723 K. Choix d'une représentation analytique des grandeurs d'excès intégrales et partielles de mélange. J. Chim. Phys. 72:83-88

1975 Ringwood AE (1975) Composition and petrology of the Earth's mantle. McGraw-Hill

1975 Suzuki I (1975) Thermal expansion of periclase and olivine, and their anharmonic properties. J. Phys. Earth 23-145-159

1975 Touloukian YS, Kirby RK, Taylor RE, Desai PD (1975) Thermophysical properties of matter: Thermal expansion, Metallic elements and alloys, IFI, Plenum, New York, Washington

1976 Angus S, Armstrong B, de Reuck K (1976) International thermodynamic tables of the fluid state – 3, Carbon dioxide. Pergamon Press, Oxford

1976 McGlashan ML, Williamson AG (1976) Isothermal liquid-vapor equilibria for system methanol – water. J. Chem. Eng. Data 21:196-199

1976 Oonk HAJ, Van Loo TJ, Vergouwen AA (1976) Empirical Gibss energy relation for immiscibility in liquid silica – metal oxide systems. Phys. Chem. Glasses 17:10-12

1976 Shannon RD (1976) Revised effective ionic radii and systematic studies of interatomic distances in halides and chalcogenides. Acta Cryst. A32:751-767

1977 Kawada K (1977) The system Mg_2SiO_4-Fe_2SiO_4 at high pressures and temperatures and the Earth's interior. Thesis, University of Tokyo

1977   Navrotsky A (1977) Calculation of effect of cation disorder on silicate spinel phase boundaries. Earth Planet Sci. Lett. 33:437-442

1977   Sumino Y, Nishizawa O, Goto T, Ohno I, Ozima M (1977) Temperature variation of elastic constants of single-crystal forsterite between -190 and 400°C. J. Phys. Earth 23:377-392

1977   Van Genderen ACG, de Kruif CG, Oonk HAJ (1977) Properties of mixed crystalline organic material prepared by zone leveling. I. Experimental determination of the EGC for the system p-dichlorobenzene + p-dibromobenzene. Z. physik. Chem. Neue Folge 107:167-173

1978   Ito E, Matsui Y (1978) Synthesis and crystal chemical characterization of $MgSiO_3$ perovskite. Earth Planet. Sci. Lett. 38:443-450

1978   Jantzen CMF, Herman H (1978) Spinodal decomposition – phase diagram representation and occurrence, In: Alper AM (Ed) Phase diagrams. Materials science and technology, Vol. V. Academic Press, New York, pp 127-184

1978   Robie RA, Hemingway BS, Fisher JR (1978) Thermodynamic properties of minerals and related substances at 298.15 K and 1 bar (10^5 pascals) pressure and at higher temperatures. U.S. Geological Survey, Bulletin 1452. U.S. Government Printing Office, Washington

1978   Spencer CF, Adler SB (1978) A critical review of equations for predicting saturated liquid density. J. Chem. Eng. Data 23:82-89

1978   Syassen K, Holzapfel WB (1978) Isothermal compression of Al and Ag to 120 kbar. J. Appl. Phys. 49:4427-4431

1979   Brouwer N, Oonk HAJ (1979) A direct method for the derivation of thermodynamic excess functions from TX phase diagrams. I. Possibilities and limitations. Z. physik. Chem. Neue Folge 105:113-123

1979   Brouwer N, Oonk HAJ (1979) A direct method for the derivation of thermodynamic excess functions from TX phase diagrams. II. Improved method applied to experimental regions of demixing. Z. physik. Chem. Neue Folge 117:55-67

1979   Hageman VBM, Oonk HAJ (1979) The region of liquid immiscibility in the system B_2O_3-BaO. Phys. Chem. Glasses 20:126-129

1979   Kieffer SW (1979) Thermodynamics and lattice vibrations of minerals: 1. Mineral heat capacities and their relationships to simple lattice vibrational models. Rev. Geophys. Space Phys. 17:1-19

1979   Kieffer SW (1979) Thermodynamics and lattice vibrations of minerals: 2. Vibrational characteristics of silicates. Rev. Geophys. Space Phys. 17:20-34

1979   Kieffer SW (1979) Thermodynamics and lattice vibrations of minerals: 3. Lattice dynamics and an approximation for minerals with application to simple substances and framework silicates. Rev. Geophys. Space Phys. 17:35-59

1979   Ohtani E (1979) Melting relation of Fe_2SiO_4 up to about 200 kbar. J. Phys. Earth 27:189-208

1980   Apelblat A, Tamir A, Wagner M (1980) Thermodynamics of acetone-chloroform mixtures. Fluid Phase Equilibria 4:229-255

1980   Bouwstra JA, Van Genderen ACG, Brouwer N, Oonk HAJ (1980) A thermodynamic method for the derivation of the solidus and liquidus curves from a set of experimental liquidus points. Thermochim. Acta 38:97-107

1980   Downie DB, Martin JF (1980) An adiabatic calorimeter for heat capacity measurements between 6 and 300 K. The molar heat capacity of aluminium. J. Chem. Thermodynamics 12:779-786

1980   Kieffer SW (1980) Thermodynamics and lattice vibrations of minerals: 4. Application to chain and sheet silicates and orthosilicates. Rev. Geophys. Space Phys. 18:862-886

1980   Lippmann F (1980) Phase diagrams depicting aqueous solubility of binary mineral systems. N. Jb. Miner. Abh. 139:1-25

1981   Dziewonski AM, Anderson DL (1981) Preliminary reference Earth model. Phys. Earth Planet. Inter. 25:297-356

1981   Göbl-Wunsch A, Heppke G, Hopf R (1981) Miscibility studies indicating a low-temperature smectic A phase in biaromatic liquid crystals with reentrant behaviour. Z. Naturforsch. A 36:1201-1204

1981   Manghnani MH, Matsui T (1981) Temperature dependence of pressure derivatives of single-crystal elastic constants of pure forsterite. ISPEI Symposium on properties of materials at high pressures and high temperatures. Toronto

1981   Ohtani E, Kumazawa M (1981) Melting of forsterite Mg_2SiO_4 up to 15 GPa. Phys. Earth Planet. Inter. 27:32-38

1981   Oonk HAJ (1981) Phase Theory – the thermodynamics of heterogeneous equilibria. Elsevier, Amsterdam

1981   Oonk H, Blok K, Van de Koot B, Brouwer N (1981) Binary common-ion alkali halide mixtures. Thermodynamic analysis of solid-liquid phase diagrams. I. Systems with negligible solid miscibility. Application of the ETXD/SIVAMIN method. Calphad 5:55-74

1981   Stacey FD, Brennan BJ, Irvine RD (1981) Finite strain theories and comparisons with seismological data. Geophys. Survey 4:189-232

1982   Bouwstra JA, Oonk HAJ (1982) Binary common-ion alkali halide mixtures. Thermodynamic analysis of solid-liquid phase diagrams. II. Systems with complete solid miscibility. Application of the LIQFIT method. Calphad 6:11-24

1982   Jamieson JC, Fritz JN, Manghnani MH (1982) Pressure measurements at high temperature in X-ray diffraction studies: gold as a primary standard. In High-Pressure Research in Geophysics. Akimoto S, Manghnani MH (Eds) Center for Academic Pusblishing Tokyo, pp 27-48

1982   Kieffer SW (1982) Thermodynamics and lattice vibrations of minerals: 5. Applications to phase equilibria, isotopic fractionation, and high-pressure thermodynamic properties. Rev. Geophys. Space Phys. 20:827-849

1982   Richet P, Bottinga Y, Deniélou L, Petitet JP, Téqui C (1982) Thermodynamic properties of quartz, cristobalite and amorphous SiO_2: drop calorimetry measurements between 1000 and 1800 K and a review from 0 to 2000 K. Geochim. Cosmochim. Acta 46:2639-2658

1982   Robie RA, Hemingway BS, Takei H (1982) Heat capacities and entropies of Mg_2SiO_4, Mn_2SiO_4 and Co_2SiO_4 between 5 and 380 K. Am. Mineral. 67:470-482

1982   Turcotte DL, Schubert G (1982) Geodynamics. p 450, John Wiley, New York.

1982   Watanabe H (1982) Thermochemical properties of synthetic high-pressure compounds relevant to the earth's mantle. In: Manghnani MH, Akimoto S (Eds) High-pressure research in geophysics. Center for Academic Publications, Japan, pp 441-464

1982   Weidner DJ, Bass JD, Ringwood AE, Sinclair W (1982) The single-crystal elastic moduli of stishovite. J. Geophys. Res. 87:4740-4746

1983   Lupis CHP (1983) Chemical thermodynamics of materials. North Holland, New York

1983   Moerkens R, Bouwstra JA, Oonk HAJ (1983) The solid-liquid equilibrium in the system p-dichlorobenzene + p-bromochlorobenzene + p-dibromobenzene. Thermodynamic assessment of binary data and calculation of ternary equilibrium. Calphad 7:219-269

1983   Oonk HAJ, Blok JG, Bouwstra JA (1983) Binary common-ion alkali halide mixtures. Thermodynamic analysis of solid-liquid phase diagrams. III. Three systems with limited and six systems with negligible solid miscibility. Application of the EXTXD/SIVAMIN method. Calphad 7:211-218

1983   Suzuki I, Anderson OL, Sumino Y (1983) Elastic properties of a single-crystal forsterite Mg_2SiO_4 up to 1200 K. Phys. Chem. Minerals 10:38-46

1984   Akaogi M, Ross NL, McMillan P, Navrotsky A (1984) The Mg_2SiO_4 polymorphs (olivine, modified spinel and spinel) – thermodynamic properties from oxide melt solution calorimetry, phase relations, and   models of lattice vibrations. Am. Mineral. 69:499-512

1984 Sawamoto H, Weidner DJ, Sasaki S, Kumazawa M (1984) Single-crystal elastic properties of the modified spinel (beta) phase of Mg_2SiO_4. Science 224:749-751

1984 Suzuki I, Takei H, Anderson OL (1984) Thermal expansion of single-crystal forsterite, Mg_2SiO_4. In: The proceedings of the eighth international thermal expansion symposium, sponsored by the national bureau of standards. Plenum, New York, pp 79-88

1985 Bauer K, Grabe D (1985) Common flagrance and flavor materials, preparation, properties and Uses. VCH Verlagsgesellschaft, Weinheim

1985 Chase Jr. MW, Davies CA, Downey Jr. JR, Frurip DJ, McDonald RA, Syverud AN (1985) JANAF Thermochemical Tables Third Edition. Part I, Al-Co; Part II, Cr-Zr. J. Phys. Chem. Ref. Data 14 Suppl. No.1

1985 Hillert M, Jansson B, Sundman B, Agren J (1985) A two-sublattice model for molten solutions with different tendency for ionization. Met. Trans. 16A:261-266

1985 Ito E, Navrotsky A (1985) Ilmenite: calorimetric, phase equilibria, and decomposition at atmospheric pressure. Am. Miner. 70:1020-1026

1985 Kato T, Kumazawa M (1985) Garnet phase of $MgSiO_3$ filling the pyroxene-ilmenite gap at very high temperature. Nature 316:803-805

1985 Krupka KM, Richard A, Hemingway BS, Ito J (1985a) Low-temperature heat capacities and derived thermodynamic properties of anthophyllite, diopside, enstatite, bronzite, and wollastonite. Am. Mineral. 70:249-260

1985 Krupka KM, Hemingway BS, Robie RA, Kerrick DM (1985b) High-temperature heat capacities and derived thermodynamic properties of anthophyllite, diopside, dolomite, enstatite, bronzite, talc, tremolite, and wollastonite. Am. Mineral. 70:261-271

1985 Kudoh Y, Takéuchi Y (1985) The crystal structure of forsterite Mg_2SiO_4 under high pressure up to 149 kb. Z. Kristallogr. 171:291-302

1985 Maroncelli M, Strauss HL, Snyder RG (1985) Structure of the n-alkane binary $n\text{-}C_{19}H_{40}$ / $n\text{-}C_{21}H_{44}$ by infrared spectroscopy and calorimetry. J. Phys. Chem. 89:5260-5267

1985 Matsui T, Manghnani MH (1985) Thermal expansion of single-crystal forsterite to 1023 K by Fizeau interferometry. Phys. Chem. Minerals 12: 201-210

1985 Okamoto H, Massalski TB (1985) The Au-Pd (gold-palladium) system. Bull. Alloy Phase Diagrams 6:229-235

1985 Van Hecke GR (1985) The Equal G Analysis. A comprehensive thermodynamics treatment for the calculation of liquid crystalline phase diagrams. J. Phys.Chem. 89:2058-2064

1985   Weidner DJ, Ito E (1985) Elasticity of MgSiO$_3$ in the ilmenite phase. Phys.Earth Planet. Inter. 40:65-70

1986   Bouwstra JA, Geels G, Kaufman L, Oonk HAJ (1986) Binary common-ion alkali halide mixtures. Thermodynamic analysis of solid-liquid phase diagrams IV. Three systems showing isodimorphism. Application of the LIQFIT method. Calphad 10:163-174

1986   Brown JM, McQueen RG (1986) Phase transitions, Grüneisen parameter and elasticity for shocked iron between 77 GPa and 400 GPa. J. Geophys. Res. 91:7485-7494

1986   Fei Y, Saxena SK (1986) A thermochemical data base for phase equilibria in the system Fe-Mg-Si-O at high pressure and temperature. Phys. Chem. Minerals 13:311-324

1986   Hageman VBM, Oonk HAJ (1986) Liquid immiscibilityin theSiO$_2$ + MgO, SiO$_2$ + SrO, SiO$_2$ + La$_2$O$_3$, and SiO$_2$ + Y$_2$O$_3$ systems. Phys. Chem. Glasses 27:194-198

1986   Kajiyoshi K (1986) High temperature equation of state for mantle minerals and their anharmonic properties. Thesis, Okayama University

1986   Knittle E, Jeanloz R, Smith GL (1986) Thermal expansion of silicate perovskite and stratification of the Earth's mantle. Nature 319:214-216

1986   Mao HK, Xu J, Bell PM (1986) Calibrations of the ruby pressure gauge to 800 kbar under quasihydrostatic conditions. J. Geophys. Res. 91:4673-4776

1986   Ming LC, Xiong D, Manghnani MH (1986) Isothermal compression of Au and Al to 20 GPa. Physica 139-140:174-176

1986   Oonk HAJ, Eisinga PJ, Brouwer N (1986) The computer program EXTXD/SIVAMIN and its application to the thermodynamic analysis of the regions of demixing in solid common-anion alkali halide mixtures. Calphad 10:1-36

1986   Oonk HAJ, Bouwstra JA, van Ekeren PJ (1986) Binary common-ion alkali halide mixtures. Correlation of thermochemical and phase-diagram data. Calphad 10:137-161

1986   Prausnitz JM, Lichtenthaler RN, Gomes de Azevedo E (1986) Molecularthermodynamics of fluid-phase equilibria. 2nd Ed., Prentice Hall, Englewood Cliffs

1986   Sawamoto H (1986) Single crystal growth of the modified spinel (β) and spinel (γ) phases of (Mg,Fe)$_2$SiO$_4$ and some geophysical implications. Phys. Chem. Miner. 13:1-10

1986   Vinet P, Ferrante J, Smith JR, Rose JH (1986) A universal equation of state for solids. J. Phys. C 19:L467-L473

1987   Akimoto S (1987) High-pressure research in geophysics: past, present and future. In Manghnani MH, Syono Y (Eds) High-pressure research in mineral physics. Terra Scientific Publishing Company (TERRAPUB), Tokyo / American Geophysical Union, Washington DC, pp 1-13

1987   Ashida T, Kume S, Ito E (1987) Thermodynamic aspects of phase boundary among α-, β-, and γ-Mg_2SiO_4. In: Manghnani MH, Syono Y   (Eds) High pressure research in mineral physics. Terra Scintific Publishing Company (TERRAPUB), Tokyo / American Geophysical Union, Washington DC, pp 269-274

1987   Bina CR, Wood BJ (1987) The olivine-spinel transitions: experimental and thermodynamic constraints and implications for the nature of the 400-km seismic discontinuity. J. Geophys. Res. 92:4853-4867

1987   Blander M, Pelton AD (1987) Thermodynamic analysis of binary liquid silicates and prediction of ternary solution properties by modified quasichemical equations. Geochim. Cosmochim. Acta 51:85-95

1987   Hageman VBM, Oonk HAJ (1987) Liquid immiscibility in the B_2O_3 – MgO, B_2O_3 – CaO, B_2O_3 – SrO, and B_2O_3 – BaO systems. Phys. Chem. Glasses 28:183-187

1987   Hofmeister AM (1987) Single-crystal absorption and reflection infrared of forsterite and fayalite. Phys. Chem. Minerals 14:499-513

1987   Horiuchi H, Ito E, Weidner DJ (1987) Perovskite-type $MgSiO_3$: single crystal X-ray diffraction study. Am. Mineral. 72:357-360

1987   Kanzaki M (1987) Ultra-high pressure phase relations in the system $Mg_4Si_4O_{12}$-$Mg_3Al_2Si_3O_{12}$. Phys. Earth Planet. Inter. 49:168-175

1987   Kudoh Y, Ito E, Takeda H (1987) Effect of pressure on the crystal structure of perovskite-type $MgSiO_3$. Phys. Chem. Minerals 14:350-354

1987   Michels MAJ, Wesker E (1987) A network model for the thermodynamics of multicomponent silicate melts. I. Binary mixtures metal oxide – silica. Calphad 11:383-393

1987   Price GD, Parker SC, Leslie M (1987) The lattice dynamics and thermodynamics of the Mg2SiO4 polymorphs. Phys. Chem. Minerals 15:181-190

1987   Reid RC, Prausnitz JM, Poling BE (1987) The Properties of Gases and Liquids. McGraw-Hill, New York

1987   Sangster J, Pelton AD (1987) Phase diagrams and thermodynamic properties of the 70 binary alkali halide systems having common ions. J. Phys. Chem. Ref. Data 16:509-561

1987   Sawamoto H (1987) Phase diagrams of $MgSiO_3$ at pressures up to 24 GPa and temperatures up to 2200 °C: Phase stability and properties of tetragonal garnet. In High Pressure Research in Mineral Physics. Manghnani MH, Syono Y (Eds), pp 209-219, Terra Scientific, Tokyo

1987 Van Ekeren PJ, Oonk HAJ (1987) Region-of-demixing calculations on HP 15C calculator; calculation of spinodal and bimodal. Calphad 11:99-99

1987 Van Ekeren PJ, Oonk HAJ (1987) Calculation of equilibria between two mixed states on HP 15C calculator. Calphad 11:101-101

1987 Van Miltenburg JC, van den Berg GJK, van Bommel MJ (1987) Construction of an adiabatic calorimeter. Measurements of the molar heat capacity of synthetic sapphire and of *n*-heptane. J. Chem. Thermodynamics 19:1129-1137

1987 Vinet P, Ferrante J, Rose JH, Smith JR (1987) Compressibility of solids. J. Geophys. Res. 92:9319-9325

1988 Ashida T, Kume S, Ito E, Navrotsky A (1988) $MgSiO_3$ Ilmenite: heat capacity, thermal expansivity and enthalpy of transformation. Phys.Chem. Miner. 16:239-245

1988 Michels MAJ, Wesker E (1988) A network model for the thermodynamics of multicomponent silicate melts. I. Binary mixtures metal oxide – silica. Calphad 12:111-126

1988 Nellis WJ, Moriarty JA, Mitchell AC, Ross M, Dandrea RG, Ashcroft NW, Holmes NC, Gathers GR (1988) Metal physics at ultrahigh pressure. Aluminum, Copper, and Lead as prototypes. Phys. Rev. Lett. 60:1414-1417

1989 Akaogi M, Ito E, Navrotsky A (1989) Olivine - modified spinel – spinel transitions in the system Mg_2SiO_4 – Fe_2SiO_4: calorimetric measurements, thermochemical calculation, and geophysical application. J. Geophys. Res. 94:15671-15685

1989 Ambrose D, Walton J (1989) Vapor pressures up to their critical temperatures of normal alkanes and 1-alkanols. Pure & Appl. Chem. 61:1395-1403

1989 Anderson OL, Isaak DL, Yamamoto S (1989) Anharmonicity and the equation of state for gold. Appl. Phys. 65:1534-1543

1989 Angel RJ, Finger LW, Hazen RM, Kanzaki M, Weidner DJ, Liebermann RC, Veblen DR (1989) Structure and twinning of single-crystal $MgSiO_3$ garnet synthesized at 17 GPa and 1800 °C. Am. Mineral. 74:509-512

1989 Barin I (1989) Thermochemical data of pure substances. VCH Verlagsgesellschaft, Weinheim

1989 Boots HMJ, De Bokx PK (1989) Theory of enthalpy-entropy compensation. J. Phys. Chem. 93:8240-8243

1989 Hillert M, Wang X (1989) A study of the thermodynamic properties of MgO-SiO_2 system. Calphad 13:253-266

1989 Holmes NC, Moriarty JA, Gathers GA (1989) The equation of state of platinum to 660 GPa (6.6 Mbar). J. Appl. Phys. 66:2962-2967

1989   Isaak DG, Anderson OL, Goto T (1989) Elasticity of single-crystal forsterite measured to 1700 K. J. Geophys. Res. 94:5895-5906

1989   Ito E, Takahashi E (1989) Postspinel transformations in the system Mg_2SiO_4-Fe_2SiO_4 and some geophysical implications. J. Geophys. Res. 94:10637-10646

1989   Katsura T, Ito E (1989) The system Mg_2SiO_4-Fe_2SiO_4 at high pressures and temperatures: precise determination of stabilities of olivine, modified spinel, and spinel. J. Geophys. Res. 94:15663-15670

1989   Ohe S (1989) Vapor-liquid equilibrium data. Elsevier, Amsterdam

1989   Ross NL, Hazen RM (1989) Single crystal X-ray diffraction study of $MgSiO_3$ perovskite from 77 to 400 K. Phys. Chem. Minerals 16:415-420

1989   Van Duijneveldt JS, Baas FSA, Oonk HAJ (1989) A program for the calculation of isobaric binary phase diagrams. Calphad 13:133-137

1989   Van Duijneveldt JS, Nguyen-Ba-Chanh, Oonk HAJ (1989) Binary mixtures of naphthalene and five of its 2-derivatives. Thermodynamic analysis of solid-liquid phase diagrams. Calphad 13:83-88

1990   Chopelas A (1990) Thermal properties of forsterite at mantle pressures derived from vibrational spectroscopy. Phys. Chem. Minerals 17:149-156

1990   Fei Y, Saxena SK, Navrotsky A (1990) Internally consistent thermodynamic data and equilibrium phase relations for compounds in the system MgO-SiO_2 at high pressure and high temperature. J. Geophys. Res. 95:6915-6928

1990   Gasparik T (1990) Phase relations in the transition zone. J. Geophys. Res. 95:15751-15769

1990   Gillet P, Le Cléach A, Madon M (1990) High-temperature Raman spectroscopy of SiO_2 and GeO_2 polymorphs: anharmonicity and thermodynamic properties at high temperatures. J. Geophys. Res. 95:21635-21655

1990   Gwanmesia GD, Rigden S, Jackson I, Liebermann RC (1990) Pressure dependence of elastic wave velocity for β-Mg_2SiO_4 and the composition of the Earth's mantle. Science 250:794-797

1990   Ito E, Akaogi M, Topor L, Navrotsky A (1990) Negative pressure-temperature slopes for reactions forming $MgSiO_3$ perovskite from calorimetry. Science 249:1275-1278

1990   Jacobs MHG (1990) The calculation of ternary phase diagrams from binary phase diagrams. Thesis, Utrecht University

1990   Königsberger E, Gamsjäger H (1990) Solid-solute phase equilibria in aqueous solution. IV Calculation of phase diagrams. Z. anorg. allg. Chem. 584:185-192

1990   Matsubara R, Toraya H, Tanaka S, Sawamoto H (1990) Precision lattice parameter determination of $(Mg,Fe)SiO_3$ tetragonal garnets. Science 247:697-699

1990 Parise JB, Wang Y, Yeganeh-Haeri A, Cox DE, Fei Y (1990) Crystal structure and thermal expansion of (Mg,Fe)SiO$_3$ perovskite. Geophys. Res. Lett., 17:2089-2092

1990 Phillips BL, Howell DA, Kirkpatrick RJ, Gasparik T (1990) Investigation of cation order in MgSiO$_3$-rich garnet using ^{29}Si and ^{27}Al MAS NMR spectroscopy. Am. Miner. 77:704-712

1990 Presnall DC, Gasparik T (1990) Melting of enstatite (MgSiO$_3$) from 10 to 16.5 GPa and the forsterite (Mg$_2$SiO$_4$) – majorite (MgSiO$_3$) eutectic at 16.5 GPa : implications for the origin of the mantle. J. Geophys. Res.95:15771-15777

1990 Ross NL, Hazen RM (1990) High pressure crystal chemistry of MgSiO$_3$ perovskite. Phys. Chem. Miner. 17:228-237

1990 Ross NL, Shu J, Hazen RM, Gasparik T (1990) High pressure crystal chemistry of stishovite. Am. Mineral. 75:739-747

1990 Tanaka T, Gokcen NA, Morita Z (1990) Relationship between enthalpy of mixing and excess entropy in liquid binary alloys. Z. Metallkd. 81:49-54

1990 Wang ZC, Lück R, Predel B (1990) New models for computing thermodynamic properties and phase diagrams of ternary systems. Part 1. Three-factor models. Calphad 14:217-234

1990 Wang ZC, Lück R, Predel B (1990) New models for computing thermodynamic properties and phase diagrams of ternary systems. Part 2. Multi-factor models. Calphad 14:235-256

1991 Boehler R, Chopelas A (1991) A new approach to laser heating in high pressure mineral physics. Geophys. Res. Lett. 18:1147-1150

1991 Bottinga Y (1991) Thermodynamic properties of silicate liquids at high pressure and their bearing on igneous petrology. In: Perchuk LL, Kushiro I (Eds) Physical chemistry of magmas. Springer, New York, pp 213-232

1991 Calvet MT, Cuevas-Diarte MA, Haget Y, van der Linde PR, Oonk HAJ (1991) Binary p-dihalobenzene systems – correlation of thermochemical and phase-diagram data. Calphad 15:225-234

1991 Chopelas A (1991) Thermal properties of β-Mg$_2$SiO$_4$ at mantle pressures derived from vibrational spectroscopy: implications for the mantle at 400 km depth. J. Geophys. Res. 96:11817-11829

1991 Fei Y, Mao HK, Mysen BO (1991) Experimental determination of element partitioning and calculation of phase relations in the MgO-FeO-SiO$_2$ system at high pressure and high temperature. J Geophys Res. 96B:2157-2169

1991 Gillet P, Richet P, Guyot F, Fiquet G (1991) High-temperature thermodynamic properties of forsterite. J. Geophys. Res. 96:11805-11816

1991   Kanzaki M (1991) Ortho/Clino enstatite transition. Phys. Chem. Minerals 17:726-730

1991   Rigden SM, Gwanmesia GD, Fitzgerald JD, Jackson I, Liebermann RC (1991) Spinel elasticity and seismic structure of the transition zone of the mantle. Nature 354:143-145

1992   Gillet P, Fiquet G, Malézieux JM, Geiger C (1992) High-pressure and high-temperature Raman spectroscopy of end-member garnets: Pyrope, grossular and andradite. Eur. J. Mineral. 4:651-664

1992   Griffen (1992) Silicate crystal chemistry, Oxford University Press, Oxford

1992   Hofmeister AM, Ito E (1992) Thermodynamic properties of $MgSiO_3$ Ilmenite from vibrational spectra. Phys. Chem. Minerals 18:423-432

1992   Tonkov EYu (1992) High pressure phase transformations. Gordon and Breach, Philadelphia

1992   Van der Kemp WJM, Blok JG, van Genderen ACG, van Ekeren PJ, Oonk HAJ (1992) Binary common-ion alkali halide mixtures; a uniform description of the liquid and solid state. Thermochim. Acta 196:301-315

1992   Winkler B, Dove MT (1992) Thermodynamic properties of $MgSiO_3$ perovskite derived from large scale molecular dynamics. Phys. Chem. Minerals 18:407-415

1992   Yagi T, Uchiyama Y, Akaogi M, Ito E (1992) Isothermal compression curve of $MgSiO_3$ tetragonal garnet. Phys. Earth Planet. Inter. 74:1-7

1993   Akaogi M, Ito E (1993a) Heat capacity of $MgSiO_3$ perovskite. Geophys. Res. Lett. 20:105-108

1993   Akaogi M, Ito E (1993b) Refinement of enthalpy measurement of $MgSiO_3$ perovskite and negative pressure-temperature slopes for perovskite-forming reactions. Geophys. Res. Lett. 20:1839-142

1993   Meng Y, Weidner DJ, Gwanmesia GD, Liebermann RC, Vaughan MT, Wang Y, Leinenweber K, Pacalo RE, Yeganeh-Haeri A, Zhao Y (1993) In situ high P-T X-Ray diffraction studies on three polymorphs (α, β, γ) of Mg_2SiO_4. J. Geophys. Res. 98:22199-22207

1993   Saxena SK, Chatterjee N, Fei Y, Shen G (1993) Thermodynamic data on oxides and silicates. Springer Verlag, Berlin

1993   Sirota EB, King HE, Singer DM, Shao HH (1993) Rotator phases of the normal alkanes: an X-ray scattering study. J. Chem. Phys. 98:5809-5824

1993   Stixrude L, Bukowinski MST (1993) Thermodynamic analysis of the system $MgO-FeO-SiO_2$ at high pressure and the structure of the lowermost mantle. In Evolution of Earth and Planets. Takahashi E, Jeanloz R, Rubie D (Eds) Geophys. Monogr. Ser. 74:131-142, AGU, Washington DC

1993 Van der Kemp WJM, Blok JG, van der Linde PR, Oonk HAJ, Schuijff A, Verdonk ML (1993) On the estimation of thermodynamic excess properties of binary solid solutions. Thermochim. Acta 225:17-30

1993 Wang SY, Sharma SK, Cooney TF (1993) Micro-Raman and infrared spectral study of forsterite under high pressure. Am. Mineral. 78:469-476

1993 Yusa H, Akaogi M, Ito E (1993), Calorimetric study of $MgSiO_3$ garnet and pyroxene: heat capacities, transition enthalpies, and equilibrium phase relations in $MgSiO_3$ at high pressures and temperatures. J. Geophys. Res. 98:6453-6460

1994 Angel RJ, Hugh-Jones DA (1994) Equation of state and thermodynamic properties of enstatite pyroxene. J. Geophys. Res. 99:19777-19783

1994 Chopelas A, Boehler R, Ko T (1994) Thermodynamics and behavior of γ-Mg_2SiO_4 at high pressure: implications for Mg_2SiO_4 phase equilibrium. Phys. Chem. Minerals 21:351-359

1994 Greene RG, Luo H, Ruoff AL (1994) Al as a simple solid: high pressure study to 220 GPa (2.2 Mbar). Phys. Rev. Lett. 73:2075-2078

1994 Hemley RJ, Prewitt CT, Kingma KJ (1994) High-pressure behavior of silica. In: Heaney RJ, Prewitt CT, Gibbs GV (Eds) Silica: physical behavior, geochemistry and materials applications. Mineral. Soc. Am. pp 41-81

1994 Lide DR, Kehiaian HV (1994) CRC Handbook of thermophysical and thermochemical data. CRC Press, Boca Raton

1994 Meng Y, Fei Y, Weidner DJ, Gwanmesia GD, Hu J (1994) Hydrostatic compression of γ-Mg_2SiO_4 to mantle pressures and 700 K: thermal equation of state and related thermoelastic properties. Phys. Chem. Minerals 21:407-412

1994 Morishima H, Kato T, Suto M, Ohtani E, Urakawa S, Utsumi W, Shimomura O, Kikegawa T (1994) The phase boundary between α- and β- Mg_2SiO_4 determined by in situ X-ray observation. Science 265:1202-1203

1994 Sorenson JM, Van Hecke GR (1994) A simple method for equal Gibbs energy analysis of phase boundaries. Calphad 18:329-333

1994 Swamy V, Saxena SK, Sundman B (1994) An assessment of the one-bar liquidus phase relations in the $MgO-SiO_2$ system. Calphad 18:157-164

1994 Van der Kemp WJM (1994) Thermodynamics of binary mixed crystals. Thesis, Utrecht University

1994 Van der Kemp WJM, Blok JG, van der Linde PR, Oonk HAJ, Schuijff A, Verdonk ML (1994) Binary alkaline earth oxide mixtures: estimation of the excess thermodynamic properties and calculation of the phase diagrams. Calphad 18:255-267

1994 Wang Y, Weidner DJ, Liebermann RC, Zhao Y (1994) PVT equation of state of (Mg,Fe)SiO₃ perovskite: constraints on composition of the lower mantle. Phys. Earth Plan. Inter. 83:13-40

1994 Yeganeh-Haeri A (1994) Synthesis and re-investigation of the elastic properties of single-crystal magnesium silicate perovskite. Phys. Earth Plan. Inter. 87:111-112

1995 Anderson OL (1995) Equations of state of solids for geophysics and ceramic science. Oxford University Press, New York

1995 Andrault D, Bouhifd MA, Itie JP, Richet P (1995) Compression and amorphization of (Mg,Fe)₂SiO₄ olivines: an X-ray diffraction study up to 70 GPa. Phys. Chem. Minerals 22:99-107

1995 Calvet T, Oonk HAJ (1995) Laevorotatory-carvoxime + dextrorotatory-carvoxime a unique binary system. Calphad 19:49-56

1995 Duffy TS, Zha CS, Downs RT, Mao HK, Hemley RJ (1995) Elasticity of forsterite to 16 GPa and the composition of the upper mantle. Nature 378:170-173

1995 Inbar I, Cohen RE (1995) High pressure effects on thermal properties of MgO. Geophys. Res. Lett. 22:1533-1536

1995 Kennett BNL, Engdahl AR, Buland R (1995) Constraints on seismic velocities in the Earth from traveltimes. Geophys. J. Int. 123:108-124

1995 Utsumi W, Funamori N, Yagi T (1995) Thermal expansivity of MgSiO₃ perovskite under high pressures up to 20 GPa. Geophys. Res. Lett. 22:1005-1008

1995 Yang H, Ghose S (1995) High-temperature single crystal X-ray diffraction studies of the ortho-proto phase transition in enstatite, Mg₂Si₂O₆ at 1360 K. Phys. Chem. Minerals 22:300-310

1995 Zhao Y, Schiferl D, Shankland TJ (1995) A high P-T single-crystal X-ray diffraction study of thermoelasticity of MgSiO₃ ortho enstatite. Phys. Chem. Minerals 22:393-398

1996 Bouhifd MA, Andrault D, Fiquet G, Richet P (1996) Thermal expansion of forsterite up to the melting point. Geophys. Res. Lett. 23:1143-1146

1996 Calvet T, Cuevas-Diarte MA, Gallis HE, Oonk HAJ (1996) Spontaneous crystallization in undercooled liquid mixtures of l-limonene + l-carvone. Recl. Trav. Chim. Pays-Bas 115:333-335

1996 Chopelas A (1996) The fluorescence sideband method for obtaining acoustic velocities at high compression: application to MgO and MgAl₂O₄. Phys. Chem. Minerals 23:25-37

1996 Chopelas A (1996) Thermal expansivity of lower mantle phases MgO and MgSiO₃ perovskite at high pressure derived from vibrational spectroscopy. Phys. Earth Plan. Inter. 98:3-15

1996 Cynn H, Carnes JD, Anderson OL (1996) Thermal properties of forsterite, including C_V, calculated from αK_T through the entropy. J. Phys. Chem. Solids 57:1593-1599

1996 Downs RT, Zha CS, Duffy TS, Finger LW (1996) The equation of state of forsterite to 17.2 GPa and effects of pressure media. Am. Mineral. 81:51-55

1996 Funamori N, Yagi T, Utsumi W, Kondo T, Uchida T, Funamori M (1996) Thermoelastic properties of $MgSiO_3$ perovskite determined by in situ X-ray observations up to 30 GPa and 2000 K. J. Geophys. Res. 101:8257-8269

1996 Gallis HE, Bougrioua F, Oonk HAJ, van Ekeren PJ, van Miltenburg JC (1996) Mixtures of d- and l-carvone: I. Differential scanning calorimetry and solid-liquid phase diagram. Thermochim. Acta 274:231-242

1996 Gallis HE, van den Berg GJK, Oonk HAJ (1996) Thermodynamic properties of crystalline d-limonene determined by adiabatic calorimetry. J. Chem. Eng. Data 41:1303-1306

1996 Gallis HE, van Miltenburg JC, Oonk HAJ, van der Eerden PJ (1996) Mixtures of d- and l-carvone. III. Thermodynamic properties of l-carvone. Thermochim. Acta 286:307-319

1996 Jacobs MHG, Jellema R, Oonk HAJ (1996) TXY-CALC, a program for the calculation of thermodynamic properties and phase equilibria in ternary systems. An application to the system (Li,Na,K)Br. Calphad 20:79-88

1996 Li B, Gwanmesia GD, Liebermann RC (1996) Sound velocities of olivine and beta polymorphs of Mg_2SiO_4 at Earth's transition zone pressures. Geophys. Res. Lett. 23:2259-2262

1996 Li B, Rigden SM, Liebermann RC (1996) Elasticity of stishovite at high pressure. Phys. Earth Planet. Inter. 96:113-127

1996 Reynard B, Rubie DC (1996) High-pressure, high-temperature Raman spectroscopic study of ilmenite-type $MgSiO_3$. Am. Mineral. 81:1092-1096

1996 Reynard B, Takir F, Guyot F, Gwanmesia GD, Liebermann RC, Gillet P (1996) High-temperature Raman spectroscopy and X-ray diffraction study of β-Mg_2SiO_4: Insights into its high-temperature thermodynamic properties and the β- to α-phase transformation mechanism and kinetics. Am. Mineral. 81:585-594

1996 Saxena SK (1996) Earth mineralogical model: Gibbs free energy minimisation computation in the system $MgO-FeO-SiO_2$. Geochim. Cosmochim. Acta 60:2379-2395

1997 Figurski G, van Ekeren PJ, Oonk HAJ (1997) Thermodynamic analysis of vapour-liquid phase diagrams I. Binary systems of nonelectrolytes with complete miscibility; application of the PXFIT- and the LIQFIT-method. Calphad 21:381-390

1997 Gillet P, Daniel I, Guyot F (1997) Anharmonic properties of Mg_2SiO_4-forsterite measured from the volume dependence of the Raman spectrum. Eur. J. Mineral. 9:255-262

1997 Hugh-Jones D (1997) Thermal expansion of $MgSiO_3$ and $FeSiO_3$ ortho- and clino pyroxenes. Am. Miner. 82:689-696

1997 McHale A (1997) Phase diagrams and ceramic processes. Chapman & Hall, New York

1997 Mondieig D, Espeau P, Robles L, Haget Y, Oonk HAJ, Cuevas-Diarte MA (1997) Mixed crystals of *n*-alkane pairs. A global view of the thermodynamic melting properties. J. Chem. Soc., Faraday Trans. 93:3343-3346

1997 Pacalo REG, Weidner DJ (1997) Elasticity of majorite, $MgSiO_3$ tetragonal garnet. Phys. Earth Planet. Inter. 99:145-154

1997 Sañé J, Rius J, Calvet T, Cuevas-Diarte MA (1997) Chiral molecular alloys: Patterson-search structure determination of L-carvone and DL-carvone from X-ray powder diffraction data at 218K. Acta Cryst. B53:702-707

1997 Sinogeikin SV, Bass JD, O'Neill B, Gasparik T (1997) Elasticity of tetragonal end-member majorite and solid solutions in the system $Mg_4Si_4O_{12}$-$Mg_3Al_2Si_3O_{12}$. Phys. Chem. Minerals 24:115-121

1997 Zha CS, Duffy TS, Mao HK, Downs RT, Hemley RJ, Weidner DJ (1997) Single-crystal elasticity of β-Mg_2SiO_4 to the pressure of the 410 km seismic discontinuity in the Earth's mantle. Earth Plan. Sci. Lett. 147:E9-E15

1998 Andrault D, Fiquet G, Guyot F, Hanfland M (1998) Pressure-induced Landau-type transition in stishovite. Science 282:720-724

1998 Fabrichnaya O (1998) The assessment of thermodynamic parameters for solid phases in the Fe-Mg-O and Fe-Mg-Si-O systems. Calphad 22:85-125

1998 Fiquet G, Andrault D, Dewaele A, Charpin T, Kunz M, Haüsermann D (1998) *P-V-T* equation of state of $MgSiO_3$ perovskite. Phys. Earth Plan. Inter. 105:21-31

1998 Flesch LM, Li B, Liebermann RC (1998) Sound velocities of polycrystalline $MgSiO_3$-orthopyroxene to 10 GPa at room temperature. Am. Miner. 83:444-450

1998 Gallis HE (1998) Time dependence of crystallization processes and polymorphism. A calorimetric study of carvone and limonene. Thesis Utrecht University

1998 Gwanmesia GD, Chen G, Liebermann RC (1998) Sound velocities in $MgSiO_3$-garnet to 8 GPa. Geophys. Res. Lett. 25:4553-4556

1998 Hillert M (1998) Phase equilibria, phase diagrams and phase transformations: their thermodynamic basis. Cambridge University Press, Cambridge

1998   Holland TJB, Powell R (1998) An internally consistent thermodynamic data set for phases of petrological interest. J. Metamorphic Geol. 16:309-343

1998   Li B, Liebermann RC, Weidner DJ (1998) Elastic moduli of wadsleyite (β-Mg_2SiO_4) to 7 Gigapascals and 873 Kelvin. Science 281:675-679

1998   López R, López DO (1998) WINIFIT. A Windows computer program for the thermodynamic assessment of TX phase diagrams using the C.I.C. (Crossed Isopolymorphism Concept). Polytechnic University of Catalonia

1998   Oonk HAJ, Mondieig D, Haget Y, Cuevas-Diarte MA. (1998) Perfect families of mixed crystals: the rotator I N-alkane case. J. Chem. Phys. 108:715-722

1999   Chopelas A (1999) Estimates of mantle relevant Clapeyron slopes in the $MgSiO_3$ system from high-pressure spectroscopic data. Am. Miner. 84:233-244

1999   Ewig CS, Thacher TS, Hagler AT (1999) Derivation of Class II force fields. 7. Nonbonded force field parameters for organic compounds. J. Chem. Phys. B 103:6998-7014

1999   Fei Y, Bertka CM (1999) Phase transformations in the Earth's mantle and mantle mineralogy. In Mantle Petrology: Field observation and High Pressure Experimentation (Eds. Y. Fei, C.M. Bertka and B.O. Myssen), vol 6, pp 189-207. The Geological Society, Washington, DC

1999   Jackson JM, Sinogeikin SV, Bass JD (1999) Elasticity of $MgSiO_3$ ortho enstatite. Am. Miner. 84:677-680

1999   Ji S, Wang Z (1999) Elastic properties of forsterite-enstatite compositions up to 3.0 GPa. Geodynamics 28:147-174

1999   Liu J, Zhang J, Flesh L, Li B, Weidner DJ, Liebermann RC (1999) Thermal equation of state of stishovite. Phys. Earth Planet. Inter. 112:257-266

1999   Metivaud V (1999) Systèmes multicomposants d'alcanes normaux dans la gamme $C_{14}H_{30} - C_{25}H_{52}$: alliances structurales et stabilité des échantillons mixtes. Applications pour la protection thermique d'installations de telecommunications et de circuits optoélectroniques. Thesis, Université de Bordeaux I

1999   Rajabalee F, Métivaud V, Mondieig D, Haget Y, Cuevas-Diarte MA (1999) New insights on the crystalline forms in binary systems of n-alkanes: characterization of the solid ordered phases in the phase diagram tricosane+pentacosane. J. Mater. Res. 14:2644-2654

1999   Rajabalee F, Métivaud V, Mondieig D, Haget Y, Oonk HAJ (1999) Thermodynamic analysis of solid-solid and solid-liquid equilibria in binary systems composed of n-alkanes: application to the system tricosane ($C_{23}H_{48}$) + Pentacosane ($C_{25}H_{52}$). Chem. Mater. 11:2788-2795

1999   Romero-Serrano A, Pelton AD (1999) Thermodynamic analysis of binary and ternary silicate systems by a structural model. ISIJ International 39:399-408

1999   Saxena SK, Dubrovinsky LS, Tutti F, Le Bihan T (1999) Equation of state of MgSiO₃ with the perovskite structure based on experimental measurement. Am. Miner. 84:226-232

1999   Shearer PM, Flanagan MP, Hedlin MA (1999) Experiments in migration processing of SS precursor data to image upper mantle discontinuity structure. J. Geophys. Res.104:7229-7242

1999   Shinmei T, Tomioka N, Fujino K, Kuroda K, Irifune T (1999) In situ X-ray diffraction study of enstatite up to 12 GPa and 1473 K and equations of state. Am. Miner. 84:1588-1594

1999   Thiéblot L, Téqui C, Richet P (1999) High-temperature heat capacity of grossular (Ca₃Al₂Si₃O₁₂), enstatite (MgSiO₃) and titanite (CaTiSiO₅). Am. Mineral. 84:848-855

2000   Chopelas A (2000) Thermal expansivity of mantle relevant magnesium silicates derived from vibrational spectroscopy at high pressure. Am. Mineral. 85:270-278

2000   Cohen RE, Gülseren O, Hemley RJ (2000) Accuracy of equation of state formulations. Am. Miner. 85:338-344

2000   Gallis HE, van Miltenburg JC, Oonk HAJ (2000) Polymorphism of mixtures of enantiomers: a thermodynamic study of mixtures of D- and L-limonene. Phys. Chem. Chem. Phys. 2:5619-5623

2000   Gemsjäger H, Königsberger E, Preis W (2000) Lippmann diagrams: theory and application to carbonate systems. Aquat. Geochem. 6:119-132

2000   Gillet P, Daniel I, Guyot F, Matas J, Chervin J-C (2000) A thermodynamic model for MgSiO₃-perovskite derived from pressure, temperature and volume dependence of the Raman mode frequencies. Phys. Earth Planet. Inter. 117:361-384

2000   Jackson JM, Sinogeikin SV, Bass JD (2000) Sound velocities and elastic properties of γ-Mg₂SiO₄ to 873 K by Brillouin spectroscopy. Am. Mineral. 85:296-303

2000   Jacobs MHG, Oonk HAJ (2000) A new equation of state based on Grover, Getting and Kennedy's empirical relation between volume and bulk modulus. The high-pressure thermodynamics of MgO. Phys. Chem. Chem. Phys. 2:2641-2646

2000   Jacobs MHG, Oonk HAJ (2000) A realistic equation of state for solids. The high-pressure and high-temperature thermodynamic properties of MgO. Calphad 24:133-148

2000   Karki BB, Wentzcovitch RM, de Gironcoli S, Baroni S (2000) High-pressure lattice dynamics and thermoelasticity of MgO. Phys. Rev. B 61:8793-8800

2000   López DO, Salud J, Tamarit Ll, Barrio M, Oonk HAJ (2000) Uniform thermodynamic description of the orientationally disordered mixed crystals of a group of neopentane derivatives. Chem. Mater. 12:1108-1114

2000   Poirier JP (2000) Itroduction to the physics of the Earth's interior. Cambridge University Press, 2nd ed. Cambridge, New York, Melbourne, Madrid

2000   Rajabalee F, Metivaud V, Oonk HAJ, Mondieig D, Waldner P (2000) Perfect families of mixed crystals: the "ordered" crystalline forms of n-alkanes. Phys. Chem. Chem. Phys. 2:1345-1350

2000   Suzuki A, Ohtani E, Morishima H, Kubo T, Kanbe Y, Kondo T, Okada T, Terasaki H, Kato T, Kikegawa T (2000) In situ determination of the boundary between wadsleyite and ringwoodite in Mg2SiO4. Geophys. Res. Lett. 27:803-806

2000   Terasaki H, Kato T, Kikegawa T (2000) In situ determination of the boundary between wadsleyite and ringwoodite in Mg$_2$SiO$_4$. Geophys. Res. Lett. 27:803-806

2000   Wang Y, Chen D, Zhang X (2000) Calculated equation of state of Al, Cu, Ta, Mo, and W to 1000 GPa. Phys. Rev. Lett. 84:3220-3223

2001   De Jong BHWS, Jacobs MHG (2001) Chemical variations affect seismic velocities less than grain size variations. Eos Trans. AGU 82(47) V51B-1010

2001   Deuss A, Woodhouse J (2001) Seismic observations of splitting of the mid transition zone in Earth's mantle. Science 294:354-357

2001   Elmsley J (2001) Nature's building blocks. An A-Z guide to the elements. Oxford University Ptress, Oxford

2001   Hirose K, Komabayashi T, Murakami M (2001) In situ measurement of the majorite-akimotoite-perovskite phase transition boundaries in MgSiO$_3$. Geophys. Res. Lett. 22:4351-4354

2001   Jacobs MHG, Oonk HAJ (2001) The Gibbs energy formulation of the α, β, and γ forms of Mg$_2$SiO$_4$ using Grover, Getting and Kennedy's empirical relation between volume and bulk modulus. Phys. Chem. Minerals 28:572-585

2001   Karki BB, Stixrude L, Wentzcovitch RM (2001) High-pressure elastic properties of major materials of Earth's mantle from first principles. Reviews of Geophysics, 39:507-534

2001   Kiefer B, Stixrude L, Hafner J, Kresse G (2001) Structure and elasticity of wadsleyite at high pressure. Am. Mineral. 86:1387-1395

2001   Koningsveld R, Stockmayer WH, Nies E (2001) Polymer phase diagrams: a textbook, Oxford University Press, Oxford

2001   Ono S, Katsura T, Ito E, Kanzaki M, Yoneda A, Walter MJ, Urakawa S, Utsumi W, Funakoshi K (2001) In situ observation of ilmenite-perovskite phase transition in MgSiO$_3$ using synchrotron radiation. Geophys. Res. Lett. 28:835-838

2001   Oonk HAJ (2001) Solid-state solubility and its limits.The alkali halide case. Pure & Appl. Chem. 73:807-823

2001 Shim S, Duffy TS, Shen G (2001) The post-spinel transformation in Mg_2SiO_4 and its relation to the 660-km seismic discontinuity. Nature 411:571-574

2001 Speziale S, Zha C, Duffy TS, Hemley RJ, Mao HK (2001) Quasi-hydrostatic compression of magnesium oxide to 52 GPa: implications for the pressure-volume-temperature equation of state. J. Geophys. Res. 106:515-528

2002 Anderson JO, Helander T, Höglund L, Shi P, Sundman B (2002) Thermo-Calc and Dictra, computational tools for materials science. Calphad 26:273-312

2002 Angel RJ, Jackson JM (2002) Elasticity and equation of state of ortho enstatite, $MgSiO_3$. Am. Miner. 87:558-561

2002 Bale CW, Chartrand P, Degterov SA, Eriksson G, Hack H, Ben Mafoud R, Melançon J, Pelton AD, Petersen S (2002) FactSage thermochemical software and databases. Calphad 26:189-228

2002 Chen SL, Daniel S, Zhang F, Chang YA, Yan XY, Xie FY, Schmid-Fetzer R, Oates WA (2002) The PANDAT software package and its applications. Calphad 26:175-188

2002 Cheynet B, Chevalier PY, Fischer E (2002) THERMOSUITE. Calphad 26:167-174

2002 Davies RH, Dinsdale AT, Gisby JA, Robinson JAJ, Martin SM (2002) MTDATA – thermodynamic and phase equilibrium software from National Physical Laboratory. Calphad 26:229-271

2002 Kaufman L (2002) Foreword. Calphad 26:141-141

2002 Lebedev S, Chevrot S, van der Hilst RD (2002) Seismic evidence for olivine phase changes at the 410- and 660 kilometer discontinuities. Science 296:1300-1302

2002 Papon P, Leblond J, Meijer PHE (2002) The Physics of Phase Transitions. Springer-Verlag, Berlin, Heidelberg

2002 Ramirez M (2002) Modelitzacio estructural de les forms ordenades en la familia dels n-alcanols. Thesis. Universitat de Barcelona

2002 Shim S, Duffy TS, Takemura K (2002) Equation of state of gold and its application to the phase boundaries near 660 km depth in the Earth's mantle. Earth Planet. Sci. Lett. 203:729-739

2002 Tomiska J (2002) ExTherm: the interactive support package of experimental thermodynamics. Calphad 26:143-154

2002 Turcotte DL, Schubert G (2002) Geodynamics, 2nd Ed., Cambridge University Press, Cambridge

2002 Van der Linde PR, Bolech M, den Besten R, Verdonk ML, van Miltenburg JC, Oonk HAJ (2002) Melting behaviour of molecular mixed crystalline materials: measurement with adiabatic calorimetry and modeling using ULTRACAL. J. Chem. Thermodynamics 34:613-629

2002 Yamanaka T, Fukuda T, Tsuchiya J (2002) Bonding character of SiO_2 stishovite under high pressures up to 30 GPa. Phys. Chem. Minerals 29:633-641

2002 Yokokawa H, Yamauchi S, Matsumoto T (2002) Thermodynamic database for Windows with gem and CHD. Calphad 26:155-166

2003 Andrault D, Angel RJ, Mosenfelder JL, Le Bihan T (2003) Equation of state of stishovite to lower mantle pressures. Am. Mineral. 88:301-307

2003 Frost DJ (2003) The structure and sharpness of $(Mg,Fe)_2SiO_4$ phase transformations in the transition zone. Earth Planet. Sci. Lett. 216:313-328

2003 Gasparik T (2003) Phase diagrams for geoscientists. Springer-Verlag, Berlin Heidelberg New York

2003 Jackson JM, Palko JW, Andrault D, Sinogeikin SV, Lakshtanov DL, Wang J, Bass JD, Zha CS (2003) Thermal expansion of natural ortho enstatite to 1473 K. Eur. J. Mineral. 15:469-473

2003 Katsura T, Yamada H, Shinmei T, Kubo A, Ono S, Kanzaki M, Yoneda A, Walter MJ, Ito E, Urakawa S, Funakoshi K, Utsumi W (2003) Post-spinel transition in Mg_2SiO_4 determined by high P-T in situ X-ray diffractometry. Phys. Earth Planet. Inter. 136:11-24

2003 Li J, Hadidiacos C, Mao HK, Fei Y, Hemley RJ (2003) Effects of pressure on thermocouples in a multi-anvil apparatus. High Pressure Res. 23: 389-401

2003 Mondieig D, Metivaud V, Oonk HAJ, Cuevas-Diarte MA (2003) Isothermal transformations in alkane alloys. Chem. Mater. 15:2552-2560

2003 Panero WR, Benedetti LR, Jeanloz R (2003) Equation of state of stishovite and interpretation of SiO_2 shock-compression data. J. Geophys. Res. 108 (B1) 2015 doi 10.1029./2001JB001663

2004 Chang YA, Chen S, Zhang F, Yan X, Xie F, Schmid-Fetzer R, Oates WA (2004) Phase diagram calculation: past, present and future. Progr. Mater. Sci. 49:313-345

2004 Chudinovskikh L, Boehler R (2004) $MgSiO_3$ phase boundaries measured in the laser-heated diamond cell. Earth Planet. Sci. Lett. 222:285-296

2004 Deschamps F, Trampert J (2004) Towards a lower mantle reference temperature and composition. Earth Planet. Sci. Lett. 222:161-175

2004 DeWaele A, Loubeyre P, Mezouar M (2004) Equations of state of six metals above 94 GPa. Phys. Rev. B. 70: 094112

2004 Fabrichnaya O, Saxena SK, Richet P, Westrum EF Jr (2004) Thermodynamic Data, Models and Phase Diagrams in Multicomponent Oxide Systems. Springer-Verlag, New York

2004 Fei Y, van Orman J, Li J, van Westrenen W, Sanloup C, Minarik W, Hirose K, Komabayashi T, Walter M (2004a) Experimentally determined postspinel transformation boundary in Mg_2SiO_4 using MgO as an internal pressure standard and its geophysical implications. J. Geophys. Res. 109:B02305

2004 Fei Y, Li J, Hirose K, Minarik W, van Orman J, Sanloup C, van Westrenen W, Komabayashi T, Funakoshi K-i (2004b) A critical evaluation of pressure scales at high temperatures by in situ X-ray diffraction measurements. Phys. Earth Planet Inter. 143-144:515-526

2004 Katsura T, Yokoshi S, Song M, Kawabe K, Tsujimura T, Kubo A, Ito E Tange Y, Tomioka N, Saito K, Nozawa A, Funakoshi KI (2004) Thermal expansion of Mg_2SiO_4 ringwoodite at high pressures. J. Geophys. Res. 109:B12209

2004 Kung J, Li B, Uchida T, Wang Y, Neuville D, Liebermann RC (2004) In situ measurements of sound velocities and densities across the orthopyroxene $\rightarrow$ high-pressure clinopyroxene transition in $MgSiO_3$ at high pressure. Phys.Earth Planet. Inter. 147:27-44

2004 Oganov AR, Dorogukupets PI (2004) Intrinsic anharmonicity in equations of state and thermodynamics of solids. J. Phys. Condens. Matter 16:1351-1360

2004 Oganov AR, Ono S (2004) Theoretical and experimental evidence for a post-perovskite phase of $MgSiO_3$ in Earth's D" layer. Nature 430:445-448

2004 Predel B, Hoch M, Pool M (2004) Phase diagrams and heterogeneous equilibria. A practical introduction. Springer-Verlag, Berlin Heidelberg

2004 Sinogeikin SV, Zhang J, Bass JD (2004) Elasticity of single crystal and polycrystalline $MgSiO_3$ perovskite by Brillouin spectroscopy. Geophys. Res.Lett. 31: L06620, doi: 10.1029/2004GL019559

2004 Tsuchiya T, Tsuchiya J, Umemoto K, Wentzcovitch RM (2004) Phase transition in $MgSiO_3$ perovskite in the Earth's lower mantle. Earth Planet. Sci. Lett. 224:241-248

2004 Wang Y, Uchida T, Zhang J, Rivers ML, Sutton SR (2004) Thermal equation of state of akimotoite $MgSiO_3$ and effects of the discontinuity. Phys. Earth Planet. Inter. 143&144:57-80

2004 Wentzcovitch RM, Stixrude L, Karki BB, Kiefer B (2004) Akimotoite to perovskite phase transition in $MgSiO_3$. Geophys. Res. Lett. 31:L10611-L10613

2005 Bovolo CI (2005) The physical and chemical composition of the lower mantle. Phys. Trans. R. Soc. 363:2811-2835

2005 Jacobs MHG, de Jong BHWS (2005) An investigation into thermodynamic consistency of data for the olivine, wadsleyite and ringwoodite form of $(Mg,Fe)_2SiO_4$. Geochim. Cosmochim. Acta 69:4361-4375

2005 Jacobs MHG, de Jong BHWS (2005) Quantum-thermodynamic treatment of anharmonicity; Wallace's theorem revisited. Phys. Chem. Minerals 32:614-626

2005 Kung J, Li B, Uchida T, Wang Y (2005) In-situ elasticity measurement for the unquenchable clinopyroxene phase: implication for the upper mantle. Geophys. Res. Lett. 32: L01307

2005 Li B, Zhang J (2005) Pressure and temperature dependence of elastic wave velocity of $MgSiO_3$ perovskite and the composition of the lower mantle. Phys. Earth Planet. Inter. 151:143-154

2005 Mattern E, Matas J, Ricard Y, Bass J (2005) Lower mantle composition and temperature from mineral physics and thermodynamic modelling. Geophys. J. Int. 160:973-990

2005 Nishihara Y, Nakayama K, Takahashi E, Iguchi T, Funakoshi K (2005) P-V-T equation of state of stishovite to the mantle transition zone conditions. Phys. Chem. Minerals 31:660-670

2005 Oonk HAJ, Tamarit JLl (2005) Condensed phases of organic materials: solid-liquid and solid-solid equilibrium. In: Weir RD, de Loos ThW (Eds) Measurement of the thermodynamic properties of multiple phases. Experimental Thermodynamics Volume VII. Elsevier, Amsterdam, pp 201-274

2005 Stixrude L, Lithgow-Bertelloni C (2005a) Thermodynamics of mantle minerals - I. Physical properties. Geophys. J. Int. 162:610-632

2005 Stixrude L, Lithgow-Bertelloni C (2005b) Mineralogy and elasticity of the oceanic upper mantle: Origin of the low-velocity zone. J. Geophys. Res. 110:B03204:1-16

2005 Tsuchiya J, Tsuchiya T, Wentzcovitch RM (2005) Vibrational and thermodynamic properties of $MgSiO_3$ postperovskite. J. Geophys. Res. 110:B02204

2006 Akahama Y, Nishimura M, Kinoshita K, Kawamura K (2006) Evidence of a fcc-hcp transition in aluminum at multimegabar pressure. Phys. Rev. Lett. 96:045505

2006 Davis JP (2006) Experimental measurement of the principal isentrope for aluminum 6061-T6 to 240 GPa. J. Appl. Phys. 99(103512):1-6

2006 Deuss A, Redfern SAT, Chambers K, Woodhouse JH (2006) The nature of the 660-kilometer discontinuity in Earth's mantle from global seismic observations of PP precursors. Science 311:198-201

2006 Inoue T, Irifune T, Higo Y, Sanchira T, Sueda Y, Yamada A, Shinmei T, Yamazaki D, Ando J, Funakoshi K, Utsumi W (2006) The phase boundary between wadsleyite and ringwoodite in Mg_2SiO_4 determined by in situ X-ray diffraction. Phys. Chem. Minerals 33:106-114

2006 Jacobs MHG, Oonk HAJ (2006) The calculation of ternary miscibility gaps using the linear contributions method: problems, benchmark systems and an application to (K,Li,Na)Br. Calphad 30:185-190

2006 Jacobs MHG, van den Berg AP, de Jong BHWS (2006) The derivation of thermo-physical properties and phase equilibria of silicate materials from lattice vibrations: application to convection in the Earth's mantle. Calphad, 30, 131-146

2006 Ono S, Kikegawa T, Ohishi Y (2006) Equation of state of $CaIrO_3$-type $MgSiO_3$ up to 144 GPa. Am. Miner. 91:475-478

2006 Sugahara M, Yoshiasa A, Komatsu Y, Yamanaka T, Bolfan-Casanova N, Nakatsuka A, Sasaki S, Tanaka M (2006) Reinvestigation of the $MgSiO_3$ perovskite structure at high pressure. Am. Mineral. 91:533-536

2006 Van Peteghem CB, Zhao J, Angel RJ, Ross NL, Bolfan-Casanova N (2006) Crystal structure and equation of state of $MgSiO_3$ perovskite. Geophys. Res. Lett. 33:L03306, doi: 10.1029/2005GL024955

2006 Wentzcovitch RM, Tsuchiya T, Tsuchiya J (2006) $MgSiO_3$ postperovskite at D" conditions. Proc. Natl. Acad. Sci. USA 103:543-546

2007 Guignot N, Andrault D, Morard G, Bolfan-Casanova N, Mezouar M (2007) Thermoelastic properties of post-perovskite phase $MgSiO_3$ determined experimentally at core-mantle boundary P-T conditions. Earth Planet. Sci. Lett. 256:162-168

2007 Jacobs MHG, de Jong BHWS (2007) Placing constraints on phase equilibria and thermophysical properties in the system $MgO-SiO_2$ by a thermodynamically consistent vibrational method. Geochim. Cosmochim. Acta 71:3630-3655

2007 Piazonni AS, Steinle-Neumann G, Bunge HP, Dolejš D (2007) A mineralogical model for density and elasticity of the Earth's mantle. Geochemistry Geophysics Geosystems 8:1-23

2008 Kresh M, Lucas M, Delaire O, Lin JYY, Fultz B (2008) Phonons in aluminum at high temperature studied by inelastic neutron scattering. Phys. Rev. B. 77(024301):1-9

2008 Oonk HAJ, Calvet MT (2008) Equilibrium between phases of matter. Phenomenology and thermodynamics. Springer, Dordrecht

2009 Deuss A (2009) Global observations of mantle discontinuities using SS and PP precursors. Surv. Geophys. 30:301-326

2009   Zhao M, Song L, Fan X (2009) The boundary theory of phase diagrams and its application. Rules for phase diagram construction with phase regions and their boundaries. Science Press Beijing and Springer-Verlag, Berlin Heidelberg

2010   Cao Q, Wang P, van der Hilst RD, de Hoop MV, Shim SH (2010) Imaging the upper mantle transition zone with a generalized Radon transform of SS precursors. Phys. Earth Planet. Inter. 180:80-91

2010   Jacobs MHG, Schmid-Fetzer R (2010) Thermodynamic properties and equation of state of fcc aluminum and bcc iron, derived from a lattice vibrational method. Phys. Chem. Minerals 37:721-739

2011   Cao Q, Van der Hilst RD, de Hoop MV, Shim S-H (2011) Seismic imaging of transition zone discontinuities suggest hot mantle west of Hawaii. Science 332:1068

2011   Jacobs MHG, van den Berg AP (2011) Complex phase distribution and seismic velocity structure of the transition zone: Convection model predictions for a magnesium-endmember olivine-pyroxene mantle. Phys. Earth Planet. Inter. 186:36-48

2011   Van den Berg AP, Yuen DA, Jacobs MHG, de Hoop MV (2011) Small scale mineralogical heterogeneity from variations in phase assemblages in the transition zone and D" layer predicted by convection modelling. J. Earth Science 22:160-168

SUBJECT INDEX

SUBSTANCES AND SYSTEMS INDEX